SEESAW25

Proceedings of the International Conference
on the Seesaw Mechanism

SEESAW25

Proceedings of the International Conference on the Seesaw Mechanism

Institut Henri Poincaré, Paris
10–11 June 2004

editors

J. Orloff
Université Blaise Pascal, France

S. Lavignac & M. Cribier
CEA–Saclay, France

World Scientific

NEW JERSEY · LONDON · SINGAPORE · BEIJING · SHANGHAI · HONG KONG · TAIPEI · CHENNAI

Published by

World Scientific Publishing Co. Pte. Ltd.

5 Toh Tuck Link, Singapore 596224

USA office: 27 Warren Street, Suite 401-402, Hackensack, NJ 07601

UK office: 57 Shelton Street, Covent Garden, London WC2H 9HE

British Library Cataloguing-in-Publication Data
A catalogue record for this book is available from the British Library.

SEESAW 25
Proceedings of the International Conference on the Seesaw Mechanism

ISBN 981-256-111-0

Printed in Singapore by Mainland Press

This book is printed on acid-free paper.

FOREWORD

Neutrino physics has undergone spectacular developments in the past 6 years, thanks to the wealthy harvest of experimental results collected mostly by SuperKamiokande (SK), the Sudbury Neutrino Observatory (SNO) and KamLAND. These have established the existence of neutrino flavour transitions, for which the most coherent interpretation relies on neutrino oscillations (or, in the case of solar neutrinos, adiabatic flavour conversion) induced by small but non-vanishing neutrino mass differences. The seesaw mechanism was devised much earlier to generically explain the smallness of possible neutrino masses by appealing to a new high scale. It is then striking that this new scale naturally falls close to the Grand Unification scale if it is to account for the observed oscillations of atmospheric neutrinos (at least for a hierarchical mass spectrum).

However, theoretical developments did not await this appealing observation, and there has been an intense activity in this field since the first mechanisms for generating Majorana masses were proposed at the turn of the eighties. The seesaw mechanism inspired a lot of works about its realizations in unified theories or about its numerous implications in particle physics and cosmology, such as the creation of the baryon asymmetry of the universe via leptogenesis or, within supersymmetric extensions, the violation of flavour and of the CP symmetry in the charged lepton sector.

These developments gained a renewed impetus with the experimental evidence for neutrino mass, and this motivated the organization of a conference in order to review the recent progress in the theoretical aspects of neutrino physics, notably in connection with the seesaw mechanism. We took the opportunity of the 25th anniversary of the talks by T. Yanagida and M. Gell-Mann, P. Ramond and R. Slansky to organize this conference at the Institut Henri Poincaré in Paris, a few days before Neutrino 2004.

Although time limitations made it impossible to give the credit they deserve to all significant contributions in the past 25 years, the participation of a large number of experts in the field allowed the conference to cover most theoretical aspects of neutrino physics related to the seesaw mechanism: the general construction of seesaw models (R. N. Mohapatra) in the GUT framework (G. Senjanović), using textures and flavour models (P. Ramond and G. G. Ross), in extra dimensions (E. Dudas) or in techni-colour theories (T. Appelquist); renormalisation group effects (M. Lindner) and the possibility of reconstructing seesaw parameters from low energy observables (S. Davidson); further effects of heavy right-handed neutrinos in cosmology via leptogenesis (T. Yanagida, M. Raidal and T. Hambye) or in the charged lepton sector in supersymmetric theories (A. Masiero and J. Hisano); alternatives to the seesaw mechanism for neutrino masses (A. Yu. Smirnov).

The key results and projects of the vast experimental programme in neutrino physics were also presented: the experimental evidence for neutrino mass (K. Heeger); the search for the absolute neutrino mass scale (Ch. Weinheimer) and cosmological implications of massive neutrinos (S. Hannestad); the quest for the third mixing angle and for CP violation in the lepton sector (C. Hagner); detectors for low energy neutrinos and cold dark matter (S. Schönert). To conclude, a round-table discussion on the question *"How to probe the origin of neutrino masses?"* was animated by the lively participation of G. Gratta, B. Kayser, S. Raby, J.W.F. Valle and Ch. Wetterich, as well as by many interventions from the audience.

The historical circumstances that lead to the idea of mixing the Standard Model neutrinos with heavier states in order to explain the smallness of their masses were addressed by several speakers. As was already known, besides the seminal works of M. Gell-Mann, P. Ramond, R. Slansky[1] and T. Yanagida[2] that are reproduced at the end of this volume for convenience, early contributions include articles by R. N. Mohapatra and G. Senjanović[3]; R. Barbieri, D.V. Nanopoulos, G. Morchio and F. Strocchi -[4]; E. Witten[5]; M. Magg and Ch. Wetterich[6]; J. Schechter and J. W. F. Valle[7]; G. Lazarides, Q. Shafi and Ch. Wetterich[8]. In contrast, the contribution of S. L. Glashow in Cargese 79 Lectures[9] resurfaced only recently, and we regret that S. Glashow could not participate in the conference. Even more recently, unfortunately after the closing of the conference, the community realized the existence of an earlier article by P. Minkowski[10], in which the seesaw mechanism was presented for the first time. To the best of our knowledge, however, the attractive idea of relating the right-handed neutrino mass scale to Grand Unification did not arise until 1979.

We cannot close this foreword without heartfully thanking the speakers for their active contribution, both to the conference and to these proceedings; the participants in the round-table discussion, who took up the delicate challenge of publicly crossing ideas; the audience, whose participation made the round-table discussion and the whole conference particularly lively; the organising committee, whose involvement has been essential for the success of the conference; and finally, the Commissariat à l'Énergie Atomique (CEA/DSM) and the Centre National de la Recherche Scientifique (départements IN2P3 et SPM), whose financial support made this conference possible.

Paris, November 2004

Michel Cribier
Stéphane Lavignac
Jean Orloff

References

1. M. Gell-Mann, P. Ramond and R. Slansky, unpublished;
   P. Ramond, *The Family Group in Grand Unified Theories*, Talk given at the 19th Sanibel Symposium, Palm Coast, Florida, Feb. 25-Mar. 2, 1979, preprint CALT-68-709 (retro-print hep-ph/9809459);
   M. Gell-Mann, P. Ramond, and R. Slansky, in *Supergravity* (P. van Nieuwenhuizen et al. eds.), North Holland, Amsterdam, 1980, p. 315.
2. T. Yanagida, *Horizontal Symmetry and Masses of Neutrinos*, in *Proceedings of the Workshop on Unified Theories and Baryon Number in the Universe* (O. Sawada and S. Sugamoto eds.), Tsukuba, Japan, Feb. 13-14, 1979, p. 95; KEK Report KEK-79-18.
3. R.N. Mohapatra and G. Senjanović, Phys. Rev. Lett. 44 (1980) 912;
   R.N. Mohapatra and G. Senjanović, Phys. Rev. D23 (1981) 165.
4. R. Barbieri, D.V. Nanopoulos, G. Morchio and F. Strocchi, Phys. Lett. B90 (1980) 91.
5. E. Witten, Phys. Lett. B91(1980) 81.
6. M. Magg and Ch. Wetterich, Phys. Lett. B94 (1980) 61;
7. J. Schechter and J.W.F. Valle, Phys. Rev. D22 (1980) 2227.
8. G. Lazarides, Q. Shafi and Ch. Wetterich, Nucl. Phys. B181 (1981) 287.
9. S. Glashow, *The Future of Elementary Particle Physics*, in *Quarks and Leptons, Cargèse 1979* (M. Lévy et. al. eds.), July 9-29, 1979, Plenum, New York, 1980, p. 687.
10. P. Minkowski, Phys. Lett. B67 (1977) 421.

SEESAW25
Seesaw Mechanism and
Neutrino Masses 25 Years Later
10-11 June 2004 - Institut Henri Poincaré, Paris

CONTENTS

SEESAW AND THE RIDDLE OF MASS

P. RAMOND*

*Institute for Fundamental Theory,
Department of Physics, University of Florida,
Gainesville FL 32611, USA
E-mail: ramond@phys.ufl.edu*

The prediction of small neutrino masses through the Seesaw Mechanism and their subsequent measurement suggests that the natural cut-off of the Standard Model is very high indeed. The recent neutrino data must be interpreted as a reflection of physics at very high energy. We examine their implications in terms of ideas of Grand Unification and Supersymmetry, and as possible hints for a unified theory of flavor.

1. Introduction

The Seesaw Mechanism[1,2], which we are here to celebrate, must be viewed in the context of the intellectual turmoil generated by the Standard Model. The renormalizability of massive Yang-Mills theories[3], the emergence of a common description of Weak and Electromagnetic Interactions[4], and the realization that the Strong Interactions weaken at shorter distances[5] established the Standard Model as the paradigm for all Fundamental Interactions except Gravity. Like all such paradigms, the Standard Model is (thankfully) incomplete, has suggested new puzzles of it own, and elicited many questions. None has been more dominating than Pati and Salam's[6] proposal that quarks and leptons are equal partners in one mathematical structure at very short distances, the idea of Grand-Unification.

To appreciate the significance of the Seesaw mechanism as the link between small neutrino masses and Physics near the Planck scale, one must first describe the great theoretical speculations which led to its creation.

---

*Work partially supported by grant DE-FG02-97ER410292-4570.5 of the US Department of Energy.

2. Triumphs of the Standard Model

The Fundamental Interactions (save for Gravity) are described by the Standard Model. It has withstood, practically unscathed, almost four decades of experiments, confirming *inter alia* its radiative structure. All of its quarks and leptons have been discovered. Its main features are

– Interactions stem from three *weakly coupled* Yang-Mills theories based on $SU(3)$, $SU(2)$ and $U(1)$.
– Quarks and leptons are needed for quantum consistency: gauge anomalies cancel between quarks and leptons.
– There are three chiral families of quarks and leptons, each with a massless neutrino.
– The gauge symmetries are spontaneously broken: the shorter the distance, the *more* the symmetry.
– It predicts a fundamental scalar particle, the Higgs boson.

Only one of these predictions has been proved wrong by experiments: neutrinos have masses. Today, only few of its parameters await measurement, the mass of the elusive Higgs particle, the strong CP-violating phase, and the mass of any of the three neutrinos.

3. Old & New Puzzles

Although the successes of the Standard Model have exceeded expectations, it has a dark side:

– It predicts CP-violation in the Strong interaction, albeit with unknown strength.
– It requires Yukawa interactions without any organizing principle.
– It fails to *explain* the values of masses and mixing patterns of quarks and charged leptons.
– It contains too many parameters to be truly fundamental.
– Without Gravitation it only describes the matter side of Einstein's equation, *sans* cosmological constant.
– It fails to account for neutrino masses.

The Standard Model presents an unfinished picture of Nature. It reminds one of the shards of a once beautifull pottery, shattered in the course of cosmological evolution.

4. Grand Unification

The quantum numbers of the three chiral families of quarks and leptons strongly suggest a more unified picture. Pati and Salam's original idea is, remarkably enough, realized by unifying the three gauge groups of the Standard Model into one. In the simplest, $SU(5)$[7], each family appears in two representations. In $SO(10)$[8], they are grouped in its fundamental spinor representation, by adding a right-handed neutrino for each family. At the next level, we find E_6[9] where each family contains several right-handed neutrinos as well as vector-like matter. Organizing the elementary particles into these beautiful structures

– Unifies the three gauge groups.
– Relates Quarks and Leptons.
– Explains anomaly cancellations.

There are indications that this idea "wants to work". When last seen, the three coupling constants of the Standard Model are perturbative. Using the renormalization group equations to continue them deep into the ultraviolet, they get closer to one another, but fail to meet at one scale: the quantum number patterns did not quite match the dynamical information. This near (thought at the time to be exact) unification introduced Planck scale physics into the realm of particle physics.

One by-product of Grand Unification is violation of baryon number. Hitherto unobserved, proton decay remains one of the most important consequences from these ideas. In a serendipitous twist, proton decay detectors now serve as the telescopes of neutrino astronomy! Other global symmetries also bite the dust: the relative lepton numbers are violated in $SU(5)$ and $SO(10)$ violated the total lepton number as well, and the extraordinary limits on these processes are consistent with the grand-unified scale.

5. Grand-unified Legacies

Grand Unification by itself does not yet have any direct experimental vindication; it is an incubator of new ideas that, even today, drive speculations on the Physics at extra-short distances.

– It linked the large grand-unified scale to tiny neutrino masses[1].
– It suggested relations between quark and charged lepton masses, although the flavor riddles of the Standard Model remain unexplained.
– It created the "gauge hierarchy" problem: why quantum corrections keep the ratio of the Higgs mass to the Unification scale small.

Moreover, two of its predictions have linked particle physics to pre-Nucleosynthesis Cosmology:

– The possibility of monopoles in our universe led to the idea of Inflationary Cosmology[10], which solves many long standing puzzles and whose prediction of a flat universe has been recently verified.
– Proton decay. This offered a framework for understanding the baryon asymmetry[11] of the Universe.

Today, only one of these predictions, tiny neutrino masses, has been borne out by experiment. On the conceptual side, it has also provided an alternative mechanism for the generation of Baryon asymmetry of the Universe through a primordial lepton asymmetry[12]. Still, Grand Unification is at most a partial theory of Nature, since it does not address Gravity (space-time is either flat or a fixed background), nor the origin of the three chiral families and its associated flavor puzzles.

6. Superstrings

At the 1973 London conference, David Olive declared Superstring Theories to be candidate "Theories of Everything", since they reproduce Einstein's gravity at large distances with no ultraviolet divergences, and also contain (some) gauge theories. This view has since gained much credence and notoriety. The matter content has gotten much closer to reality[13], although this unification of the gravitational and gauge forces takes place in a somewhat unsettling background:

– Fermions and Bosons are related by a new type of symmetry: *Supersymmetry*[14].
– Ultimate Unification takes place in nine or ten space dimensions!

Nature at the millifermi displays neither Supersymmetry nor extra space dimensions. Yet, the lesson of the Standard Model of more symmetries at shorter distances provide an argument for these to be fabrics of the Ultimate Theory; these symmetries are somehow destroyed in the process of cosmological evolution. To compare the highly symmetric superstring theories to Nature, a dynamical understanding of their breakdown is required, an understanding that still eludes us.

To relate to Nature, experiments at energies at which these symmetries appear must be carried out. All could be just around the energy corner, although circumstantial evidence lends more credence to low- energy Supersymmetry than to low-energy extra dimensions. The collapse of the extra space dimensions occurs first, while Supersymmetry hangs on to later times (lower energies). It is a challenge to theory to find a dynamical reason which triggers the breakdown of higher-dimensional space (perhaps through brane formation), while leaving Supersymmetry nearly intact.

7. Supersymmetry

Supersymmetry is an attractive theoretical concept; it is required by the unification of gravity and gauge interactions, and links fermions and bosons. Also, the mass of the spinless superpartner of a Weyl fermion, inherits quantum-naturality[15] through the chiral symmetry of its partner.

Morever, when applied to the Standard Model, it yields quantitative predictions that fit remarkably well with Gauge Unification. With Supersymmetry,

– The Gauge hierarchy problem is managed: the Higgs mass is stabilized even in the presence of a large (grand-unification) scale
– The three gauge couplings of the Standard Model run to a single value in the deep ultraviolet with the addition of superpartners in the TeV range. Thus naturally emerges a new scale using the renormalization group, a scale that matches the quantum number patterns of the elementary particles.
– With supersymmetry the renormalization group displays an infrared fixed point that predicts[16] the top quark mass, in agreement with experiment.
– Under a large class of ultraviolet initial conditions, the same renormalization group shows that the breaking of supersymmetry triggers electroweak breaking[17].

Supersymmetry at low energy is the leading theory for physics beyond the Standard Model, although many puzzles remain unanswered and new ones are created as well.

For one, there are almost as many theories of supersymmetry breaking as there are theorists, and none, theories and theorists alike, are convincing. It is an experimental question.

In addition, Supersymmetry deepens the flavor riddles of the Standard Model by predicting new scalar particles which generically produce flavor-changing neutral processes. Even if the breaking mechanism is flavor-blind (tasteless), non-trivial effects are expected: supersymmetry-breaking is already highly constrained by the existing data set.

The existence of low-energy Supersymmetry will soon be tested at the LHC. May the supersymmetry-breaking mechanism parameters prove to be so unique as to allow intellectually-challenged theorists (the author included) to infer its origin from the LHC data alone!

8. Minute Neutrino Masses

The only solid experimental evidence to date for physics beyond the Standard Model is the observation of oscillation among neutrino species. Thirty five years of experiments on solar neutrinos, Homestake[18], GALLEX[19], SAGE[20], SUPERK[21] and SNO[22], yield

$$\Delta m_\odot^2 \ = \ |\ m_{\nu_1}^2 - m_{\nu_2}^2\ | \ \sim \ 7. \times 10^{-5} \ \mathrm{eV}^2 \ ,$$

with corroborating evidence on antineutrinos[23]. Neutrinos born in Cosmic ray collisions[24], and on earth[25] give

$$\Delta m_\oplus^2 \ = \ |\ m_{\nu_2}^2 - m_{\nu_3}^2\ | \ \sim \ 3. \times 10^{-3} \ \mathrm{eV}^2 \ .$$

The best bound to their absolute value of the masses comes from WMAP[26]

$$\sum_i m_{\nu_i} \ < \ .71 \ \mathrm{eV} \ .$$

These experimental findings are not sufficient to determine fully the mass patterns. One oscillates between three patterns, *hierarchy,*

$$|m_{\nu_1}| < |m_{\nu_2}| \ll |m_{\nu_3}| \ ,$$

inverse hierarchy

$$|m_{\nu_1}| \simeq |m_{\nu_2}| \gg |m_{\nu_3}| \ ,$$

and *hyperfine*

$$|m_{\nu_1}| \simeq |m_{\nu_2}| \simeq |m_{\nu_3}| \ .$$

The mixing patterns provide some surprises, since it contains one small angle and two large angles. In terms of the MNS mixing matrix,

$$\begin{pmatrix} \cos\theta_\odot & \sin\theta_\odot & \epsilon \\ -\cos\theta_\oplus \sin\theta_\odot & \cos\theta_\oplus \cos\theta_\odot & \sin\theta_\oplus \\ \sin\theta_\oplus \sin\theta_\odot & -\sin\theta_\oplus \cos\theta_\odot & \cos\theta_\oplus \end{pmatrix} \ ,$$

the various experiments yield

$$\sin^2 2\theta_\oplus > 0.85 , \qquad 0.30 < \tan^2 \theta_\odot < 0.65 ,$$

while there is a only a limit[27] on the third angle

$$| \epsilon |^2 < 0.05 .$$

Spectacular as they are, these results generate new questions for experimenters:

- Are the masses Majorana-like (i.e. lepton number violating)?
- What are their absolute values?
- Can one measure the sign of Δm^2?
- What is the value of the CHOOZ angle?
- Is CP-violation in the lepton sector observable?

They also generate new theoretical questions

- Are there right-handed neutrinos?
- How many? How heavy, and with what hierarchy?
- Where do they live? Brane or bulk?
- Do their decays trigger leptogenesis[12]?

9. Standard Model Analysis

In the context of Grand Unification, one needs to discuss both quark and lepton mass matrices. To that effect, recall that the masses and mixings of the quarks are determined from the diagonalization of Yukawa matrices generated by the $\Delta I_\mathrm{W} = \frac{1}{2}$ breaking of electroweak symmetry, for charge $2/3$

$$\mathcal{U}_{2/3} \begin{pmatrix} m_u & 0 & 0 \\ 0 & m_c & 0 \\ 0 & 0 & m_t \end{pmatrix} \mathcal{V}_{2/3}^\dagger ,$$

and charge $-1/3$

$$\mathcal{U}_{-1/3} \begin{pmatrix} m_d & 0 & 0 \\ 0 & m_s & 0 \\ 0 & 0 & m_b \end{pmatrix} \mathcal{V}_{-1/3}^\dagger ,$$

resulting in the observable CKM matrix

$$\mathcal{U}_{CKM} \;\equiv\; \mathcal{U}^{\dagger}_{2/3}\,\mathcal{U}_{-1/3} \;,$$

which, up to corrections of the order of the Cabibbo angle, $\theta_c \sim 13°$, is equal to the unit matrix. This implies similar family mixings for up-like and down-like quarks. Their masses are of course highly hierarchical. The charged lepton Yukawa matrix

$$\mathcal{U}_{-1}\begin{pmatrix} m_e & 0 & 0 \\ 0 & m_\mu & 0 \\ 0 & 0 & m_\tau \end{pmatrix}\mathcal{V}^{\dagger}_{-1}$$

also stems from $\Delta I_{\mathrm{W}} = \frac{1}{2}$ electroweak breaking, and has hierarchical eigenvalues.

To obtain neutrino masses in the Standard Model, it is simplest to add one right-handed neutrino for each family. This yields another $\Delta I_{\mathrm{W}} = \frac{1}{2}$ Yukawa matrix

$$\mathcal{U}_0\begin{pmatrix} m_1 & 0 & 0 \\ 0 & m_2 & 0 \\ 0 & 0 & m_3 \end{pmatrix}\mathcal{V}^{\dagger}_0 \;,$$

but does not explain the extraordinary gap between charged and neutral leptons.

The right-handed neutrino masses are of Majorana type, since they have no gauge quantum numbers to forbid it (unlike electrons, say), and necessarily violate total lepton number.

In the context of effective field theories, one expects their masses to be of the order of lepton number breaking. Total lepton number-violating processes have never been seen resulting in a bound from neutrinoless double β decay experiments. So either they are very large or zero.

If they are zero, the analysis proceeds as in the quark sector, and the observable MNS lepton mixing matrix is just

$$\mathcal{U}_{MNS} \;\equiv\; \mathcal{U}^{\dagger}_{-1}\,\mathcal{U}_0 \;.$$

As for the quarks, it would be generated solely from the isospinor breaking of electroweak symmetry, even though the mixing patterns are so different.

In the belief that global symmetries are an endangered species (for one, black holes eat them up), we expect their masses to set the scale of the Standard model's cut-off, since they are unprotected by gauge symmetries. This yields the seesaw where large right-handed masses engender tiny neutrino masses, the latter being suppressed over that of the charged particles by the ratio of the two scales

$$\frac{\Delta I_{\mathrm{W}} = \frac{1}{2}}{\Delta I_{\mathrm{W}} = 0} \ ,$$

thus introducing a large electroweak-singlet scale in the Standard Model. The neutrino mass matrix is then

$$\mathcal{M}^{(0)}_{Seesaw} \ = \ \mathcal{M}^{(0)}_{Dirac} \ \frac{1}{\mathcal{M}^{(0)}_{Majorana}} \ \mathcal{M}^{(0)\,T}_{Dirac} \ ,$$

which we can rewrite as

$$\mathcal{M}^{(0)}_{Seesaw} \ = \ \mathcal{U}_0 \, \mathcal{C} \, \mathcal{U}_0^T \ ,$$

in terms of the central matrix[28]

$$\mathcal{C} \ = \ \mathcal{D}_0 \, \mathcal{V}_0^\dagger \ \frac{1}{\mathcal{M}^{(0)}_{Majorana}} \ \mathcal{V}_0^* \, \mathcal{D}_0 \ .$$

It is diagonalized by the unitary matrix $\mathcal{F}$

$$\mathcal{C} \ = \ \mathcal{F} \mathcal{D}_\nu \mathcal{F}^T \ ,$$

where the mass eigenstates produced in β-decay are (unimaginatively labelled as "1", "2", "3")

$$\mathcal{D}_\nu \ = \ \begin{pmatrix} m_{\nu_1} & 0 & 0 \\ 0 & m_{\nu_2} & 0 \\ 0 & 0 & m_{\nu_3} \end{pmatrix} \ .$$

The effect of the Seesaw is to add the unitary $\mathcal{F}$ matrix to the MNS lepton matrix

$$\mathcal{U}_{MNS} \ = \ \mathcal{U}_{-1}^\dagger \mathcal{U}_0 \, \mathcal{F} \ .$$

This framework enables us to recast theoretical questions in terms of $\mathcal{F}$. In particular, where do the large angles come from? We catalog models in terms of the number of large angles contained in $\mathcal{F}$, none, one or two?

10. A Modicum of Grand Unification

To answer that question, we must turn to Grand Unification ideas for guidance, where relations between the $\Delta I_{\mathrm{W}} = \frac{1}{2}$ quark and lepton Yukawa matrices appear naturally.

In $SU(5)$, the charge $-1/3$ and charge -1 Yukawa matrices are family-transposes of one another.

$$\mathcal{M}^{(-1/3)} \sim \mathcal{M}^{(-1)\,T} \, .$$

In $SO(10)$, it is the charge $2/3$ Yukawa matrix that is related to the Dirac charge 0 matrix

$$\mathcal{M}^{(2/3)} \sim \mathcal{M}^{(0)}_{Dirac} \, .$$

These result in naive expectations for the unitary matrices that yield observable mixings

$$\mathcal{U}_{-1/3} \sim \mathcal{V}^{*}_{-1} \, ; \qquad \mathcal{U}_{2/3} \sim \mathcal{U}_0 \, .$$

Assuming this pinch of grand-unification, we can relate the CKM and MNS matrices

$$
\begin{aligned}
\mathcal{U}_{MNS} &= \mathcal{U}^{\dagger}_{-1} \mathcal{U}_0 \, \mathcal{F} \\
&\sim \mathcal{U}^{\dagger}_{-1} \mathcal{U}_{-1/3} \mathcal{U}^{\dagger}_{CKM} \, \mathcal{F} \\
&\sim \left(\mathcal{V}^{T}_{-1/3} \mathcal{U}_{-1/3} \right) \mathcal{U}^{\dagger}_{CKM} \, \mathcal{F}
\end{aligned}
$$

Hence two wide classes of models:

I-) Family-Symmetric $\mathcal{M}_{-1/3}$ Yukawa matrices. In these we have

$$\mathcal{U}_{-1/3} = \mathcal{V}^{*}_{-1/3} \, ,$$

so that

$$\mathcal{U}_{MNS} = \mathcal{U}^{\dagger}_{CKM} \, \mathcal{F} \, .$$

In these models, $\mathcal{F}$ necessarily contains two large angles. In the absence of any symmetry acting on $\mathcal{F}$, these models require a highly structured $\mathcal{F}$ matrix, which could even be non-Abelian.

Interestingly, these models provide a testable prediction for the size of the CHOOZ angle. With a family-symmetric charge $-1/3$ matrix, the MNS matrix reads

$$
\mathcal{U}_{MNS} = \mathcal{U}_{CKM}^{\dagger} \times
$$

$$
\begin{pmatrix}
\cos\theta_{\odot} & \sin\theta_{\odot} & \lambda^{\gamma} \\
-\cos\theta_{\oplus}\sin\theta_{\odot} & \cos\theta_{\oplus}\cos\theta_{\odot} & \sin\theta_{\oplus} \\
\sin\theta_{\oplus}\sin\theta_{\odot} & -\sin\theta_{\oplus}\cos\theta_{\odot} & \cos\theta_{\oplus}
\end{pmatrix},
$$

where we have chosen to fill the zero in the $\mathcal{F}$ matrix by a Cabibbo effect, with γ presumably greater than one. It follows that

$$
\theta_{13} \sim \lambda\sin\theta_{\oplus} \sim \frac{1}{\sqrt{2}}\lambda \, .
$$

It will be interesting to see if this definite prediction of type I models, $\theta_{13} \sim 7-9°$, is borne out in future experiments.

II-) Family-Skewed $\mathcal{M}_{-1/3}$ Yukawa matrices. One can make a compelling arguments for at least one large angle to reside in $\mathcal{U}_{-1}$. If we extend the Wolfenstein[29] expansion of the CKM matrix in powers of the Cabibbo angle λ to include quark mass ratios

$$
\frac{m_s}{m_b} \sim \lambda^2 \qquad \frac{m_d}{m_b} \sim \lambda^4 \, ,
$$

we find the charge $-1/3$ Yukawa matrix

$$
\mathcal{M}^{(-1/3)} = \begin{pmatrix}
\lambda^4 & \lambda^3 & \lambda^3 \\
\lambda^? & \lambda^2 & \lambda^2 \\
\lambda^? & \lambda^? & 1
\end{pmatrix} \, .
$$

If the exponents are related to charges, as in the Froggatt-Nielsen[30] schemes, the lower diagonal exponents are known, and we get the orders of magnitude

$$\mathcal{M}^{(-1/3)} = \begin{pmatrix} \lambda^4 & \lambda^3 & \lambda^3 \\ \lambda^3 & \lambda^2 & \lambda^2 \\ \lambda^1 & 1 & 1 \end{pmatrix} ,$$

which is not family-symmetric. In the limit of no Cabibbo mixing,

$$\mathcal{M}^{(-1/3)} \approx \begin{pmatrix} 0 & 0 & 0 \\ 0 & 0 & 0 \\ 0 & a & b \end{pmatrix} + \mathcal{O}(\lambda) ,$$

and

$$\mathcal{U}_{MNS} = \begin{pmatrix} 1 & 0 & 0 \\ 0 & \cos\theta_\oplus & \sin\theta_\oplus \\ 0 & -\sin\theta_\oplus & \cos\theta_\oplus \end{pmatrix} \mathcal{F} ,$$

where

$$\tan\theta_\oplus = \frac{a}{b} ,$$

is of order one[31]. In these models, $\mathcal{F}$ need contain only one large angle, which is very natural, although they give no generic prediction for the CHOOZ angle.

11. Right-Handed Hierarchy

In most models, $\mathcal{F}$ must contain at least one large angle to accomodate the data. This presents a puzzle since $\mathcal{F}$ diagonalizes a matrix which contains the neutral Dirac Yukawa matrix which is presumably hierarchical, coming from the isospinor electroweak breaking. This suggests special restrictions put upon the Majorana mass matrix of the right-handed neutrinos. We want to illustrate this point by looking at a 2×2 two-families case[28], and write

$$\mathcal{D}_0 = m \begin{pmatrix} a\lambda^\beta & 0 \\ 0 & 1 \end{pmatrix} ,$$

and define M_1 , M_2 to be the eigenvalues of the right-handed neutrino's Majorana mass matrix. This matrix can be diagonalized by a large mixing angle in one of two cases:

– Its matrix elements have similar orders of magnitude $\mathcal{C}_{11} \sim \mathcal{C}_{22} \sim \mathcal{C}_{12}$, in which case we find that

$$\frac{M_1}{M_2} \sim \lambda^{2\beta} \,,$$

suggesting a doubly *correlated hierarchy* betwen the $\Delta I_{\mathrm{W}} = 0$ and $\Delta I_{\mathrm{W}} = \frac{1}{2}$ Sectors. This agrees well with grand-unified models such as $SO(10)$ and E_6, where each right-handed neutrinos is part of a family.

–A large mixing angle can occur if the diagonal elements are much smaller than the diagonal ones, that is $\mathcal{C}_{11}$, $\mathcal{C}_{22} \ll \mathcal{C}_{12}$. Then we find

$$\frac{\lambda^{\alpha} \, m^2}{\sqrt{-M_1 M_2}} \begin{pmatrix} 0 & a \\ a & 0 \end{pmatrix} .$$

Hence maximal mixing may infer that some of the right-handed neutrinos are Dirac partners of one another, leading to conservation in the right-handed mass matrix, of a relative lepton number $L_1 - L_2$.

12. Cabibbo Flop

As we have seen, Grand-Unification, even in its simplest form, implies Cabibbo-sized effects in the MNS matrix. In the quark sector, Cabibbo mixing is the strongest between the first and second family. Applied to the lepton sector, the solar angle may be maximal, with a Cabibbo correction of $13°$[32,33].

Recently, we[34] have been exploring possible Wolfenstein parametrizations of the MNS matrix, in the hope that some regularity might emerge from the data, once Cabibbo effects are taken into account.

Since we do not know how the Cabibbo angle is generated in flavor theories. So we start by asking if the limit $\theta_c \to 0$ makes any theoretical sense. To simplify matters, assume there is only one small parameter in the flavor sector; then the quark and charged lepton masses of the first two families are zero. In the same limit, $\mathcal{U}_{CKM} = 1$, and there are no neutral flavor changes. Of course the mixing between the first two families is undetermined.

We do not know $\mathcal{U}_{MNS}$ in that limit, the starting point of a Wolfenstein parametrization for the lepton mixing matrix.

The measured values of the lepton mixing angles are

$$\theta_{\odot} = 32.5^{\circ}\,{}^{+\,2.4^{\circ}}_{-\,2.3^{\circ}} \;\; ; \;\;\; \theta_{\oplus} = 45.00^{\circ}\,{}^{+10^{\circ}}_{-10^{\circ}} \;\; ; \;\;\; \theta_{CHOOZ} < 13^{\circ} \; .$$

The solar angle is well measured, but the atmospheric angle is not, and could very well be non-maximal. Furthermore, their values could be affected by Cabibbo flop of $\pm\,13^{\circ}$, and the CHOOZ angle could well be a Cabibbo effect. Our starting point is

$$\mathcal{U}_{MNS} = \begin{pmatrix} \cos\eta_{\odot} & \sin\eta_{\odot} & 0 \\ -\cos\eta_{\oplus}\sin\eta_{\odot} & \cos\eta_{\oplus}\cos\eta_{\odot} & \sin\eta_{\oplus} \\ \sin\eta_{\oplus}\sin\eta_{\odot} & -\sin\eta_{\oplus}\cos\eta_{\odot} & \cos\eta_{\oplus} \end{pmatrix} + \cdots \; ,$$

with a range of initial angles

$$15^{\circ} \; < \; \eta_{\odot} \; < \; 45^{\circ} \; ; \;\;\;\;\; 30^{\circ} \; < \; \eta_{\oplus} \; < \; 60^{\circ} \; .$$

We write the Wolfenstein expansion of the MNS Matrix in the form

$$\mathcal{U}_{MNS} \; \equiv \; \mathcal{W} \; + \; \mathcal{O}(\lambda) \; ,$$

where the starting matrix is split in two parts, showing the large angles

$$\mathcal{W} \; = \; \mathcal{W}_{\oplus}\,\mathcal{W}_{\odot} \; ,$$

with

$$\mathcal{W}_{\oplus} \; = \; \begin{pmatrix} 1 & 0 & 0 \\ 0 & \cos\eta_{\oplus} & -\sin\eta_{\oplus} \\ 0 & \sin\eta_{\oplus} & \cos\eta_{\oplus} \end{pmatrix} \; ,$$

$$\mathcal{W}_{\odot} \; = \; \begin{pmatrix} \cos\eta_{\odot} & \sin\eta_{\odot} & 0 \\ -\sin\eta_{\odot} & \cos\eta_{\odot} & 0 \\ 0 & 0 & 1 \end{pmatrix} \; .$$

We introduce Cabibbo flop through the unitary matrix

$$\mathcal{V} \; = \; I + \Delta(\lambda) \; ,$$

with $\Delta(0) \; = \; 1$. Unlike the quark sector it does not commute with the starting matrix

$$[\mathcal{W}, \mathcal{V}(\lambda)] \neq 0 \, .$$

This means that Cabibbo effects from the left and from the right or even in between the two starting matrices are not equivalent. Hence we consider basic flops

- Left $\quad \mathcal{U}_{MNS} = \mathcal{V}(\lambda)\,\mathcal{W}$
- Right $\quad \mathcal{U}_{MNS} = \mathcal{W}\,\mathcal{V}(\lambda)$
- Middle $\quad \mathcal{U}_{MNS} = \mathcal{W}_\oplus\mathcal{V}(\lambda)\,\mathcal{W}_\odot$

and we can have one $\mathcal{O}(\lambda)$ correction (single flop), or two (double flop). The present data is not sufficient to single out a particular Wolfenstein parametrization, but the hope is that by considering possible Cabibbo effects on various starting matrices, generic features suggestive of flavor patterns might become obvious. In particular, they would restrict the size of the CHOOZ angle and of the CP-violation.

To illustrate these points, consider the effect of flop matrices, shown here to $\mathcal{O}(\lambda^3)$,

$$\mathcal{V}_{12}(\lambda) \;=\; \begin{pmatrix} 1 - \frac{a^2}{2}\lambda^2 & a\,\lambda & b\,\lambda^2 \\ -a\,\lambda & 1 - \frac{a^2}{2}\lambda^2 & 0 \\ -b\,\lambda^2 & 0 & 1 \end{pmatrix}$$

$$\mathcal{V}_{23}(\lambda) \;=\; \begin{pmatrix} 1 & 0 & b\,\lambda^2 \\ 0 & 1 - \frac{a^2}{2}\lambda^2 & a\,\lambda \\ -b\,\lambda^2 & -a\,\lambda & 1 - \frac{a^2}{2}\lambda^2 \end{pmatrix}$$

$$\mathcal{V}_{\text{double}}(\lambda) \;=\; \begin{pmatrix} 1 - \frac{a^2}{2}\lambda^2 & a\,\lambda & (b + \frac{aa'}{2})\,\lambda^2 \\ -a\,\lambda & 1 - \frac{a^2+a'^2}{2}\lambda^2 & a'\,\lambda \\ (\frac{aa'}{2} - b)\,\lambda^2 & -a'\,\lambda & 1 - \frac{a'^2}{2}\lambda^2 \end{pmatrix} \, ,$$

where we have limited ourselves to $a = \pm 1$; $\quad a' = \pm 1$; $\quad 0.8 < b < 1.2$. For instance, a left single flop $\mathcal{V}_{23}$, yields values for the starting angles that are different from the data,

$\eta^\circ_\odot$	$\eta^\circ_\oplus$	$\theta^\circ_\odot$	$\theta^\circ_\oplus$	θ°_{13}
30	30	~ 31	43	$.06 - .4$
30	60	$31 - 32$	~ 47	$.6 - 2.5$

Right single flops with $\mathcal{V}_{23}$ and $\mathcal{V}_{12}$ produce:

$\eta^\circ_\odot$	$\eta^\circ_\oplus$	$\theta^\circ_\odot$	$\theta^\circ_\oplus$	θ°_{13}
30	60	$30 - 31$	$48 - 50$	$3 - 10$
15	45	~ 30.3	$44 - 45$	$2 - 4$
45	45	~ 32	$44 - 46$	$1 - 3$

A right double flop with $\mathcal{V}_{\text{double}}$:

$\eta^\circ_\odot$	$\eta^\circ_\oplus$	$\theta^\circ_\odot$	$\theta^\circ_\oplus$	θ°_{13}
15	60	~ 30.3	$45 - 51$	$1 - 6$
45	60	32	$48 - 53$	$6 - 12$

Finally a left double flop with $\mathcal{V}_{\text{double}}$:

$\eta^\circ_\odot$	$\eta^\circ_\oplus$	$\theta^\circ_\odot$	$\theta^\circ_\oplus$	θ°_{13}
45	30	~ 34	$40 - 46$	$5 - 8$

We see that double flops can produce a larger CHOOZ angle. Also a left flop from a family-symmetric Yukawa,

$$\mathcal{U}_{MNS} = \begin{pmatrix} 1 & \lambda & \lambda^3 \\ \lambda & 1 & \lambda^2 \\ \lambda^3 & \lambda^2 & 1 \end{pmatrix} \begin{pmatrix} \cos\eta_\odot & \sin\eta_\odot & 0 \\ -\cos\eta_\oplus \sin\eta_\odot & \cos\eta_\oplus \cos\eta_\odot & \sin\eta_\oplus \\ \sin\eta_\oplus \sin\eta_\odot & -\sin\eta_\oplus \cos\eta_\odot & \cos\eta_\oplus \end{pmatrix} ,$$

yields $\eta_\odot \sim 40° \; ; \; \eta_\oplus \sim 45° \; ; \; \theta_{13} \sim 0.7\lambda \sim 9°$, which we have already seen. Finally we note that CP-violation effects can be much larger than in the quark sector. This is because the CP-violating lepton invariant[35,36] is

$$J \sim (\lambda - \lambda^3) \sin\delta ,$$

to be compared with that in the quark sector which is of order λ^6. If the limit of zero Cabibbo mixing is meaningful for theory, analyses of the type we have just presented will assume some importance. One important remark emerges: precision measurements of the MNS matrix is quite important for theory.

13. Conclusions

We are beginning to read the new lepton data, but there is much work to do before a credible theory of flavor is proposed. The Seesaw Mechanism links static neutrino to physics that can never be reached by accelerators, creating a new era of the physics which centers around right-handed neutrinos. With no electroweak quantum numbers, they could hold the key to the flavor puzzles. The second large neutrino mixing angle suggests that hierarchy is independent of electroweak breaking, and occurs at grand-unified scales.

I would like to express my gratitude to P. Binétruy, M. Cribier, J. Orloff, S. Lavignac and D. Vignaud for organizing this conference *très sympathique* in the heart of Paris, at the Institut Henri Poincaré. I also wish to thank my collaborators A. Datta and L. Everett for many useful insights.

References

1. M. Gell-Mann, P. Ramond, and R. Slansky in Sanibel Talk, CALT-68-709, Feb 1979, hep-ph/9809459 (retroprint), and in *Supergravity* (North Holland, Amsterdam 1979). T. Yanagida, in *Proceedings of the Workshop on Unified Theory and Baryon Number of the Universe*, KEK, Japan, Feb 1979.
2. I became aware after this conference of a prescient paper by P. Minkowski, *Phys. Lett. B67, 421(1977)*, in which the seesaw matrix is proposed. It predates our contribution, but is presented in a limited context that does not establish the link to Planck scale physics, the heart of the seesaw mechanism as we know it.
3. Gerard 't Hooft and M.J.G. Veltman *Nucl.Phys.* B44:189-213,1972.
4. S.L. Glashow, *Nucl.Phys.*22:579-588,1961; Steven Weinberg, *Phys. Rev. Lett.*19:1264-1266,1967; Abdus Salam, "Svartholm: Elementary Particle Theory, Proceedings Of The Nobel Symposium" (1968), Lerum, Sweden.
5. H.David Politzer, *Phys.Rev.Lett.* 30:1346-1349,1973; D.J. Gross, Frank Wilczek *Phys.Rev.*D8:3633-3652,1973.
6. J. C. Pati and A. Salam, *Phys. Rev.* D10, 275, 1974.
7. H. Georgi and S.L. Glashow, *Phys.Rev.Lett.* 32:438-441,1974.
8. H. Fritzsch and P. Minkowski, *Annals Phys.* 93:193-266,1975 ; H. Georgi, Invited Talk at Williamsburg Conference, 1975.
9. F. Gürsey, P. Ramond and P. Sikivie, *Phys.Lett.* B60:177,1976.
10. Alan H. Guth, *Phys.Rev.* D 23:347-356,1981.
11. M. Yoshimura, *Phys. Rev. Lett.* 41:281-284,1978, Erratum-ibid.42:746,1979.
12. M. Fukugita, T. Yanagida, *Phys. Lett. B* 174:45,1986.
13. David J. Gross, Jeffrey A. Harvey, Emil J. Martinec, Ryan Rohm, *Phys.Rev.Lett.* 54:502-505,1985; P. Candelas, Gary T. Horowitz, Andrew Strominger, Edward Witten, *Nucl.Phys.* B258:46-74,1985.
14. P. Ramond *Phys. Rev. D3*:2415-2418,1971.
15. E. Gildener, S. Weinberg, *Phys.Rev.* D 13:3333,1976.
16. B. Pendleton, Graham G. Ross, *Phys.Lett. B* 98:291,1981.

17. Luis E. Ibañez, Graham G. Ross, *Phys.Lett.*B110:215-220,1982; Luis Alvarez-Gaumé, Mark Claudson, Mark B. Wise, *Nucl.Phys.*B207:96,1982.

18. Raymond Davis, Jr., Don S. Harmer, Kenneth C. Hoffman, *Phys.Rev.Lett.*20:1205-1209,1968.

19. W. Hampel et al., GALLEX collaboration, *Phys. Lett.* **B447** (1999) 127.

20. J.N. Abdurashitov et al., SAGE collaboration, *Phys. Rev.* **C60** (1999) 055801.

21. The Super-Kamiokande Collaboration, *Phys. Lett.* *B* **539** (2002) 179.

22. The SNO Collaboration, *Phys. Rev. Lett.* **89** (2002) 011301.

23. The KamLAND Collaboration, *Phys. Rev. Lett.* **90** (2003) 021802.

24. The Super-Kamiokande Collaboration, *Phys. Rev. Lett.* **85** (2000) 3999.

25. The K2K Collaboration, *Phys. Rev. Lett.* **90** (2003) 041801.

26. C.L. Bennett et al., *Astrophys.J.Suppl.*148:1,2003; D.N. Spergel et al., *Astrophys.J.Suppl.*148:175,2003.

27. M. Apollonio *et al.*, *Phys. Lett.* *B* **338** (1998) 383; *Phys. Lett.* *B* **420** (1998) 397.

28. Aseshkrishna Datta, Fu-Sin Ling, Pierre Ramond,*Nucl.Phys.* B 671:383-400,2003.

29. L. Wolfenstein, *Phys. Rev. Lett.* **51** (1983) 1945.

30. C. Froggatt and H.B. Nielsen, *Nucl. Phys.* *B* **147** (1979) 277.

31. N. Irges, S. Lavignac, P. Ramond, *Phys. Rev.* *D* 58:035003,1998.

32. P. Ramond, Invited talk at the Fujihara Seminar "Neutrino Mass and Seesaw Mechanism", KEK, Japan, February 2004.

33. Hisakazu Minakata and Alexei Yu. Smirnov, hep-ph/0405088. S. Petcov and A. Smirnov, *Phys. Lett.* *B* 322:109-118,1994

34. Aseshkrishna Datta, L. Everett, Pierre Ramond, in preparation.

35. C. Jarlskog, *Phys. Rev. Lett.* 55:1039,1985.

36. I. Dunietz, O.W. Greenberg, Dan-di Wu, *Phys. Rev. Lett.* 55:2935,1985.

LEPTOGENESIS IN SUSY THEORIES

TSUTOMU YANAGIDA[1,2]

[1] *Department of Physics, University of Tokyo,*
Tokyo 113-0033, Japan
[2] *Research Center for the Early Universe, University of Tokyo,*
Tokyo 113-0033, Japan

The seesaw mechanism is very attractive since it relates small neutrino masses to ultrahigh-energy physics beyond the standard model. This new physics predicts superheavy Majorana neutrinos, whose decays produce a lepton-number asymmetry in the universe (leptogenesis). The lepton-number asymmetry is converted into a baryon-number asymmetry by nonperturbative effects in the electroweak gauge theory. Thus, the seesaw mechanism connects small neutrino masses to the baryon-number asymmetry in the universe observed today. We show, in this talk, that a nonthermal leptogenesis via an inflaton decay is an interesting alternative to the thermal leptogenesis. In this scenario, the reheating temperature of inflation T_R can be as low as $T_R \simeq 10^{6-7}$ GeV, which may avoid an overproduction of gravitinos.

1. Introduction

Heavy Majorana neutrinos is a prediction of a large class of unified theories, which naturally generate small neutrino masses via the seesaw mechanism. Thus, very tiny neutrino masses is a window to a new physics at ultrahigh-energy scales beyond the standard model [1]. This is the prime reason why we are interested in neutrino masses and mixings. Experimental observation on the neutrino masses suggests masses of the heavy Majorana neutrinos N_i $(i = 1-3)$ to be in a range of $M_i \simeq 10^{10-15}$ GeV. It seems, unfortunately, infeasible to produce the heavy neutrinos directly in laboratory experiments in order to confirm the seesaw mechanism. Therefore, it is very important to search for other possible evidences of the superheavy Majorana neutrinos.

The early universe may be an interesting place to consider, since the heavy neutrinos N_i could be produced through scattering processes of the thermal particles in the early universe if the universe's temperature T is higher than their masses M_i. The heavy Majorana neutrinos start to decay when the temperature of the universe cools down to their masses. They

20

have two distinct decay channels; one is to a Higgs boson + a lepton and the other is to a Higgs boson + an antilepton. If CP is violated in the decay processes, the decay rates of each channels become different from each other, yielding the lepton-number asymmetry in the final state. This lepton-number asymmetry is converted to the baryon-number asymmetry [2] in the early universe through so called "shpaleron processes" [3]. This is called leptogenesis [2]. The universe's baryon asymmetry is, therefore, a quite natural product of the decays of superheavy Majorana neutrinos. Detailed calculations [4] show that the baryon-number asymmetry in the present universe can be naturally explained with the aid of neutrino masses suggested from the solar and atmospheric neutrino oscillation experiments. It is also shown that the thermal leptogenesis requires the temperature of the early universe to be $T \gtrsim 2 \times 10^9$ GeV.

Supersymmetry (SUSY) is a basic ingredient of unification of all interactions and all matters. Thus, the supergravity (SUGRA) is the framework for low-energy effective field theories. The seesaw mechanism is easily incorporated in this framework. However, there is a serious problem in the SUGRA, that is the gravitino problem. If the gravitino is unstable, it has a long lifetime and decays during or after the big-bang nucleosynthesis (BBN) for an interesting range of the gravitino mass, $m_{3/2} \simeq 100\text{GeV} - 10\text{TeV}$. The decay products destroy the light elements produced by the BBN and hence the relic abundance of the gravitino is constrained from above to keep the success of the BBN. This leads to the upper bound of the reheating temperature T_R after inflation, since the abundance of the gravitino is proportional to the reheating temperature. The recent detailed analyses derived a stringent upper bound $T_R \lesssim 10^{6-7}$ GeV when the gravitino decay has hadronic modes [5]. This upper bound is much lower than the temperature for the leptogenesis, $T_R \gtrsim 2 \times 10^9$ GeV. Therefore, the thermal leptogenesis seems hard to work in SUGRA for $m_{3/2} \simeq 100\text{GeV} - 10\text{TeV}$, unless some enhancement mechanism operates on the lepton-asymmetry production [6].

There have been proposed various solutions to the above problem [7,8,9,10,11]. We consider that the nonthermal leptogenesis via inflaton decay [7] is the most interesting. In the next section, we show that a general argument on the inflaton-decay scenario gives us a lower bound on a mass of the light neutrinos as $m_{\nu_3} \gtrsim 0.01$ eV. In section 3, we adopt a chaotic inflation model, since it is free from the initial condition problem [12] and also from so called η problem. Here, a shift symmetry [13] plays a crucial role on generating a flat potential required for the chaotic inflation in SUGRA.

In section 4, we discuss the nonthermal leptogenesis in the chaotic inflation model and show how naturally the universe's baryon asymmetry is explained in the present scenario. Here, the reheating temperature is estimated as low as 10^{6-7} GeV. The last section is devoted to conclusions and a brief review on possible solutions to the gravitino problem in the thermal leptogenesis.

2. Nonthermal leptogenesis from inflaton decay

Let us consider that the inflaton Φ decays dominantly into a pair of the lightest heavy Majorana neutrinos, that is $\Phi \to N_1 + N_1$, assuming the other decay modes including N_2 and N_3 are energetically forbidden, for simplicity. Then, the produced N_1's decay successively into $H + \ell$ or $H^* + \ell^*$. If the reheating temperature T_R is lower than the mass M_1 of the heavy neutrino N_1, the out-of-equilibrium condition [14] is automatically satisfied.

The above two channels of the N_1 decay have different branching ratios when CP conservation is violated. Interference between tree-level and one-loop diagrams generates lepton-number asymmetry [15] as,

$$\epsilon \equiv \frac{\Gamma(N_1 \to H + \ell) - \Gamma(N_1 \to H^* + \ell^*)}{\Gamma_{N_1}}$$

$$\simeq -\frac{3}{8\pi(hh^\dagger)_{11}}\left[\mathrm{Im}(hh^\dagger)_{13}^2\frac{M_1}{M_3} + \mathrm{Im}(hh^\dagger)_{12}^2\frac{M_1}{M_2}\right], \tag{1}$$

where h is a 3×3 matrix of Yukawa coupling of the Higgs H. The lepton asymmetry parameter ϵ is written as [16]

$$\epsilon = -\frac{3}{8\pi}\frac{M_1}{\langle H \rangle^2}m_{\nu_3}\delta_{\mathrm{eff}}. \tag{2}$$

Here, the effective CP-violating phase δ_{eff} is given by

$$\delta_{\mathrm{eff}} = \frac{\mathrm{Im}\left[h_{13}^2 + \frac{m_{\nu_2}}{m_{\nu_3}}h_{12}^2 + \frac{m_{\nu_1}}{m_{\nu_3}}h_{11}^2\right]}{|h_{13}|^2 + |h_{12}|^2 + |h_{11}|^2}, \tag{3}$$

and the seesaw mass formula $m_\nu \simeq m^2/M$ with $m = h\langle H \rangle$ [1,17] has been used. Then, we obtain the ϵ parameter as

$$\epsilon \simeq -2 \times 10^{-6}\left(\frac{M_1}{10^{10}\mathrm{GeV}}\right)\left(\frac{m_{\nu_3}}{0.05\mathrm{eV}}\right)\delta_{\mathrm{eff}}, \tag{4}$$

where we have taken $\langle H \rangle \simeq 174$ GeV.

The chain decays of $\Phi \to N_1 + N_1$ and $N_1 \to H + \ell$ or $H^* + \ell^*$ reheat the universe and produce not only the lepton-number asymmetry but also entropy of the thermal bath. Then, the ratio of the lepton number to entropy density after the reheating is estimated as [7]

$$\frac{n_L}{s} \simeq -\frac{3}{2}\epsilon\frac{T_R}{m_\Phi}$$

$$\simeq 3 \times 10^{-10} \left(\frac{T_R}{10^6 \mathrm{GeV}}\right) \left(\frac{M_1}{m_\Phi}\right) \left(\frac{m_{\nu_3}}{0.05\mathrm{eV}}\right), \qquad (5)$$

where m_Φ is the inflaton mass and we have taken $\delta_{\mathrm{eff}} = 1$. This lepton-number asymmetry is converted into the baryon-number asymmetry by the sphaleron effects and we obtain [18]

$$\frac{n_B}{s} \simeq -\frac{8}{23}\frac{n_L}{s}. \qquad (6)$$

From the observation [19]

$$\frac{n_B}{s} \simeq 0.9 \times 10^{-10}, \qquad (7)$$

we derive a constraint

$$m_{\nu_3} \gtrsim 0.01\mathrm{eV}. \qquad (8)$$

Here we have assumed $T_R \lesssim 10^7$ GeV to satisfy the cosmological constraint on the gravitino abundance [5] discussed in the introduction. It is very interesting that the neutrino mass suggested from the atmospheric neutrino oscillation experiments, $\sqrt{m_{23}^2} \simeq 0.05$ eV, just satisfies the above constraint.

3. A chaotic inflation model in SUGRA

The inflationary early universe is the most attractive hypothesis in modern cosmology, since it not only solves longstanding problems in cosmology, that is the horizon and flatness problems [20], but also accounts for the origin of density fluctuations [21]. Among various types of inflation models proposed so far, chaotic inflation model is the most attractive since it can realize an inflationary expansion of the universe even at the Planck time and hence it is free from the initial condition problem [12]. However, the chaotic inflation is not easily realized in SUGRA [22]. The reason is that the minimal supergravity potential has an exponential factor, $\exp(\phi^*\phi/M_G^2)$, which prevents any scalar field ϕ from having a value larger than the reduced Planck scale $M_G \simeq 2.4 \times 10^{18}$ GeV. However, the inflaton φ should have an initial value much larger than M_G at the Planck time to cause the chaotic inflation.

The above problem can be solved by making a use of a shift symmetry of the inflaton chiral multiplet $\Phi(x,\theta)$. We assume [13] that the theory is invariant under a shift of Φ;

$$\Phi \to \Phi + iCM_G, \tag{9}$$

where C is a real constant parameter. Hereafter, we take the unit where $M_G = 1$. The Kähler potential is, thus, a function of $\Phi + \Phi^*$, that is $K(\Phi, \Phi^*) = K(\Phi + \Phi^*)$. Now it is clear that the exponential factor $\exp(K(\Phi + \Phi^*))$ discussed above does not contain the imaginary part of the scalar components of Φ, that is called as φ. Thus, the φ may have an initial value much larger than the unity and plays a role of the inflaton.

As long as the shift symmetry is exact, however, the inflaton φ never has a potential and the inflation is never generated. Therefore, we must introduce a breaking of the shift symmetry. For this purpose we introduce a spurion chiral field Ξ and extend the shift symmetry as [23]

$$\Phi \to \Phi + iC,$$
$$\Xi \to \frac{\Phi}{\Phi + iC}\Xi. \tag{10}$$

Notice that the product of $\Xi\Phi$ is invariant of the shift symmetry and the vacuum value $\langle\Xi\rangle = m \ll 1$ represents the breaking of the symmetry. In addition to the shift symmetry we impose R symmetry to suppress the constant term in the superpotential. This compels us to introduce another chiral multiplet $X(x,\theta)$ with the R charge 2 to get a superpotential. Then, we have a superpotential invariant under the shift and R symmetries[a],

$$W = X\left(c_1\Xi\Phi + c_2(\Xi\Phi)^2 + c_3(\Xi\Phi)^3 + \cdots\right), \tag{11}$$

where we have considered that Ξ and Φ carry vanishing R charges and the coefficients c_i are constants of order 1. We see that the first term in the superpotential dominates unless $\langle\Xi\rangle\Phi = m\Phi \geq 1$. Thus, we assume the dominance of the first term in the following analysis, that is[b] ,

$$W \simeq mX\Phi. \tag{12}$$

The Kähler potential is given by[c]

$$K = \frac{1}{2}(\Phi + \Phi^*)^2 + XX^* + \cdots. \tag{13}$$

---

[a]Absence of other possible terms such as $W \propto X$ is explained below.
[b]We take $c_1 = 1$, for simplicity.
[c]The term $(\Phi + \Phi^*)$ in the Kähler potential does not affect the inflation dynamics [23].

Then, the potential for the inflaton φ is given by

$$V(\varphi) \simeq \frac{1}{2}m^2\varphi^2. \tag{14}$$

This is nothing but the inflaton potential of the chaotic inflation model and the normalization of the density fluctuations at the COBE scale gives

$$m \simeq 10^{13}\text{GeV} \simeq 10^{-5}. \tag{15}$$

We should note here that we have discarded a linear term of X in the superpotential, since it disturbs the inflation dynamics. However, it is easy to forbid such an unwanted term in the superpotential by imposing an extra U(1) symmetry, in which X and Ξ have the U(1) charges $-n$ and $+n$. We assume this U(1) symmetry is broken by $\langle\Theta\rangle = \vartheta$ where the spurion field Θ carries the U(1) charge -1. Then, we see that the linear term of X in the superpotential is completely forbidden for $n > 0$, while the term $X\Xi\Phi$ is allowed.

4. Leptogenesis in the chaotic inflation model

Let us consider the inflaton decay. The inflaton chiral multiplet Φ may couple to a pair of the heavy Majorana neutrinos N_i, since the Φ is a gauge singlet and has a vanishing R charge[d]. The superpotential contributes to the inflaton φ decay is given by,

$$\begin{aligned}
W &= \lambda_i\langle\Theta\rangle^{n+2\gamma_i}\langle\Xi\rangle\Phi N_i N_i, \\
&= \lambda_i\vartheta^{n+2\gamma_i}m\Phi N_i N_i, \tag{16}
\end{aligned}$$

where λ_i are constants of order 1 and we have assumed the U(1) charge for N_i to be $+\gamma_i$. If $M_i < m_\Phi/2$ the inflaton φ can decay into the N_i pair and the decay rate is given by

$$\Gamma_\varphi \simeq (\lambda\vartheta^{n+2\gamma_i})^2\frac{m^3}{32\pi} \simeq 10\vartheta^{2(n+2\gamma_i)}\text{GeV}, \tag{17}$$

for $m \simeq 10^{13}$ GeV. Then, the reheating temperature is estimated as

$$T_R \simeq \lambda_i\vartheta^{n+2\gamma_i} \times 10^9\text{GeV}. \tag{18}$$

---

[d]The coupling to a pair of Higgs multiplets is forbidden by the R symmetry or a discrete $B - L$ symmetry [23].

For a given reheating temperature we estimate the baryon-number asymmetry produced via the inflaton decay as

$$\frac{n_B}{s} \simeq 1 \times 10^{-10} \left(\frac{T_R}{10^6\,\mathrm{GeV}}\right) \left(\frac{M_1}{m_\Phi}\right) \left(\frac{m_{\nu_3}}{0.05\mathrm{eV}}\right),$$

$$\simeq 1 \times 10^{-7} \vartheta^{n+2\gamma_i} \left(\frac{M_i}{m_\Phi}\right) \left(\frac{m_{\nu_3}}{0.05\mathrm{eV}}\right), \tag{19}$$

for $\lambda_i = 1$.

It is an intriguing possibility to identify the above extra U(1) symmetry with the Froggatt and Nielsen U(1) [24]. If it is the case, the mass of the heavy Majorana neutrinos, M_i, is given by

$$M_i \simeq \vartheta^{2\gamma_i} M_0, \tag{20}$$

where[e]

$$M_0 \simeq \vartheta^{2\beta_i} \times 10^{15}\,\mathrm{GeV}. \tag{21}$$

Here, β_i is the U(1) charge for lepton-doublet chiral multiplets ℓ_i, and we have assumed an hierarchy $m_{\nu_3} > m_{\nu_{1,2}}$ and used $m_{\nu_3} \simeq 0.05$ eV. It is clear that the dominant contribution to n_B/s comes from the decay of N_i whose mass is near the inflaton mass $\lesssim m_\Phi/2 \simeq 10^{13}$ GeV. The phenomenological analyses give us $\vartheta \simeq 1/17$ [25]. Thus, we consider the case of $\gamma_i + \beta_i = 1$ which provides $M_i \simeq \vartheta^2 \times 10^{15}\mathrm{GeV} \simeq 3 \times 10^{12}$ GeV. Then, we find

$$\frac{n_B}{s} \simeq \frac{1}{3} \times 10^{-7} \vartheta^{n+2-2\beta_i}. \tag{22}$$

The observation [19] $n_B/s \simeq 0.9 \times 10^{-10}$ suggests the exponent $n+2-2\beta_i = 2$. For $n = 1$ we get $\beta_i = 1/2$ leading to a prediction, $\tan\beta \equiv \langle H \rangle/\langle \bar{H} \rangle \simeq 10 - 15$. Notice, however, that the prediction on $\tan\beta$ depends on the unknown Froggatt and Nielsen charge n.

5. Conclusions

The seesaw mechanism relates small neutrino masses to ultrahigh-energy new physics [1]. Thus, the small neutrino masses and mixing angles are a window to the new physics beyond the standard model. The superheavy Majorana neutrinos is a prediction of the seesaw mechanism. In this talk, we have shown that the decays of the superheavy Majorana neutrinos N_i produce the lepton-number asymmetry, which is converted into the baryon-number symmetry in our universe (leptogenesis).

---

[e] M_0 represents the discrete $B - L$ breaking scale.

The thermal leptogenesis is very attractive, since we do not need any extra assumption except for the presence of the superheavy Majorana neutrinos and the reheating temperature $T_R \gtrsim 2 \times 10^9$ GeV. This required reheating temperature, however, causes a serious cosmological problem in SUGRA, namely too many gravitinos are produced to keep the success of the BBN with such high reheating temperatures, if the gravitino has a mass $m_{3/2} \simeq 100\text{GeV} - 10\text{TeV}$ and it is unstable. This problem forces us to consider nonthermal leptogeneses. We have found that a nonthermal leptogenesis via an inflaton decay is an interesting alternative to the thermal leptogenesis and it works very well to explain the baryon-number asymmetry $n_B/s \simeq 0.9 \times 10^{-10}$ in the present universe.

However, in this last section, we briefly review on attempts to solve the gravitino problem in the thermal leptogenesis.

The first one is the proposal by Pilaftsis who considers a quasi-degenerate heavy Majorana neutrinos $(M_1 \simeq M_2)$ [6]. In this model the lepton-asymmetry parameter ϵ is enhanced by a factor of $M_1/(M_1 - M_2)$ and hence the decays of both N_1 and N_2 may produce enough asymmetry even for $T_R \lesssim 10^{6-7}$ GeV.

The second is the proposal by Bolz, Buchmuller and Plumacher [26] who consider the case where the gravitino is the stable lightest SUSY particle (LSP). In this case the next LSP is the subject of the cosmological constraint, since it's decay products may destroy the light elements created by the BBN as like the unstable gravitino. The detailed analyses show that this scenario survives marginally only for a small parameter region of $m_{3/2} \simeq 10 - 100\text{GeV}$ [27]. It should be noted here that the next LSP decay to the gravitino may provide us with an independent measurement of the Planck scale [28].

The third solution is given by gauge-mediation model in which the gravitino is the stable LSP of mass $m_{3/2} \lesssim 1$ GeV. If the gravitino mass is $m_{3/2} \simeq 1 - 30\text{eV}$, we have no gravitino problem. The dark matter may be the axion. For $m_{3/2} \simeq 100\text{keV} - 1\text{GeV}$, there is an interesting possibility that a late-time entropy production in a class of gauge mediation models may render naturally the gravitino to be a dominant component of the dark matter [29]. In this scenario the reheating temperature may be as high as $T_R \simeq 10^{13}$ GeV.

The last solution is to assume an anomaly mediation with the gravitino mass $\gtrsim 100$ TeV. In this case the gravitino decays before the BBN and hence there is no cosmological problem. However, the gravitino decay mode contains always one LSP and hence the relic abundance of the gravitino must be constrained from above so that the density of the nonthermal LSP produced by the gravitino decay do not exceed the dark-matter density. This condition leads to $T_R \lesssim 10^{11}$ GeV [30] which is well consistent with the thermal leptogenesis.

We stress here that the above each solutions predict distinct particle spectra at TeV scale, which may be testable in future collider experiments such as LHC. If all of them are excluded, we will consider the nonthermal leptogenesis seriously. On the other hand, if one of them is confirmed the thermal leptogenesis will become more convincing.

Acknowledgements

The author is grateful to the organizers of *SEESAW25* in Paris for the hospitality during the stay.

References

1. T. Yanagida, *in Proc. of the Workshop on "the Unified Theory and the Baryon Number in the Universe"*, Tsukuba, Japan, Feb. 13-14, 1979, eds. O. Sawada and S. Sugamoto, (KEK Report KEK-79-18, 1979, Tsukuba) p.95; Progr. Theor. Phys. **64** (1980) 1103; P. Ramond, *in a Talk given at Sanibel Symposium*, Palm Coast, Fla., Feb. 25-Mar. 2, 1979, preprint CALT-68-709. See also S. Glashow, *in Proc. of the Cargèse Summer Institute on "Quarks and Leptons"*, Cargèse, July 9-29, 1979, eds. M. Lévy et. al, (Plenum, 1980, New York), p.707.
2. M. Fukugita and T. Yanagida, Phys. Lett. B **174**, 45 (1986).
3. V. A. Kuzmin, V. A. Rubakov and M. E. Shaposhnikov, Phys. Lett. **155**, 36 (1985).
4. W. Buchmuller, P. Di Bari and M. Plumacher, hep-ph/0302092; hep-ph/0406014; G. F. Giudice, A. Notari, M. Raidal, A. Riotto and A. Strumia, hep-ph/0310123.
5. M. Kawasaki, K. Kohri and T. Moroi, astro-ph/0408426.
6. A. Pilafsis, Phys. Rev. D**56**, 5431 (1997); J. Ellis, M. Raidal and T. Yanagida, Phys. Lett. B**546**, 228 (2002).
7. K. Kumekawa, T. Moroi and T. Yanagida, Progr. Theor. Phys. **92**, 437 (1994); G. Lazarides, hep-ph/9904428 and references therein; G. F. Giudice, M. Peloso, A. Riotto and I. Tkachev, J. High Energy Phys. **08**, 014 (1999); T. Asaka, K. Hamaguchi, M. Kawasaki and T. Yanagida, Phys. Lett. B**464**, 12 (1999); Phys. Rev. D**61**, 083512 (2000).
8. H. Murayama, H. Suzuki, T. Yanagida and J. Yokoyama, Phys. Rev. Lett. **70**, 1912 (1993); J. R. Ellis, M. Raidal and T. Yanagida, Phys. Lett. B**581**, 9 (2004).
9. H. Murayama and T. Yanagida, Phys. Lett. B**322**, 349 (1994); K. Hamaguchi, H. Murayama and T. Yanagida, Phys. Rev. D **65**, 043512 (2002).

10. H. Murayama and T. Yanagida, in ref. [9]; M. Dine, L. Randall and S. Thomas, Nucl.Phys. B**458**, 291 (1996); T. Asaka, M. Fujii, K. Hamaguchi and T. Yanagida, Phys. Rev. D**62**, 123514 (2000).

11. Y. Grossman, T. Kashti, Y. Nir and E. Roulet, Phys. Rev. Lett. **91**, 251801 (2003); G. D'Ambrosio, G. F. Giudice and M. Raidal, Phys. Lett. B**575**, 75 (2003).

12. A. Linde, Phys. Lett. **129**B, 177 (1983); See also A. Linde, *in a Talk given at Nobel Symposium 2003* "Cosmology and String Theory", Sigtunastiftelsen, Sweden; hep-th/0402051.

13. M. Kawasaki, M. Yamaguchi and T. Yanagida, Phys. Rev. Lett. **85**, 3572 (2000).

14. A. D. Sakharov, ZhETF Pis'ma **5**, 32 (1967).

15. M. Fukugita and T. Yanagida, in ref. [2]; M. Flanz, E. A. Paschos and U. Sakar, Phys. Lett. B **345**, 248 (1995); L. Covi, E. Roulet and F. Vissani, Phys. Lett. B **384**, 169 (1996); W. Buchmuller and M. Plumacher, Phys. Lett. B **431**, 354 (1998).

16. K. Hamaguchi, H. Murayama and T. Yanagida, in ref. [9]; S. Davidson and A. Ibarra, Phys. Lett. B**535**,25 (2002).

17. After this conference, I became aware of a paper by P. Minkowski, Phys. Letter **67**B, 421 (1977), in which the seesaw matrix is discussed in a calculation of $\mu \to e + \gamma$ decay amplitude. I thank Paul Frampton for sending an information on it to me.

18. J. A. Harvey and M. S. Turner, Phys. Rev. D**42**, 3344 (1990).

19. D. N. Spergel et al. Astrophy. J. Suppl. **148**, 175 (2003).

20. A. H. Guth, Phys. Rev. D**23**, 347 (1981).

21. A. H. Guth and So-Y. Pi, Phys. Rev. Lett. **49**, 1110 (1982); S. Hawking, Phys. Lett. **115**B, 295 (1982); A. A. Starobinsky, Phys. Lett. **117**B, 175 (1982).

22. A. S. Goncharov and A. D. Linde, Phys. Lett. **139**B, 27 (1984); H. Murayama, H. Suzuki, T. Yanagida and J. Yokoyama, Phys. Rev. D**50**, R2356 (1994).

23. M. Kawasaki, M. Yamaguchi and T. Yanagida, Phys. Rev. D**63**, 103514 (2001).

24. C. D. Froggatt and H. B. Nielsen, Nucl. Phys. B**147**, 277 (1979).

25. J. Sato and T. Yanagida, Phys. Lett. B**430**, 127 (1998); W. Buchmuller and T. Yanagida, Phys. Lett. B**445**, 399 (1999).

26. M. Bolz, W. Buchmuller and M. Plumacher, Phys. Lett. B**443**, 209 (1998).

27. M. Fujii, M. Ibe and T. Yanagida, Phys. Lett. B**579**, 6 (2004) [hep-ph/0310142]; J. R. Ellis, K. A. Olive, Y. Santoso and V. C. Spanos, Phys.Lett. B**588** , 7 (2004) [hep-ph/0312262]; J. L. Feng, S. Su and F. Takayama, hep-ph/0404198 ; L. Roszkowski and R. R. de Austri, hep-ph/0408227.

28. W. Buchmuller, K. Hamaguchi, M. Ratz and T. Yanagida, Phys. Lett. B**588**, 90 (2004) [hep-ph/0402179].

29. M. Fujii, T. Yanagida, Phys.Lett. B**549**,273 (2002) [hep-ph/0208191]; M. Fujii, M. Ibe and T. Yanagida, Phys.Lett. B**579**, 6 (2004) [hep-ph/0310142].

30. See, for an explicit model, M. Ibe, R. Kitano, H. Murayama and T. Yanagida, hep-ph/0403198.

SEESAW MECHANISM AND ITS IMPLICATIONS

R.N. MOHAPATRA

Department of Physics, University of Maryland, College Park, MD-20742, USA

The seesaw mechanism is introduced and some of its different realizations and applications are discussed. It is pointed out how they can be used to understand the bi-large mixing patterns among neutrinos in combination with the assumptions about high scale physics such as grand unification or quasi-degeneracy.

1. Introduction

The discovery of neutrino masses and mixings has been an important milestone in the history of particle physics and rightly qualifies as the first evidence for new physics beyond the standard model. The amount of new information on neutrinos already established from various neutrino oscillation searches has provided very strong clues to new symmetries of particles and new directions for unification. Enough puzzles have emerged making this field a hotbed for theory research with implications ranging all the way from supersymmetry and grand unification to cosmology and astrophysics.

A major cornerstone for the theory research in this field has been the seesaw mechanism introduced 25 years ago in four independently written papers[1] to understand why neutrino masses are so much smaller than the masses of other fermions of the standard model. Even though there was no solid evidence for neutrino masses then, there were very well motivated extensions of the standard models that led to nonzero masses for neutrinos. It was therefore incumbent on those models that they have a mechanism for understanding why upper limits on neutrino masses known at that time were so small and the seesaw mechanism was introduced in the context of specific such models in the year 1979 e.g. horizontal, left-right and

SO(10) models to achieve this goal. A general operator description of small neutrino mass without any specific model was written down the same year[2]. A very minimal nonsupersymmetric SO(10) model was constructed soon after as an application[3]. It was clear from this early enthusiasm about the idea that if the experimental evidence for neutrino masses ever appeared then, seesaw mechanism would be a major tool in understanding its various ramifications. As we see below, this has indeed turned out to be the case.

2. Seesaw mechanism

To appreciate the simplicity and beauty of the seesaw mechanism, let us start with a discussion of neutrino mass in the standard model. It is based on the gauge group $SU(3)_c \times SU(2)_L \times U(1)_Y$ group under which the quarks and leptons transform as follows: Quarks $\mathbf{Q}_L^T \equiv (u_L, d_L)(3, 2, \frac{1}{3})$; $\mathbf{u}_R(3, 1, \frac{4}{3})$; $\mathbf{d}_R(3, 1, -\frac{2}{3})$; leptons $\mathbf{L}^T \equiv (\nu_L, e_L)(1, 2, -1)$; $\mathbf{e}_R(1, 1, -2)$; Higgs Boson $\mathbf{H}(1, 2, +1)$; Color Gauge Fields $\mathbf{G}_a(8, 1, 0)$; Weak Gauge Fields $\mathbf{W}^\pm, \mathbf{Z}, \gamma(1, 3 + 1, 0)$.

The electroweak symmetry $SU(2)_L \times U(1)_Y$ is broken by the vacuum expectation of the Higgs doublet $< H^0 > = v_{wk} \simeq 246$ GeV, which gives mass to the gauge bosons and the fermions, all fermions except the neutrino. The model had been a complete success in describing all known low energy phenomena, until the evidence for neutrino masses appeared.

Note that there is no right handed neutrino in the standard model and this directly leads to the fact that neutrinos are massless at the tree level. This result holds not only to all orders in perturbation theory but also when nonperturbative effects are taken into account due to the existence of an exact B-L symmetry of the standard model. It would therefore appear that nonzero neutrino mass ought to be connected to breaking of B-L symmetry.

A simple way to generate neutrino masses is to introduce right handed neutrinos N_R, one per family into the standard model. The standard model Lagrangian now allows for a new Yukawa coupling of the form $h_\nu \bar{L} H N_R$ which after electroweak symmetry breaking leads to a neutrino mass $\sim h_\nu v_{wk}$. Since h_ν is expected to be of same order as the charged fermion couplings in the model, this mass is much too large to describe neutrino oscillations. Luckily, since the N_R's are singlets under the standard model gauge group, they are allowed to have Majorana masses unlike the charged fermions. We denote them by $M_R N_R^T C^{-1} N_R$ (where C is the Dirac charge conjugation matrix). The masses M_R are not constrained by the gauge symmetry and can therefore be arbitrarily large (i.e. $M_R \gg h_\nu v_{wk}$). This together with mass induced by Yukawa couplings (called the Dirac mass) leads to a the mass matrix for the neutrinos (left and right handed neutrinos together) which has the form

$$M_\nu = \begin{pmatrix} 0 & M_D \\ M_D^T & M_R \end{pmatrix} \tag{1}$$

where M_D and M_R are 3×3 matrices. Diagonalizing this mass matrix, one gets the mass matrix for the light neutrino masses to be as follows:

$$M_\nu = -M_D^T M_R^{-1} M_D \tag{2}$$

Since as already noted M_R can be much larger than M_D which is likely to be of order $= h_\nu v_{wk}$, one finds that $m_\nu \ll m_{e,u,d}$ very naturally. This is known as the seesaw mechanism[1] and it provides a natural explanation of why neutrino masses are small.

Seesaw mechanism of course raises its own questions:

- what is the scale of M_R and what determines it ?
- Is there a natural reason for the existence of the right handed neutrinos ?
- Is the seesaw mechanism by itself enough to explain all aspects of neutrino masses and mixings.

Below, we try to answer some of these questions and discuss how far one can go towards explaining observations.

3. Why seesaw mechanism is so appealing ?

Even though the standard model has enjoyed incredible success in explaining all low energy observations, it has long been recognized that it cannot be a complete theory due to many issues it leaves unaddressed, e.g. the gauge hierarchy problem, strong CP problem as well as the fermion mass and mixing problem etc. Besides, it also has some aesthetic inadequacies that cry out for new physics. In the latter category are such questions as (i) an omission of the right handed neutrino which simple quark lepton symmetry observed in weak interaction would have demanded; (ii) a somewhat adhoc definition of the electric charge in terms of an U(1) charge called "weak hypercharge Y" i.e. $Q = I_{3L} + \frac{Y}{2}$ and (iii) the origin of parity violation. It turns out that inclusion of the three right handed neutrinos helps to remove the aesthetic cloud that hangs over the standard model while maintaining all its phenomenological successes.

The fact that the addition of one right handed neutrino per generation to the standard model restores quark lepton symmetry is obvious. But in 1973, Pati and Salam pointed out that there is a compelling symmetry reason for including the right handed neutrino if one considers leptons as a fourth color in Nature[4], the other three being associated with quarks. In fact in the presence of the N_R's, the minimal anomaly free gauge group of weak interactions expands beyond the standard model and becomes the left-right symmetric group $SU(2)_L \times SU(2)_R \times U(1)_{B-L}{}^5$ which is a subgroup of the $SU(2)_L \times SU(2)_R \times SU(4)_c$ group introduced by Pati and Salam. We will see below that both these gauge groups lead to a picture of weak interaction which is fundamentally different from that envisaged in the standard model in that *weak interactions like the strong and electromagnetic ones are parity conserving at very high energies and observed maximally parity violating V-A structure of low energy weak processes is a consequence of gauge symmetry breaking.*

To see this explicitly, we consider the left-right symmetric theory of weak interactions under which fermions and Higgs bosons transform as follows:

$$\mathbf{Q}_L \equiv \begin{pmatrix} u_L \\ d_L \end{pmatrix} (2,1,+\tfrac{1}{3}); \quad \mathbf{Q}_R \equiv \begin{pmatrix} u_R \\ d_R \end{pmatrix} (1,2,\tfrac{1}{3}); \quad \mathbf{L}_L \equiv \begin{pmatrix} \nu_L \\ e_L \end{pmatrix} (2,1,-1);$$

$\mathbf{L}_R \equiv \begin{pmatrix} \nu_R \\ e_R \end{pmatrix} (1,2,-1)$. The Higgs fields transform as ϕ (2,2,0) ; $\mathbf{\Delta}_L$ (3,1,+ 2); $\mathbf{\Delta}_R$ (1,3,+ 2).

It is clear that this theory leads to a weak interaction Lagrangian of the form

$$L_{wk} = \frac{g}{2} \left(\vec{j}_L^{\mu} \cdot \vec{W}_{L,\mu} + \vec{j}_R^{\mu} \cdot \vec{W}_{R,\mu} \right) \tag{3}$$

which is parity conserving prior to symmetry breaking. Furthermore, in this theory, the elctric charge formula is given by[6]:

$$Q = I_{3L} + I_{3R} + \frac{B-L}{2}, \tag{4}$$

where each term has a physical meaning unlike the case of the standard model. When only the gauge symmetry $SU(2)_R \times U(1)_{B-L}$ is broken down, one finds the relation $\Delta I_{3R} = -\Delta\left(\frac{B-L}{2}\right)$. This connects $B-L$ breaking i.e. $\Delta(B-L) \neq 0$ to the breakdown of parity symmetry i.e. $\Delta I_{3R} \neq 0$. It also reveals the true meaning of the standard model hypercharge as $\frac{Y}{2} = I_{3,R} + \frac{B-L}{2}$.

To discuss the implications of these observations, note that in stage I, the gauge symmetry is broken by the Higgs multiplets $\Delta_L(3,1,2) \oplus \Delta_R(1,3,2)$ to the standard model and in stage II by the bidoublet $\phi(2,2,0)$ as in the standard model. In the first stage, the right handed neutrino picks up a mass of order $f < \Delta_R^0 > \equiv f v_R$. Denoting the left and right handed neutrino by (ν, N) (in a two component notation), the mass matrix for neutrinos at this stage looks like

$$M_\nu^0 = \begin{pmatrix} 0 & 0 \\ 0 & f v_R \end{pmatrix} \tag{5}$$

At this stage, familiar standard model particles are all massless. As soon as the standard model symmetry is broken by the bidoublet ϕ i.e. $< \phi > \equiv diag(\kappa, \kappa')$, the W and Z boson as well as the fermions pick up mass. I will generically denote κ, κ' by a common symbol v_{wk}. The contribution to neutrino mass at this stage look like

$$M_\nu^0 = \begin{pmatrix} f v_L & h v_{wk} \\ h v_{wk} & f v_R \end{pmatrix} \tag{6}$$

where $v_L = \frac{v_{wk}^2}{v_R}$. The appearance of the $f v_L$ term is a reflection of parity invariance of the model. Note that except for the $\nu\nu$ entry, the neutrino mass matrix in Eq.(6) is exactly in the same form as in Eq. (1). Diagonalizing this matrix, we get a modified seesaw formula for the light neutrino mass matrix

$$M_\nu = f v_L - h_\nu^T f_R^{-1} h_\nu \left(\frac{v_{wk}^2}{v_R} \right) \tag{7}$$

The important point to note is that v_L is suppressed by the same factor as the second term in Eq. (7) so that despite the new contribution to neutrino masses, seesaw suppression remains[7]. This is called the type II seesaw whereas the formula in Eq. (1) is called type I seesaw formula[a].

An important physical meaning of the seesaw formula is brought out when it is viewed in the context of left-right models. Note that $m_\nu \to 0$ when v_R goes to infinity. In the same limit the weak interactions become pure V-A type. Therefore, left-right model derivation of the seesaw formula smoothly connects smallness of neutrino mass with suppression of V+A part of the weak interactions providing an important clarification of a major puzzle of the standard model i.e. why are weak interactions are near maximally parity violating ? The answer is that they are near maximally parity violating because the neutrino mass happens to be small. This point was emphasized in the fourth paper in Ref. [1].

---

[a]The triplet contribution to neutrino masses without the seesaw suppression was considered in [8] and triplet contribution by itself outside the framework of parity symmetric models have been considered in [9].

In a subsequent section, we will discuss the connection of the seesaw mass scale with the scale of grand unification, which is suggested by the value of the atmospheric Δm_A^2. SO(10) is the simplest gauge group that contains the right handed neutrino needed to implement the seesaw mechanism and also it is important to note that the left-right symmetric gauge group is a subgroup of the SO(10) group, which therefore provides an attractive over all grand unified framework for the discussion of neutrino masses. The extra bonus one may expect is that since bigger symmetries tend to relate different parameters of a theory, one may be able to predict neutrino masses and mixings. We will present a model where indeed this happens.

We also note that since type I seesaw involves the Dirac mass of the neutrino, which is likely to scale with generation the same way as the charged fermions of the standard model, unless there is extreme hierarchy among the right handed neutrinos, one would expect the ν spectrum to be hierarchical. On the other hand, it has been realized for a long time[11] that if neutrino masses are quasi-degenerate, it is a tell-tale sign of type II seesaw with the triplet vev term being the dominant one. However, a normal hierarchy can also arise with type II seesaw as we discuss in the example below.

4. Seesaw and large neutrino mixings

While seesaw mechanism provides a simple framework for understanding the smallness of neutrino masses, it does not throw any light on the question of why neutrino mixings are large. The point is that mixings are a consequence of the structure of the light neutrino mass matrix and the seesaw mechanism is only statement about the scale of new physics. This can also be understood by doing a simple parameter counting. If we work in a basis where the right handed neutrino masses are diagonal, there are 18 parameters describing the seesaw formula for neutrino masses - three RH neutrino masses and 15 parameters in the Dirac mass matrix. On the other hand, there are only nine observables (three masses, three mixing angle snd three phases) describing low energy neutrino sector. Thus there are twice as many parameters as observables. As a result, the neutrino mass matrix needs inputs beyond the simple seesaw mechanism to fix the neutrino mass matrix.

In order to understand large mixings, one has to go beyond the simple seesaw mechanism to particular models. This is any way necessary to limit the scale of the right handed neutrino far below the Planck scale as seems to be the case. Many such models have been considered that use horizontal symmetry, grand unification, discrete symmetries, assumption of single right handed neutrino dominance etc.[10] to derive large mixings. In the following section, I will focus on a recently discussed minimal SO(10) model, where without any assumption other than SO(10) grand unification, one can indeed predict all but one neutrino parameters. I will then consider a case where assumption of quasi-degeneracy in the neutrino spectrum at high scale leads in a natural way via radiative corrections to large mixings at low energies.

To understand the fundamental physics behind neutrino mixings, we first write down the neutrino mass matrix that leads to maximal solar and atmospheric mixing. We consider the case of normal hierarchy where we have

$$M_\nu = \frac{\sqrt{\Delta m_A^2}}{2} \begin{pmatrix} c\epsilon & b\epsilon & d\epsilon \\ b\epsilon & 1 + a\epsilon & -1 \\ d\epsilon & -1 & 1 + \epsilon \end{pmatrix} \tag{8}$$

where $\epsilon \simeq \sqrt{\frac{\Delta m_\odot^2}{\Delta m_A^2}}$ and parameters a, b, c, d are of order one. Any theory of neutrino which attempts to explain the observed mixing pattern for the case of normal hierarchy must strive to get a mass matrix of this form. In the next section, we give a simple example of a minimal SO(10) grand unified theory that gives this mass matrix without any extra assumptions.

5. A predictive minimal SO(10) theory for neutrinos

The main reason for considering SO(10) for neutrino masses is that its **16** dimensional spinor representation consists of all fifteen standard model fermions plus the right handed neutrino arranged according to the it $SU(2)_L \times SU(2)_R \times SU(4)_c$ subgroup[4] as follows:

$$\Psi = \begin{pmatrix} u_1 \; u_2 \; u_3 \; \nu \\ d_1 \; d_2 \; d_3 \; e \end{pmatrix} \tag{9}$$

There are three such spinors for three fermion families.

In order to implement the seesaw mechanism in the SO(10) model, one must break the B-L symmetry, since the right handed neutrino mass breaks this symmetry. One implication of this is that the seesaw scale is at or below the GUT scale. Secondly in the context of supersymmetric SO(10) models, the way B-L breaks has profound consequences for low energy physics. For instance, if B-L is broken by a Higgs field belonging to the **16** dimensional Higgs field (to be denoted by Ψ_H), then the field that acquires a nonzero vev has the quantum numbers of the ν_R field i.e. B-L breaks by one unit. In this case higher dimensional operators of the form $\Psi\Psi\Psi\Psi_H$ will lead to R-parity violating operators in the effective low energy MSSM theory such as $QLd^c, u^c d^c d^c$ etc which can lead to large breaking of lepton and baryon number symmetry and hence unacceptable rates for proton decay. This theory also has no dark matter candidate without making additional assumptions.

On the other hand, one may break B-L by a **126** dimensional Higgs field. The member of this multiplet that acquires vev has $B - L = 2$ and leaves R-parity as an automatic symmetry of the low energy Lagrangian. There is a naturally stable dark matter in this case. It has recently been shown that this class of models lead to a very predictive scenario for neutrino mixings[12,13,14,15]. We summarize this model below.

As already noted earlier, any theory with asymptotic parity symmetry leads to type II seesaw formula. It turns out that if the B-L symmetry is broken by **16** Higgs fields, the first term in the type II seesaw (effective triplet vev induced term) becomes very small compared to the type I term. On the other hand, if B-L is broken by a **126** field, then the first term in the type II seesaw formula is not necessarily small and can in principle dominate in the seesaw formula. We will discuss a model of this type below.

The basic ingredients of this model are that one considers only two Higgs multiplets that contribute to fermion masses i.e. one **10** and one **126**. A unique property of the **126** multiplet is that it not only breaks the B-L symmetry and therefore contributes to right handed neutrino masses, but it also contributes to charged fermion masses by virtue of the fact that it contains MSSM doublets which mix with those from the **10** dimensional multiplets and survive down to the MSSM scale. This leads to a tremendous reduction of the number of arbitrary parameters, as we will see below.

There are only two Yukawa coupling matrices in this model: (i) h for the **10** Higgs and (ii) f for the **126** Higgs. SO(10) has the property that the Yukawa couplings involving the **10** and **126** Higgs representations are symmetric. Therefore if we assume that CP violation arises from other sectors of the theory (e.g. squark masses) and work in a basis where one of these two sets of Yukawa coupling matrices is diagonal, then it will have only nine parameters. Noting the fact that the (2,2,15) submultiplet of **126** has a pair of standard model doublets that contributes to charged fermion masses, one can write the quark and lepton mass matrices as follows[12]:

$$M_u = h\kappa_u + f v_u \tag{10}$$
$$M_d = h\kappa_d + f v_d$$
$$M_\ell = h\kappa_d - 3 f v_d$$
$$M_{\nu_D} = h\kappa_u - 3 f v_u$$

where $\kappa_{u,d}$ arc thc vcv's of the up and down standard model type Higgs fields in the **10** multiplet and $v_{u,d}$ are the corresponding vevs for the same doublets in **126**. Note that there are 13 parameters in the above equations and there are 13 inputs (six quark masses, three lepton masses and three quark mixing angles and weak scale). Thus all parameters of the model that go into fermion masses are determined.

To determine the light neutrino masses, we use the seesaw formula in Eq. (7), i.e.

$$M_\nu = f v_L - h_\nu^T f_R^{-1} h_\nu \left(\frac{v_{wk}^2}{v_R} \right). \tag{11}$$

The coupling matrix f is nothing but the **126** Yukawa coupling that appears in Eq. (10). Thus all parameters that give neutrino mixings except an overall scale are determined. These models were extensively discussed in the last decade[13] using type I seesaw formula. Their predictions for neutrino masses and mixings are either ruled out or are at best marginal.

There has been a revival of these models due to an observation for a two generation version of it[14]. It was pointed out in Ref.[14] that if the direct triplet term in type II seesaw dominates, then it provides a very natural understanding of the large atmospheric mixing angle for the case of two generations without invoking any symmetries. Subsequently it was shown[15] that the same $b - \tau$ mass convergence also provides an explanation of large solar mixing as well as small θ_{13} making the model realistic and experimentally interesting.

A simple way to see how large mixings arise in this model is to note that when the triplet term dominates the seesaw formula, we have the neutrino mass matrix $M_\nu \propto f$, where f matrix is the **126** coupling to fermions discussed earlier. Using the above equations, one can derive the following sumrule (sumrule was already noted in the second reference of [13]):

$$M_\nu = c(M_d - M_\ell) \tag{12}$$

38

where numerically $c \approx 10^{-9}$ GeV. To see how this leads to large atmospheric and solar mixing, let us work in the basis where the down quark mass matrix is diagonal. All the quark mixing effects are then in the up quark mass matrix i.e. $M_u = U_{CKM}^T M_u^d U_{CKM}$. Note further that the minimality of the Higgs content leads to the following sumrule among the mass matrices:

$$k\tilde{M}_\ell = r\tilde{M}_d + \tilde{M}_u \qquad (13)$$

where the tilde denotes the fact that we have made the mass matrices dimensionless by dividing them by the heaviest mass of the species i.e. up quark mass matrix by m_t, down quark mass matrix by m_b etc. k, r are functions of the symmetry breaking parameters of the model. Using the Wolfenstein parameterization for quark mixings, we can conclude that that we have

$$M_{d,\ell} \approx m_{b,\tau} \begin{pmatrix} \lambda^3 & \lambda^3 & \lambda^3 \\ \lambda^3 & \lambda^2 & \lambda^2 \\ \lambda^3 & \lambda^2 & 1 \end{pmatrix} \qquad (14)$$

where $\lambda \sim 0.22$ and the matrix elements are supposed to give only the approximate order of magnitude.

An important consequence of the relation between the charged lepton and the quark mass matrices in Eq. (13) is that the charged lepton contribution to the neutrino mixing matrix i.e. $U_\ell \simeq 1 + O(\lambda)$ or close to identity matrix. As a result the neutrino mixing matrix is given by $U_{PMNS} = U_\ell^\dagger U_\nu \simeq U_\nu$, since in U_ℓ, all mixing angles are small. Thus the dominant contribution to large mixings will come from U_ν, which in turn will be dictated by the sum rule in Eq. (12). Let us now see how how this comes about.

As we extrapolate the quark masses to the GUT scale, due to the fact that $m_b - m_\tau \approx m_\tau \lambda^2$ for a wide range of values of $\tan\beta$, the neutrino mass matrix $M_\nu = c(M_d - M_\ell)$ takes roughly the form

$$M_\nu = c(M_d - M_\ell) \approx m_0 \begin{pmatrix} \lambda^3 & \lambda^3 & \lambda^3 \\ \lambda^3 & \lambda^2 & \lambda^2 \\ \lambda^3 & \lambda^2 & \lambda^2 \end{pmatrix} \qquad (15)$$

This mass matrix is in the form discussed in Eq. (8) and it is easy to see that both the θ_{12} (solar angle) and θ_{23} (the atmospheric angle) are now large. The detailed magnitudes of these angles of course depend on the details of the quark masses at the GUT scale. Using the extrapolated values of the quark masses and mixing angles to the GUT scale, the predictions of this model for various oscillation parameters are given in Ref.[15]. The predictions for the solar and atmospheric mixing angles fall within 3 σ range of the present central values. Specifically the prediction for U_{e3} (see Fig. 1) can be tested in MINOS as well as other planned Long Base Line neutrino experiments such as Numi-Off-Axis, JPARC etc.

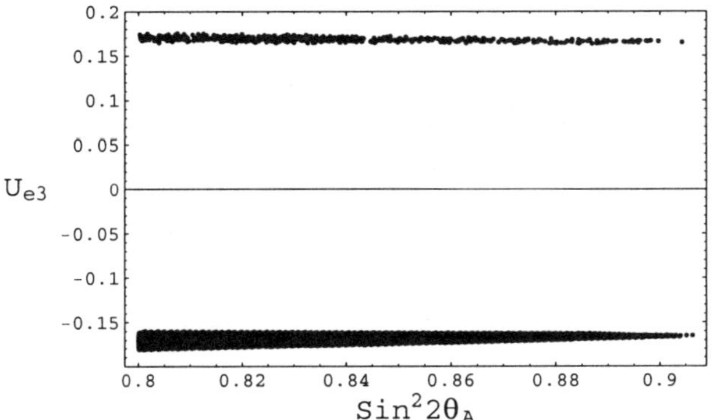

Figure 1. The figure shows the predictions of the minimal SO(10) model for $sin^2 2\theta_A$ and U_{e3} for the allowed range of parameters in the model. Note that U_{e3} is very close to the upper limit allowed by the existing reactor experiments.

There is a simple explanation of why the U_{e3} comes out to be large. This can also be seen from the mass sumrule in Eq.12. Roughly, for a matrix with hierarchical eigen values as is the case here, the mixing angle $tan2\theta_{13} \sim \frac{M_{\nu,13}}{M_{\nu,33}} \simeq \frac{\lambda^3 m_\tau}{m_b(M_U) - m_\tau(M_U)}$. Since to get large mixings, we need $m_b(M_U) - m_\tau(M_U) \simeq m_\tau \lambda^2$, we see that $U_{e3} \simeq \lambda$ upto a factor of order one. Indeed the detailed calculations lead to 0.16 which is not far from this value.

6. CP violation in the minimal SO(10) model

In the discussion given above, it was assumed that CP violation is non-CKM type and resides in the soft SUSY breaking terms of the Lagrangian. The overwhelming evidence from experiments seem to be that CP violation is perhaps is of CKM type. It has recently been pointed out that with slight modification, one can include CKM CP violation in the model[16]. The basic idea is to include all higher dimensional operators of type $h'\Psi\Psi\bar{\Delta}\Sigma/M$ where $\bar{\Delta}$ and Σ denote respectively the **126** and the **210** dimensional representation. It is then clear that those operators transforming as **10** and

126 representations will simply redefine the h, f coupling matrices and add no new physics. On the other hand the higher dimensional operator that transforms like an effective **120** representation will add a new piece to all fermion masses. Now suppose we introduce a parity symmetry into the theory which transforms Ψ to Ψ^{c*}, then it turns out that the couplings h and f become real and symmetric matrices whereas the **120** coupling (denoted by h') becomes imaginary and antisymmetric. This process introduces three new parameters into the theory and the charged fermion masses are related to the fundamental couplings in the theory as follows:

$$M_u = h\kappa_u + fv_u + h'v_u \tag{16}$$

$$\begin{aligned}
M_d &= h\kappa_d + fv_d + h'v_d \\
M_\ell &= h\kappa_d - 3fv_d - 3h'v_d \\
M_{\nu_D} &= h\kappa_u - 3fv_u - 3h'v_u
\end{aligned}$$

$$\tag{17}$$

Note that the extra contribution compared to Eq. (10) is antisymmetric which therefore does not interfere with the mechanism that lead to $M_{\nu,33}$ becoming small as a result of $b - \tau$ convergence. Hence the natural way that θ_A became large in the CP conserving case remains.

Let us discuss if the new model is still predictive in the neutrino sector. Of the three new parameters, one is determined by the CP violating quark phase. the two others are determined by the solar mixing angle and the solar mass difference squared. Therefore we lose the prediction for these parameters. However, we can predict in addition to θ_A (see above), θ_{13} and the Dirac phase for the neutrinos.

7. Radiative generation of large mixings: another application of type II seesaw

As alluded before, type II seesaw liberates the neutrinos from obeying normal generational hierarchy and instead could easily be quasi-degenerate in mass. This raises a new way to understand the large mixings instead of having to generate them in the original seesaw theory as is normally done. The basic idea is that at the seesaw scale, all mixings angles are small. Since the observed neutrino mixings are the weak scale observables, one must extrapolate[17] the seesaw scale mass matrices to the weak scale and recalculate the mixing angles. The extrapolation formula is $M_\nu(M_Z) = \mathbf{I}M_\nu(v_R)\mathbf{I}$ where $\mathbf{I}_{\alpha\alpha} = \left(1 - \frac{h_\alpha^2}{16\pi^2}\right)$. Note that since $h_\alpha = \sqrt{2}m_\alpha/v_{wk}$ (α being the charged lepton index), in the extrapolation only the τ-lepton makes a difference. In the MSSM, this increases the $M_{\tau\tau}$ entry of the neutrino mass matrix and essentially leaves the others unchanged. It was shown[18] that if the muon and the tau neutrinos are nearly degenerate but not degenerate enough in mass at the seesaw scale, the radiative corrections can become large enough so that at the weak scale the two diagonal elements of M_ν become much more degenerate. This leads to an enhancement of the mixing angle to become almost maximal value. This can also be seen from the renormalization group equations when they are written in the mass basis[19]. Denoting the mixing angles as θ_{ij} where i, j stand for generations, the equations are:

$$\frac{ds_{23}}{dt} = -F_\tau c_{23}{}^2 \left(-s_{12} U_{\tau 1} D_{31} + c_{12} U_{\tau 2} D_{32}\right), \tag{18}$$

$$\frac{ds_{13}}{dt} = -F_\tau c_{23} c_{13}{}^2 \left(c_{12} U_{\tau 1} D_{31} + s_{12} U_{\tau 2} D_{32}\right), \tag{19}$$

$$\frac{ds_{12}}{dt} = -F_\tau c_{12} \left(c_{23} s_{13} s_{12} U_{\tau 1} D_{31} - c_{23} s_{13} c_{12} U_{\tau 2} D_{32} \right.$$
$$\left. + U_{\tau 1} U_{\tau 2} D_{21}\right). \tag{20}$$

where $D_{ij} = (m_i + m_j)) / (m_i - m_j)$ and $U_{\tau 1,2,3}$ are functions of the neutrino mixings angles. The presence of $(m_i - m_j)$ in the denominator makes it clear that as $m_i \simeq m_j$, that particular coefficient becomes large and as we extrapolate from the GUT scale to the weak scale, small mixing angles at GUT scale become large at the weak scale. It has been shown recently that indeed such a mechanism for understanding large mixings can work for three generations[20]. It was shown that if we identify the seesaw scale neutrino mixing angles with the corresponding quark mixings and assume quasi-degenerate neutrinos, the weak scale solar and atmospheric angles get magnified to the desired level while due to the extreme smallness of V_{ub}, the magnified value of U_{e3} remains within its present upper limit. In figure 2, we show the evolution of the mixing angles to the weak scale. A requirement for this scenario to work is that the common mass of neutrinos must be larger than 0.1 eV, a result that can be tested in neutrinoless double beta experiments.

8. Other realizations of seesaw

As we saw from the previous discussion, the conventional seesaw mechanism requires rather high scale for the B-L symmetry breaking and the corresponding right handed neutrino mass (of order 10^{15} GeV). There is

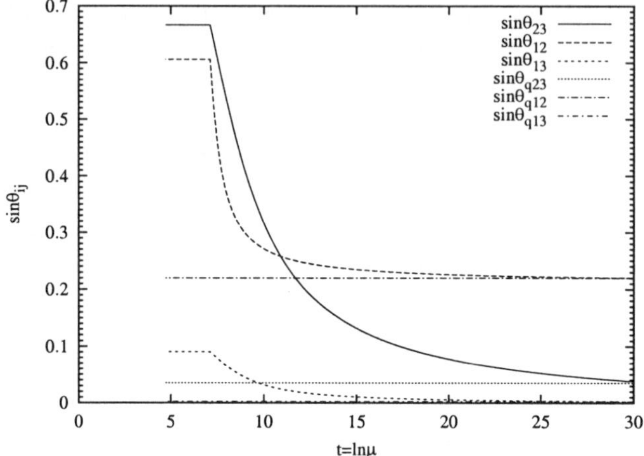

Figure 2.   Radiative magnification of small quark-like neutrino mixings at the see-saw scale to bilarge values at low energies. The solid, dashed and dotted lines represent $\sin\theta_{23}$, $\sin\theta_{13}$, and $\sin\theta_{12}$, respectively.

however no way at present to know what the scale of B-L symmetry breaking is. There are for example models bases on string compactification[22] where the $B - L$ scale is quite possibly in the TeV range. In this case small neutrino mass can be implemented by a double seesaw mechanism suggested in Ref.[21]. The idea is to take a right handed neutrino N whose Majorana mass is forbidden by some symmetry and a singlet neutrino S which has extra quantum numbers which prevent it from coupling to the left handed neutrino but which is allowed to couple to the right handed neutrino. One can then write a three by three neutrino mass matrix in the basis (ν, N, S) of the form:

$$M = \begin{pmatrix} 0 & m_D & 0 \\ m_D & 0 & M \\ 0 & M & \mu \end{pmatrix} \tag{21}$$

For the case $\mu \ll M \approx M_{B-L}$, (where M_{B-L} is the $B - L$ breaking scale), this matrix has one light and two heavy states. The lightest eigenvalue is given by $m_\nu \sim m_d M^{-1} \mu M^{-1} m_D$. There is a double suppression by the heavy mass compared to the usual seesaw mechanism and hence the name double seesaw. A generalization of this mechanism to the case of three generations is straightforward. One important point here is that to keep $\mu \sim m_D$, one also needs some additional gauge symmetries, which often are a part of the string models. In fact this mechanism is sometimes invoked in string models with TeV scale Z' to understand neutrino masses[22].

It has recently been noted[23] that if there is parity symmetry in these models, the 13 and 31 entries of the above neutrino mass matrix get filled by a small seesaw suppressed entry. This has interesting applications in neutrino model building.

There are also other alternatives to seesaw, [24] discussed in literature, that we do not discuss.

9. Conclusion

In summary, the seesaw mechanism is by far the simplest and most appealing way to understand neutrino masses. It not only improves the aesthetic appeal of the standard model by restoring quark-lepton symmetry but it also makes weak interactions asymptotically parity conserving. Further more it connects neutrino masses with the hypothesis of grand unification. In this talk I have discussed three different realizations of this mechanism : type I, type II and double seesaw. I have also noted how using the type II seesaw mechanism, one can have simple understanding of large neutrino mixings among neutrinos. A particularly interesting framework is provided by the minimal SO(10) model with **126** Higgs fields, which provides a very predictive framework for neutrinos.

This work is supported by the National Science Foundation Grant No. PHY-0354401.

References

1. M. Gell-Mann, P. Ramond, and R. Slansky, *Supergravity* (P. van Nieuwenhuizen et al. eds.), North Holland, Amsterdam, 1980, p. 315; T. Yanagida, in *Proceedings of the Workshop on the Unified Theory and the Baryon Number in the Universe* (O. Sawada and A. Sugamoto, eds.), KEK, Tsukuba, Japan, 1979, p. 95; S. L. Glashow, *The future of elementary particle physics*, in *Proceedings of the 1979 Cargèse Summer Institute on Quarks and Leptons* (M. Lévy et al. eds.), Plenum Press, New York, 1980, pp. 687–713; R. N. Mohapatra and G. Senjanović, Phys. Rev. Lett. **44**, 912 (1980); P. Ramond, hep-ph/9809459.
2. S. Weinberg, Phys. Rev. Lett. **43**, 1566 (1979)
3. E. Witten, Phys. Lett. **B 91**, 81 (1980).
4. J. C. Pati and A. salam, Phys. Rev. **D10**, 275 (1974).
5. J. C. Pati and A. Salam, Ref.[4]; R. N. Mohapatra and J. C. Pati, Phys. Rev. **D 11**, 566, 2558 (1975); G. Senjanović and R. N. Mohapatra, Phys. Rev. **D 12**, 1502 (1975).
6. R. N. Mohapatra and R. E. Marshak, Phys. Lett. **B 91**, 222 (1980); A. Davidson, Phys. Rev. **D20**, 776 (1979).

7. G. Lazarides, Q. Shafi and C. Wetterich, Nucl.Phys.**B181**, 287 (1981); R. N. Mohapatra and G. Senjanović, Phys. Rev. **D 23**, 165 (1981);

8. J. Schechter and J. W. F. Valle, Phys. Rev. **D 22**, 2227 (1980).

9. E. Ma and U. Sarkar, Phys. Rev. Lett. **80**, 5716 (1998).

10. A. Smirnov, hep-ph/0311259; S. F. King, Rept.Prog.Phys. **67**, 107 (2004); G. Altarelli and F. Feruglio, hep-ph/0405048; R. N. Mohapatra, hep-ph/0211252; New J. Phys., **6**, 82 (2004).

11. See for instance D. O. Caldwell and R. N. Mohapatra, Phys. Rev. D **48**, 3259 (1993); D. G. Lee and R. N. Mohapatra, Phys. Lett. B **329**, 463 (1994); S. Antusch and S. F. King, hep-ph/0402121.

12. K. S. Babu and R. N. Mohapatra, Phys. Rev. Lett. **70**, 2845 (1993).

13. D. G. Lee and R. N. Mohapatra, Phys. Rev. **D 51**, 1353 (1995); L. Lavoura, Phys. Rev. **D 48**, 5440 (1993); B. Brahmachari and R. N. Mohapatra, Phys. Rev. **D58**, 015001 (1998); K. Oda, E. Takasugi, M. Tanaka and M. Yoshimura, Phys. Rev. **D 58**, 055001 (1999); K. Matsuda, Y. Koide, T. Fukuyama and N. Okada, Phys. Rev. **D 65**, 033008 (2002); T. Fukuyama and N. Okada, hep-ph/0205066; N. Oshimo, hep-ph/0305166.

14. B. Bajc, G. Senjanovic and F. Vissani, Phys. Rev. Lett. **90**, 051802 (2003) [arXiv:hep-ph/0210207]

15. H. S. Goh, R. N. Mohapatra and S. P. Ng, Phys. Lett. B **570**, 215 (2003) [arXiv:hep-ph/0303055].

16. B. Dutta, Y. Mimura and R. N. Mohapatra, hep-ph/0406262.

17. K.S. Babu, C.N. Leung and J. Pantaleone, Phys. Lett. **B319**, 191 (1993); P. Chankowski and Z. Pluciennik, Phys. Lett. **B316**, 312 (1993).

18. K. S. Balaji, A. Dighe, R. N. Mohapatra and M. K. Parida, Phys. Rev. Lett. **84**, 5034 (2000); Phys. Lett. **B481** , 33 (2000).

19. J. Casas, J. Espinoza, J. Ibara and S. Navaro, hep-ph/9905381; Nucl. Phys. **B 573**, 652 (2000).

20. R. N. Mohapatra, M. K. Parida and G. Rajasekaran, Phys.Rev. **D69**, 053007 (2004); hep-ph/0301234.

21. R. N. Mohapatra, Phys. Rev. Lett. **56**, 561 (1986); R. N. Mohapatra and J. W. F. Valle, Phys. Rev. **D 34**, 1642 (1986).

22. Junhai Kang, Paul Langacker, and Tianjun Li, UPR-1010T

23. S. M. Barr, Phys. Rev. Lett. **92**, 101601 (2004)

24. Talk by A. Smirnov at this workshop; P. Langacker, Int. J. Mod. Phys. A **18**, 4015 (2003).

SEE-SAW AND GRAND UNIFICATION

GORAN SENJANOVIĆ

International Centre for Theoretical Physics, 34100 Trieste, Italy

I review the profound connection between the see-saw mechanism for neutrino masses and grand unification. This connections points naturally towards SO(10) grand unified theory. The emphasis will be in the supersymmetric theory, but I also discuss salient features of the ordinary SO(10). Particular attention is paid to the crucial issue of the minimal such theory, i.e. the question of the Higgs sector needed to break SO(10) down to the Minimal Supersymmetric Standard Model or the Standard Model. Some essential features of the see-saw mechanism are clarified, in particular its precise origin at the high scale.

Prelude

I have been asked by the organizers of the SEESAW25 to review the see-saw mechanism in connection with grand unification. Due to an enormous body of work in the field, neither time nor space allow me to do a complete job. Instead I focus here on the work regarding pure grand unification, the work connected with the search for the minimal grand unified theory, both supersymmetric and not.

I am forced thus to omit some important issues such as doublet triplet splitting, grand unification in extra dimension, non minimal models, fermion mass textures, and more. For some complementary reviews of these topics (and not only) see: [1,2,3,4].

1. The parameters

We know today [5] that neutrinos are massive and we know the two mass differences that correspond to the atmospheric and solar neutrino oscillations

$$\Delta m_A^2 \simeq (2.5 \pm 0.6)10^{-3} eV^2 ; \quad \Delta m_\odot^2 \simeq (8.2 \pm 0.6)10^{-5} eV^2 \qquad (1)$$

The corresponding mixing angles are

$$\theta_A = 45^o \pm 6^o ; \quad \theta_\odot = 32.5^o \pm 2.5^o \qquad (2)$$

From (1), the mass of the heaviest neutrino has a lower limit

$$m_\nu^{\max} \geq 5 \times 10^{-2} eV \tag{3}$$

We also know from β decay that

$$m_{\nu_e} \leq 2.2eV \quad (95\% c.l.) \tag{4}$$

and from cosmological data we know that the sum of neutrino masses is small

$$\sum m_\nu \lesssim 0.4eV - 1.7eV \tag{5}$$

Thus, even if degenerate, neutrinos are very light, $m_\nu \lesssim (0.15eV - 0.6eV)$. How to understand so small neutrino masses? A simple answer lies in the see-saw mechanism.

2. See-saw in the Standard Model

The see-saw mechanism [6] of neutrino masses in the context of the Standard Model (SM), is obtained by adding a right-handed neutrino and writing the most general $SU(2)_L \times U(1)_Y$ Yukawa Lagrangean

$$L_y(\nu) = y_D \bar{\nu}_L \, \phi \, \nu_R + M_R \, \nu_R^T C \, \nu_R + h.c. \tag{6}$$

In our symbolic notation, y_D is a Yukawa matrix in flavor space. From (1), at least two light neutrinos are massive, and thus one needs at least two right-handed neutrinos. Having in mind grand unification with quark-lepton symmetry *a la* Pati-Salam as a natural framework to study neutrino masses, in what follows I assume three right-handed neutrinos and suppress generation indices.

Now, (6) creates a mess, unless $M_R \gg y_D\langle\phi\rangle$. This is quite natural though, since M_R is a gauge invariant quantity and thus expected to be very large: $M_R \gg M_W$. The principle that gauge invariant quantities lie much above the scale of the breaking of the symmetry in question lies at the heart of the see-saw mechanism. In what follows, I will stick to it consistently.

Since $y_D \leq 1$, with $M_R \gg \langle\phi\rangle$ one gets automatically small neutrino masses

$$M_\nu^I = m_D^T M_R^{-1} m_D \tag{7}$$

where $m_D \equiv y_D\langle\phi\rangle$ and I stands for the type I see-saw, which has become the common name for this realization of small neutrino masses.

Alternatively, you could add to the SM a $SU(2)_L$ triplet Δ, with $B - L(\Delta) = 2$, and the Yukawa couplings

$$L_y(\nu) = y_\nu \ell_L^T C \Delta \ell_L + h.c. \tag{8}$$

where ℓ_L stands for the leptonic doublets. The triplet gets a non-vanishing vev $\langle \Delta \rangle \simeq \frac{M_W^2}{M_\Delta}$ if the triplet mass $M_\Delta \gg M_W$. The same principle as before ensures small neutrino masses [7]

$$M_\nu^{II} \simeq y_\nu \frac{M_W^2}{M_\Delta} \tag{9}$$

where the superscript II stands for the type II see-saw as is commonly called.

In the SM we cannot distinguish the two mechanisms for $M_R, M_\Delta \gg M_W$. They are both simply the expression of the effective operator analysis which tells us that the leading $SU(2)_L \times U(1)_Y$ Yukawa coupling is of dimension 5 [8]

$$L_y^{\text{eff}}(\nu) = f \frac{1}{M} \left(\ell^T \sigma_2 \Phi \right) C \left(\phi^T \sigma_2 \ell \right) \tag{10}$$

where $M \gg M_W$ and f is a matrix in generation space. Hence small, see-saw like suppressed neutrino masses

$$M_\nu \simeq f \frac{M_W^2}{M} \tag{11}$$

Obviously, both type I and type II see-saw are of the form (11), as they have to be. There is no sense in trying to distinguish type I form type II in the SM, not without any new physics being invoked. After all, if $y_D \propto y_\nu \propto M_R$, we will even have the same flavor structure for neutrino mass matrices. In order to study this issue, we must go beyond the SM.

Our task is highly nontrivial. In order that the see-saw mechanism be tested, in order that it be a theory, we need first of all to know the origin of the mechanism: is it type I, type II or something else? [a] Since M_R has to be very large (unless Yukawa couplings are extremely small), it is natural to consider grand unification as a framework of new large mass scales and of quark-lepton unification which sheds light on y_D (and/or y_ν).

---

[a] It is interesting that the only other alternative is the fermionic triplet (actually triplets, at least two are needed similar to right-handed neutrinos)[9].

3. How to incorporate see-saw in GUTs ?

This question is intimately tied up with the fundamental issue of the choice of *the* grand unified theory. After three decades of grand unification, there is no consensus today of what the theory is.

First, the symmetry reasoning. In the SM, $B - L$ is an accidental, anomaly free $U(1)$ global symmetry. It is of course broken by M_R (or M in the effective operator language). If you believe in an accidental global $B - L$, then the SU(5) direction is the natural one, since this phenomenon persists.

On the other hand, the fact that $B - L$ is anomaly free makes a strong case for its gauging. This would in turn induce $(B - L)^3$ anomaly; the simplest remedy is to introduce the right-handed neutrinos (one per generation). The natural setting is then provided by Left-Right $(L - R)$ symmetric theories [10] where right-handed neutrinos are a must and $B - L$ has a simple physical interpretation from the electric charge formula

$$Q = T_{3L} + T_{3R} + \frac{B - L}{2} \tag{12}$$

The new scale M_R is then the scale of $SU(2)_R$, or better to say $L - R$ (parity) symmetry breaking. It is important to stress that both type I and type II see-saw emerge naturally in this case and are deeply connected. This route points towards Pati-Salam quark-lepton unification and SO(10) as a grand unified theory (GUT).

In the next two sections, I go through both SU(5) and SO(10) supersymmetric theories. Of course, SU(5) needs low energy supersymmetry (or split supersymmetry [11]), whereas the same cannot be said of the SO(10) theory. I will thus discuss both supersymmetric and ordinary SO(10).

4. Grand unification: SU(5)

As remarked before, SU(5) is a natural theory if you give up gauging $B - L$. The minimal SU(5) theory fails, for the gauge couplings do not unify. With low energy supersymmetry the couplings unify as predicted more than two decades ago [12].

The minimal theory with only 5_H (and $\overline{5_H}$ in SUSY), predicts $m_d = m_e$ at M_{GUT}, generation by generation [13]. Whereas $m_b = m_\tau$ works well, for other generations $|m_\mu| \simeq 3|m_s|$, $|m_e| \simeq 1/3|m_d|$ is needed (again, at M_{GUT}). This is easily achieved without any change in the theory, by adding higher dimensional operators suppressed by $1/M_{Pl}$. The theory loses then its predictivity in determining precisely M_{GUT} and τ_p, but is saved from being ruled out by a too fast $d = 5$ proton decay [14]. Alternatively, one could add more Higgs superfields, say 45_H as in the Georgi-Jarlskog [15] approach.

What about neutrino masses in $SU(5)$? You could choose from three simple possibilities, none of them very appealing:

(1) Add right-handed neutrinos, SU(5) singlets. In this case M_R is a gauge invariant quantity, and by the principle of naturalness $M_R \gg M_{GUT}$. This is no good, since then $m_\nu \ll M_W^2/M_{GUT} \simeq 10^{-3}eV$, which is too small to explain the solar and especially the atmospheric neutrino data.

(2) Add a $SU(2)_L$ triplet as before, with Δ contained in 15_H, a two index symmetric Higgs superfield. With the same principle as above, you reach the same conclusion.

(3) You could write a higher-dimensional operator *a la* Weinberg [8]

$$O_5 = f\frac{1}{M_\nu}\bar{5}_F\bar{5}_F 5_H 5_H \tag{13}$$

with $M_\nu \gg M_{GUT}$, say $M_\nu \sim M_{Pl}$ [16]. Strictly speaking, (1) and (2) correspond to this, as in the SM case discussed before. Again, the Planck scale suppression is too large to account for the atmospheric or solar neutrino data. Such terms can be relevant though for small splittings in the case of degenerate neutrinos [17].

All this does not prove that we must give up on SU(5). After all, we can fine-tune M_R, the trouble is that the Yukawas are arbitrary, as much as in the SM. Such a theory, with all the tree-level and $1/M_{Pl}$ corrections to Yukawa couplings needed to correct fermionic mass relations, has too many parameters. In order to pursue this direction, one should go beyond SU(5) and invoke extra family horizontal symmetries, discrete, global or local (for a review and references see [2]). Instead, we turn to SO(10) which is tailor fit for a theory of fermion masses.

5. SO(10): the minimal theory of matter and gauge coupling unification

There are a number of features that make SO(10) special:

(1) a family of fermions is unified in a 16-dimensional spinorial representation; this in turn predicts the existence of right-handed neutrinos

(2) $L - R$ symmetry is a finite gauge transformation in the form of charge conjugation. This is a consequence of both left-handed

fermions f_L and its charged conjugated counterparts $(f^c)_L \equiv C\overline{f}_R^T$ residing in the same representation 16_F.

(3) in the supersymmetric version, matter parity $M = (-1)^{3(B-L)}$, equivalent to the R-parity $R = M(-1)^{2S}$, is a gauge transformation [18], a part of the center Z_4 of SO(10). It simply reads $16 \rightarrow -16$, $10 \rightarrow 10$. Its fate depends then on the pattern of symmetry breaking (or the choice of Higgs fields); it turns out that in the renormalizable version of the theory R-parity remains exact at all energies [19,20]. The lightest supersymmetric partner (LSP) is then stable and is a natural candidate for the dark matter of the universe.

(4) its other maximal subgroup, besides $SU(5) \times U(1)$, is $SO(4) \times SO(6) = SU(2)_L \times SU(2)_R \times SU(4)_c$ symmetry of Pati and Salam. It explains immediately the somewhat mysterious relations $m_d = m_e$ (or $m_d = 1/3 m_e$) of SU(5).

(5) the unification of gauge couplings can be achieved with or without supersymmetry.

(6) the minimal renormalizable version (with no higher dimensional $1/M_{Pl}$ terms) offers a simple and deep connection between $b - \tau$ unification and a large atmospheric mixing angle in the context of the type II see-saw [21].

In order to understand some of these results, and in order to address the issue of construction of the theory, we turn now to the Yukawa sector.

5.1. Yukawa sector

Fermions belong to the spinor representation 16_F [22]. From

$$16 \times 16 = 10 + 120 + \overline{126} \tag{14}$$

the most general Yukawa sector in general contains 10_H, 120_H and $\overline{126}_H$, respectively the fundamental vector representation, the three-index antisymmetric representation and the five-index antisymmetric and anti-self-dual representation. $\overline{126}_H$ is necessarily complex, supersymmetric or not; 10_H and $\overline{126}_H$ Yukawa matrices are symmetric in generation space, while the 120_H one is antisymmetric.

Understanding fermion masses is easier in the Pati-Salam language of one of the two maximal subgroups of SO(10), $G_{PS} = SU(4)_c \times SU(2)_L \times SU(2)_R$ (the other being $SU(5) \times U(1)$). Let us decompose the relevant representations under G_{PS}

$$16 = (4, 2, 1) + (\bar{4}, 1, 2) \tag{15}$$

$$10 = (1, 2, 2) + (6, 1, 1) \tag{16}$$

$$120 = (1, 2, 2) + (6, 3, 1) + (6, 1, 3) + (15, 2, 2)$$
$$+ (10, 1, 1) + (\overline{10}, 1, 1) \tag{17}$$

$$\overline{126} = (\overline{10}, 3, 1) + (10, 1, 3) + (15, 2, 2) + (6, 1, 1) \tag{18}$$

Clearly, the see-saw mechanism, whether type I or II, requires $\overline{126}$: it contains both $(10, 1, 3)$ whose vev gives a mass to ν_R (type I), and $(\overline{10}, 3, 1)$, which contains a color singlet, $B - L = 2$ field Δ_L, that can give directly a small mass to ν_L (type II). A reader familiar with the SU(5) language sees this immediately from the decomposition under this group

$$\overline{126} = 1 + 5 + 15 + \overline{45} + 50 \tag{19}$$

The 1 of SU(5) belongs to the $(10, 1, 3)$ of G_{PS} and gives a mass for ν_R, while 15 corresponds to the $(\overline{10}, 3, 1)$ and gives the direct mass to ν_L.

Of course, $\overline{126}_H$ can be a fundamental field, or a composite of two $\overline{16_H}$ fields, or can even be induced as a two-loop effective representation built out of a 10_H and two gauge 45-dim representations. In what follows I shall discuss carefully all three possibilities.

Normally the light Higgs is chosen to be the smallest one, 10_H. Since $\langle 10_H \rangle = \langle (1, 2, 2) \rangle_{PS}$ is a $SU(4)_c$ singlet, $m_d = m_e$ follows immediately, independently of the number of 10_H you wish to have. Thus we must add either 120_H or $\overline{126}_H$ or both in order to correct the bad mass relations. Both of these fields contain $(15, 2, 2)_{PS}$, and its vev gives the relation $m_e = -3m_d$.

As $\overline{126}_H$ is needed anyway for the see-saw, it is natural to take this first. The crucial point here is that in general $(1, 2, 2)$ and $(15, 2, 2)$ mix through $\langle (10, 1, 3) \rangle$ [23] and thus the light Higgs is a mixture if the two. In other words, $\langle (15, 2, 2) \rangle$ in $\overline{126}_H$ is in general non-vanishing [b]. It is rather appealing that 10_H and $\overline{126}_H$ may be sufficient for all the fermion masses, with only two sets of symmetric Yukawa coupling matrices.

5.2. *An instructive failure*

Before proceeding, let me emphasize the crucial point of the necessity of 120_H or $\overline{126}_H$ in the charged fermion sector on an instructive failure: a

---

[b]In supersymmetry this is not automatic, but depends on the Higgs superfields needed to break SO(10) at M_{GUT}.

simple and beautiful model by Witten [24]. The model is non-supersymmetric and the SUSY lovers may place the blame for the failure here. It uses $\langle 16_H \rangle$ in order to break $B - L$, and the "light" Higgs is 10_H. Witten noticed an ingenious and simple way of generating an effective mass for the right-handed neutrino, through a two-loop effect which gives

$$M_{\nu_R} \simeq y_{up} \left(\frac{\alpha}{\pi}\right)^2 M_{GUT} \tag{20}$$

where one takes all the large mass scales, together with $\langle 16_H \rangle$, of the order M_{GUT}. Since $\langle 10_H \rangle = \langle (1, 2, 2)_{PS} \rangle$ preserves quark-lepton symmetry, it is easy to see that

$$M_\nu \propto M_u$$
$$M_e = M_d$$
$$M_u \propto M_d \tag{21}$$

so that $V_{\text{lepton}} = V_{\text{quark}} = 1$. The model fails badly. Is it yet another example of beautiful theories killed by the ugly facts of nature?

The original motivation of Witten was a desire to know the scale of M_{ν_R} and increase M_ν, at that time neutrino masses were expected to be larger. But the real achievement of this simple, elegant, minimal SO(10) theory is the predictivity of the structure of M_{ν_R} and thus M_ν. It is an example of a good, albeit wrong theory: it fails because it predicts.

What is the moral behind the failure? Not easy to answer. The main problem, in my opinion, was to ignore the fact that with only 10_H already charged fermion masses fail. As I have argued repeatedly, one needs a $\overline{126}_H$ (or a 120_H), one way or another, a lesson we keep in what follows.

6. Supersymmetric SO(10) GUT

In supersymmetry 10_H is necessarily complex and the bidoublet $(1, 2, 2)$ in 10_H contains the two Higgs doublets of the MSSM, with the vevs v^u and v^d in general different: $\tan\beta \equiv v^u/v^d \neq 1$ in general. In order to study the physics of SO(10), we need to know what the theory is, i.e. its Higgs content. There are two orthogonal approaches to the issue, as we discuss now

6.1. *Small representations*

The idea: take the smallest Higgs fields (least number of fields, not of representations) that can break SO(10) down to the MSSM and give realistic fermion masses and mixings. The following fields are both necessary and sufficient

$$45_H, 16_H + \overline{16}_H, 10_H \tag{22}$$

It all looks simple and easy to deal with, but the superpotential becomes extremely complicated. First, at the renormalizable level it is too simple. The pure Higgs and the Yukawa superpotential at the renormalizable level take the form

$$W_H = m_{45}45_H^2 + m_{16}16_H\overline{16}_H + \lambda_1 16_H\Gamma^2\overline{16}_H 45_H$$
$$m_{10}10_H^2 + \lambda_2 16_H\Gamma 16_F 10_H + \lambda_3 \overline{16}_H\Gamma\overline{16}_H 10_H \tag{23}$$

$$W_y = y_{10}16_F\Gamma 16_F 10_H \tag{24}$$

where Γ stands for the Clifford algebra matrices of $SO(10)$, $\Gamma_1...\Gamma_{10}$, and the products of Γ's are written in a symbolic notation (both internal and Lorentz charge conjugation are omitted).

Clearly, both W_H and W_y are insufficient. The fermion mass matrices would be completely unrealistic and the vevs $\langle 45_H\rangle$, $\langle 6_H\rangle$, $\langle\overline{16}_H\rangle$ would all point in the $SU(5)$ direction. Thus, one adds non-renormalizable operators

$$\Delta W_H = \frac{1}{M_{Pl}}\left[(45_H^2)^2 + 45_H^4 + (16_H\overline{16}_H)^2 + (16_H\Gamma^2\overline{16}_H)^2 + (16_H\Gamma^4\overline{16}_H)^2\right.$$
$$+(16_H\Gamma 16_H)^2 + (16_H\Gamma^5 16_H)^2 + \{16_H \to \overline{16}_H\}$$
$$\left.+16_H\Gamma^4\overline{16}_H 45_H^2 + 16_H\Gamma^3\overline{16}_H 45_H 10_H + \{16_H \to \overline{16}_H\}\right] \tag{25}$$

$$\Delta W_y = \frac{1}{M_{Pl}}\left[16_F\Gamma 16_F\, 16_H\Gamma 16_H + \{16_H \to \overline{16}_H\}\right.$$
$$\left.16_F\Gamma^3 16_F 45_H 10_H + 16_F\Gamma^5 16_F\overline{16}_H\Gamma^5\overline{16}_H\right] \tag{26}$$

where I take for simplicity all the couplings to be unity; there are simply too many of them. The large number of Yukawa couplings means very little predictivity.

The way out is to add flavor symmetries and to play the texture game and thus reduce the number of couplings. This in a sense goes beyond grand unification and appeals to new physics at M_{Pl} and/or new symmetries (see e.g. [25]).

To me, maybe the least appealing aspect of this approach is the loss of R (matter) parity due to 16_H and $\overline{16}_H$; it must be postulated by hand as much as in the MSSM.

On the positive side, it is an asymptotically free theory and one can work in the perturbative regime all the way up to M_{Pl}. While this sounds nice, I am not sure what it means in practice. It would be crucial if you were able to make high precision determination of M_{GUT} or m_T, the mass of colored triplets responsible for $d = 5$ proton decay. The trouble is that the lack of knowledge of the superpotential couplings is sufficient even in the minimal SU(5) theory to prevent this task; in SO(10) it gets even worse.

Maybe more relevant is the fact that in this scenario $M_R \simeq M_{GUT}^2/M_{Pl} \simeq 10^{13} - 10^{14} GeV$, which fits nicely with the neutrino masses via see-saw. Furthermore, see-saw can be considered "clean", of the pure type I, since the type II effect is suppressed by $1/M_{Pl}$. Most important, the $m_b \simeq m_\tau$ relation from (24) is maintained due to small $1/M_{Pl}$ effects relevant only for the first two generations.

Now, the higher dimensional operators can be mimicked by the inclusion of singlets when they are integrating out (assuming them much heavier that M_{GUT}, as expected by the gauge principle). This paves the way for model building if one is willing to fine-tune their masses to lie below M_{GUT}, and this way one can get the double type I see-saw formula (see e.g.[26]). Recently Barr [27] has shown how in a particular case one can obtain the see-saw formula linear in y_D (and not quadratic as usual). Although obtained in a completely different manner, this is what happens in the Witten's model, and thus in my opinion does not really represent a new type of see-saw. Of course, this allows for different models of neutrino masses and mixings. In order to stick to minimal theories, I refrain here from dicussing this and similiar proposals; this does not imply that they are without merit.

6.2. *Big is Better approach*

The non-renormalizable operators in reality mean invoking new physics beyond grand unification. This may be necessary, but still, one should be more ambitious and try to use the renormalizable theory only. This means large representations necessarily: at least $\overline{126}_H$ is needed in order to give the mass to ν_R (in supersymmetry, one must add 126_H). The consequence is the loss of asymptotic freedom above M_{GUT}, the coupling constants grow large at the scale $\Lambda_F \simeq 10 M_{GUT}$. To me this is a priori neither good nor bad, but if it bothers you, you should skip the rest of the section.

Once we accept large representations, we should minimize their number. The minimal theory contains, on top of 10_H, 126_H and $\overline{126}_H$, also 210_H

[28,29,30,31] with the decomposition

$$210_H = (1,1,1)_- + (15,1,1)_+ + (15,1,3) + (15,3,1) + (6,2,2) + (10,2,2) + (\overline{10},2,2)$$

(27)

where the -(+) subscript denotes the properties of the color singlets under charge conjugation.

The Higgs superpotential is remarkably simple

$$W_H = m_{210}(210_H)^2 + m_{126}\overline{126_H}126_H + m_{10}(10_H)^2 + \lambda(210_H)^3$$
$$+ \eta 126_H\overline{126}_H 210_H + \alpha 10_H 126_H 210_H + \overline{\alpha}10_H\overline{126}_H 210_H \quad (28)$$

and the Yukawa one even simpler

$$W_Y = y_{10}16_F\Gamma 16_F 10_H + y_{126}16_F\Gamma^5 16_F\overline{126}_H \tag{29}$$

Remarkably enough, this may be sufficient, without any higher dimensional operators; however, the situation is not completely clear.

There is a small number of parameters: $3 + 6\text{x}2 = 15$ real Yukawa couplings, and 11 real parameters in the Higgs sector. In this sense the theory can be considered as the minimal supersymmetric GUT in general [31]. As usual, I am not counting the parameters associated with the SUSY breaking terms.

The nicest feature of this program (and the best justification for the use of large representations) is the following. Besides the $\langle(10,1,3)\rangle$ which gives masses to the ν_R's, also the $\langle(15,2,2)\rangle$ in $\overline{126}_H$ gets a vev [29,23]. Approximately

$$\langle 15,2,2\rangle_{\overline{126}} \simeq \frac{M_{PS}}{M_{GUT}}\langle 1,2,2\rangle \tag{30}$$

with $M_{PS} = \langle 15,2,2\rangle$ being the scale of $SU(4)_c$ symmetry breaking. In SUSY, $M_{PS} \leq M_{GUT}$ and thus one can have correct mass relations for the charged fermions.

What is lost, though, is the $b-\tau$ unification, i.e. with $\langle(15,2,2)\rangle_{\overline{126}} \neq 0$, $m_b = m_\tau$ at M_{GUT} becomes an accident. However, in the case of type II see-saw, there is a profound connection between $b-\tau$ unification and a large atmospheric mixing angle. The fermionic mass matrices are obtained

56

from (29)

$$M_u = v_{10}y_{10} + v^u_{126}y_{126} \; ,$$
$$M_d = v_{10}y_{10} + v^d_{126}y_{126} \; ,$$
$$M_e = v_{10}y_{10} - 3v^d_{126}y_{126} \; ,$$
$$M_{\nu_D} = v_{10}y_{10} - 3v^u_{126}y_{126} \; , \tag{31}$$
$$M_{\nu_R} = y_{126}\langle(10,1,3)\rangle \; , \tag{32}$$
$$M_{\nu_L} = y_{126}\langle(\overline{10},3,1)\rangle \; , \tag{33}$$
$$\tag{34}$$

where $\langle(\overline{10},3,1)\rangle \simeq M_W^2/M_{GUT}$ provides a direct (type II) see-saw mass for light neutrinos. The form in (31) is readily understandable, if you notice that $\langle(1,2,2)\rangle$ is a $SU(4)_c$ singlet with $m_q = m_\ell$, and $\langle(15,2,2)\rangle$ is a $SU(4)_c$ adjoint, with $m_\ell = -3m_q$ The vevs of the bidoublets are denoted by v^u and v^d as usual.

Now, suppose that type II dominates, or $M_\nu \propto y_{126} \propto M_e - M_d$, so that

$$M_\nu \propto M_e - M_d \tag{35}$$

Let us now look at the 2nd and 3rd generations first. In the basis of diagonal M_e, and for the small mixing ϵ_{de}

$$M_\nu \propto \begin{pmatrix} m_\mu - m_s & \epsilon_{de} \\ \epsilon_{de} & m_\tau - m_b \end{pmatrix} \tag{36}$$

obviously, large atmospheric mixing can only be obtained for $m_b \simeq m_\tau$ [21].

Of course, there was no reason whatsoever to assume type II see-saw. Actually, we should reverse the argument: the experimental fact of $m_b \simeq m_\tau$ at M_{GUT}, and large $\theta_{\rm atm}$ seem to favor the type type II see-saw. It can be shown, in the same approximation of 2-3 generations, that type I cannot dominate: it gives a small $\theta_{\rm atm}$ [32]. This gives hope to disentangle the nature of the see-saw in this theory. As a check, it can be shown that the two types of see-saw are really inequivalent [32].

The three generation numerical studies supported a type II see-saw [c] with the interesting prediction of a large θ_{13} and a hierarchical neutrino mass spectrum [35]. Somewhat better fits are obtained with a small contribution of 120_H [36] or higher dimensional operators [37].

---

[c]Type I can apparently be saved with CP phases, see [33]. For earlier work on type I see [34]

I wish to stress an important feature of this programme. Since 126 ($\overline{126}$) is invariant under matter parity, R parity remains exact at all energies and thus the lightest supersymmetric particle is stable and a natural candidate for the dark matter.

6.2.1. *Mass scales*

In SO(10) we have in principle more than one scale above M_W (and Λ_{SUSY}): the GUT scale, the Pati-Salam scale where $SU(4)_c$ is broken, the L-R scale where parity (charge conjugation) is broken, the scales of the breaking of $SU(2)_R$ and $U(1)_{B-L}$. Of course, these may be one and the same scale, as expected with low-energy supersymmetry. This solution is certainly there, since the gauge couplings of the MSSM unify successfully and encourage the single step breaking of SO(10).

Is there any room for intermediate mass scales in SUSY SO(10)? It is certainly appealing to have an intermediate see-saw mass scale M_R, between $10^{12} - 10^{15} GeV$ or so. In the non-renormalizable case, with 16_H and $\overline{16}_H$, this is precisely what happens: $M_R \simeq cM_{GUT}^2/M_{Pl} \simeq c(10^{13} - 10^{14})GeV$. In the renormalizable case, with 126_H and $\overline{126}_H$, one needs to perform a renormalization group study using unification constraints. While this is in principle possible, in practice it is hard due to the large number of fields. The stage has recently been set, for all the particle masses were computed [38,39], and the preliminary studies show that the situation may be under control [40]. It is interesting that the existence of intermediate mass scales lowers the GUT scale [38,41] (as was found before in models with 54_H and 45_H [20]), allowing for a possibly observable $d = 6$ proton decay.

Notice that a complete study is basically impossible. In order to perform the running, you need to know particle masses precisely. Now, suppose you stick to the principle of minimal fine-tuning. As an example, you fine-tune the mass of the W and Z in the SM, then you know that the Higgs mass and the fermion masses are at the same scale

$$m_H = \frac{\sqrt{\lambda}}{g}m_W\,, \quad m_f = \frac{y_f}{g}m_W \tag{37}$$

where λ is a ϕ^4 coupling, and y_f an appropriate fermionic Yukawa coupling. Of course, you know the fermion masses in the SM model, and you know $m_H \simeq m_W$.

In an analogous manner, at some large scale m_G a group G is broken and there are usually a number of states that lie at m_G, with masses

$$m_i = \alpha_i m_G \tag{38}$$

where α_i is an approximate dimensionless coupling. Most renormalization group studies typically argue that $\alpha_i \simeq O(1)$ is natural, and rely on that heavily. In the SM, you could then take $m_H \simeq m_W$, $m_f \simeq m_W$; while reasonable for the Higgs, it is nonsense for the fermions (except for the top quark).

In supersymmetry *all* the couplings are of Yukawa type, i.e. self-renormalizable, and thus taking $\alpha_i \simeq O(1)$ may be as wrong as taking all $y_f \simeq O(1)$. While a possibly reasonable approach when trying to get a qualitative idea of a theory, it is clearly unacceptable when a high-precision study of M_{GUT} is called for.

6.2.2. *Proton decay*

As you know, $d = 6$ proton decay gives $\tau_p(d = 6) \propto M_{GUT}^4$, while $(d = 5)$ gives $\tau_p(d = 5) \propto M_{GUT}^2$. In view of the discussion above, the high-precision determination of τ_p appears almost impossible in SO(10) (and even in SU(5)). Preliminary studies [42] indicate fast $d = 5$ decay as expected.

You may wonder if our renormalizable theory makes sense at all. After all, we are ignoring the higher dimensional operators of order $M_{GUT}/M_{Pl} \simeq 10^{-2} - 10^{-3}$. If they are present with the coefficients of order one, we can forget almost everything we said about the predictions, especially in the Yukawa sector. However, we actually know that the presence of $1/M_{Pl}$ operators is not automatic (at least not with the coefficients of order 1). Operators of the type (in symbolic notation)

$$O_5^p = \frac{c}{M_{Pl}} 16_F^4 \tag{39}$$

are allowed by SO(10) and they give

$$O_5^p = \frac{c}{M_{Pl}} [(QQQL) + (Q^c Q^c Q^c L^c)] \tag{40}$$

These are the well-known $d = 5$ proton decay operators, and for $c \simeq O(1)$ they give $\tau_p \simeq 10^{23} yr$. Agreement with experiment requires

$$c \leq 10^{-6} \tag{41}$$

Could this be a signal that $1/M_{Pl}$ operators are small in general? Alternatively, you need to understand why just this one is to be so small. It is appealing to assume that this may be generic; if so, neglecting $1/M_{Pl}$ contributions in the study of fermion masses and mixings is fully justified.

6.2.3. *Leptogenesis*

The see-saw mechanism provides a natural framework for baryogenesis through leptogenesis, obtained by the out-of-equilibrium decay of heavy right-handed neutrinos [43]. This works nicely for large M_R, in a sense too nicely. Already type I see-saw works by itself, but the presence of the type II term makes things more complicated [44]. One cannot be a priori sure whether the decay of right-handed neutrinos or the heavy Higgs triplets is responsible for the asymmetry, although the hierarchy of Yukawa couplings points towards ν_R decay. In the type II see-saw, the most natural scenario is the ν_R decay, but with the triplets running in the loops [45]. This and related issues are now under investigation [46].

7. SUSY? Who needs her?

In the last two decades, and especially after its success with gauge coupling unification, grand unification by an large got tied up with low energy supersymmetry. This is certainly well motivated, since supersymmetry is the only mechanism in field theory which controls the gauge hierarchy. On the other hand, I hope to have convinced you that the right grand unified theory should be based on SO(10), not SU(5). If so, gauge coupling unification needs no supersymmetry whatsoever. It only says that there must be intermediate scales [47], such as Pati-Salam $SU(4)_c \times SU(2)_L \times SU(2)_R$ or Left-Right $SU(3)_c \times SU(2)_L \times SU(2)_R \times U(1)_{B-L}$ symmetry, between M_W and M_{GUT} (the $SU(5)$ route is ruled out). An oasis or two in the desert is always welcome.

Thus if we accept the fine-tuning, as we seem to be forced in the case of the cosmological constant, we can as well study the ordinary, non-supersymmetric version of the theory. In this context the idea of the cosmic attractors [48] as the solution to the gauge hierarchy becomes extremelly appealing. It needs no supersymmetry whatsoever, and enhances the motivation for ordinary grand unified theories. In what follows I discuss some essential features of a possible minimal such theory.

Let me stick to a purely renormalizable theory for the sake of simplicity and predictivity. The minimal such theory is based on 210_H, $\overline{126}_H$ (no need for 126_H as in SUSY) and a "light" 10_H. In this case, the theory is asymptotically free, and thus there is no advantage with small representations. A purely renormalizable theory can alternatively be built with 45_H and 54_H instead of 210_H. Notice that 45_H would not suffice: it turns out to preserve SU(5) [49], and $\overline{126}_H$ must preserve it in order not to break the

SM symmetry.

Intermediate mass scales help lower the masses of ν_R, but create potential problems for the charged fermions on the other hand. We have seen in the supersymmetric version that the light Higgs is a mixture of $(1, 2, 2)$ in 10_H and $(15, 2, 2)$ in $\overline{126}_H$. This is crucial if one is to get correct mass relations between down quarks and charged leptons. These fields can mix through $(15, 1, 1)$ in 45_H or 210_H, and $(10, 1, 3)$ in $\overline{126}_H$, either by the trilinear couplings $(1, 2, 2), (15, 2, 2), (15, 1, 1)$ or the quartic ones $(1, 2, 2), (15, 2, 2), (15, 1, 1)^2; ; (1, 2, 2), (15, 2, 2), (10, 1, 3)^2$ In other words

$$\langle(15, 2, 2)\rangle \simeq \left(\frac{M_I}{M_{GUT}}\right)^n \langle(1, 2, 2)\rangle \tag{42}$$

where $n = 1, 2$ for trilinear and quartic mixings respectively (which depends on what the GUT scale fields are). Since $\langle(15, 2, 2)\rangle$ is needed for the second generation, with $m_\mu/M_W \simeq 10^{-3}$, we have the constraint

$$\left(\frac{M_I}{M_{GUT}}\right)^n \geq 10^{-3} \tag{43}$$

This can be used to eliminate the single intermediate scale chain of symmetry breaking [50].

Another important difference with the SUSY situation lies in the Yukawa sector where now, in the minimal theory, 10_H is real. This implies $v_{10}^u = v_{10}^d$ and fitting fermionic masses and mixings becomes impossible [50].

If it does fail, one could just add another 10_H; this means unfortunately another y_{10}. You can avoid it by postulating a Peccei-Quinn symmetry

$$16_F \to e^{i\alpha}16_F, \quad 10_H \to e^{-2i\alpha}10_H, \quad \overline{126}_H \to e^{-2i\alpha}\overline{126}_H \tag{44}$$

with 10_H now complex. This can give you naturally the axionic dark matter, at the expense of introducing additional $\overline{126}$ (or 126) in order to break both $B - L$ and $U(1)_{PQ}$. Although somewhat unappealing and against the rules of sticking to the pure $SO(10)$, the loss of neutralinos as the dark matter may necessitate this. If you dislike this fact, simply work with two 10_H's and two y_{10}'s. Adding another 10_H is especially mild in the type II see-saw since the relation $M_d - M_e \propto M_\nu$ is independent of the number of 10_H's. Thus $b - \tau$ unification is still connected with a large θ_{atm} as in the supersymmetric case. In recent years not much attention was devoted to the ordinary $SO(10)$, except for the work of the Napoli group (see e.g. Ref. [51]).

8. Summary and Outlook

The see-saw mechanism has emerged in recent years as the simplest and the most natural way of explaining small neutrino masses. It simply means adding right-handed neutrinos to the Standard Model, allowing them to have large gauge invariant masses M_R which break $B - L$ symmetry and through Dirac Yukawa couplings y_D with left-handed neutrinos give the latter non-vanishing, but small masses. As appealing as this may be, as useless it is in practice, The lack of knowledge of M_R and y_D leaves neutrino masses arbitrary.

We have argued in this talk that the most natural framework for see-saw is grand unification, a theory of large mass scale and the stage for the $q - \ell$ symmetry which can hopefully connect y_D with quark Yukawa couplings.

If you accept this, then by now you should be convinced that the right GUT is based on SO(10) gauge group. Basically all you ever wanted is there: right-handed neutrinos, Pati-Salam quark-lepton symmetry, charge conjugation as a L-R symmetry and much more. The trouble is that physics depends not only on the gauge symmetry, but almost as much on the choice of the Higgs sector. It is here that the practitioners cannot agree yet and most attention is devoted to two rather orthogonal approaches. One insists on perturbativity all the way to the Planck scale and chooses small representations: 16_H ($+ \overline{16}_H$ in SUSY) and 45_H. This program then uses $1/M_{Pl}$ operators to generate the physically acceptable superpotential; uses textures to simplify the theory and thus appeal to physics beyond grand unification.

The other approach sticks to the pure SO(10) theory with no $1/M_{Pl}$ operators. This means large representation $\overline{126}_H$ ($+126_H$ in SUSY), 210_H; strong couplings in SUSY at $\lambda_F = 10M_{GUT}$, but is blessed with a small number of couplings and is a complete theory of matter and non-gravitational interactions. Although I prefer this direction, I will refrain from taking a stand here. Suffice it to say that this program is good enough to be testable: it may be only a question of time before proven wrong.

I guess the main message of this short review is a caution when discussing the see-saw mechanism. By itself, it is only an aesthetically appealing scenario devoid of practical use. It makes sense to discuss it only in the context of a well-defined theory based on firm physical principles.

Acknowledgments

I am grateful to the organizers of this excellent meeting for the warm hospitality and for the seesaw birthday party. I wish to acknowledge the great fun Rabi Mohapatra and I always had doing physics, and especially while working on the seesaw. I thank my friends Charan Aulakh, Borut Bajc, Pavel Fileviez-Pérez, Thomas Hambye, Alejandra Melfo and Francesco Vissani for enjoyable ongoing collaboration. Thanks are also due to the members of the CMS group at FESB, Univ. of Split, Croatia for their hospitality during the write-up of this talk. I am especially indebted to Alejandra Melfo for her great help in the preparation of this review. This work is supported by EEC (TMR contracts ERBFMRX-CT960090 and HPRN-CT-2000-00152).

References

1. R. N. Mohapatra, arXiv:hep-ph/0211252.
2. G. Altarelli and F. Feruglio, arXiv:hep-ph/0405048.
3. C. H. Albright, Int. J. Mod. Phys. A **18** (2003) 3947.
4. J. C. Pati, arXiv:hep-ph/0407220.
5. For some recent reviews see e.g. S. M. Bilenky, C. Giunti, J. A. Grifols and E. Masso, Phys. Rept. **379** (2003) 69; V. Barger, D. Marfatia and K. Whisnant, Int. J. Mod. Phys. E **12** (2003) 569; A. Y. Smirnov, Int. J. Mod. Phys. A **19** (2004) 1180.
6. P. Minkowski, Phys. Lett. B **67** (1977) 421; T. Yanagida, proceedings of the *Workshop on Unified Theories and Baryon Number in the Universe*, Tsukuba, 1979, eds. A. Sawada, A. Sugamoto, KEK Report No. 79-18, Tsukuba; S. Glashow, in *Quarks and Leptons, Cargèse 1979*, eds. M. Lévy. et al., (Plenum, 1980, N ew York); M. Gell-Mann, P. Ramond, R. Slansky, proceedings of the *Supergravity Stony Brook Workshop*, New York, 1979, eds. P. Van Niewenhuizen, D. Freeman (North-Holland, Amsterdam); R. Mohapatra, G. Senjanović, Phys.Rev.Lett. **44** (1980) 912
7. G. Lazarides, Q. Shafi and C. Wetterich, Nucl. Phys. B **181** (1981) 287; R. N. Mohapatra and G. Senjanović, Phys. Rev. D **23** (1981) 165.
8. S. Weinberg, Phys. Rev. Lett. **43** (1979) 1566.
9. T. Hambye, Y. Lin, A. Notari, M. Papucci and A. Strumia, Nucl. Phys. B **695** (2004) 169.
10. J. C. Pati and A. Salam, Phys. Rev. D **10** (1974) 275. R. N. Mohapatra and J. C. Pati, Phys. Rev. D **11** (1975) 2558; G. Senjanović and R. N. Mohapatra, Phys. Rev. D **12** (1975) 1502; G. Senjanović, Nucl. Phys. B **153** (1979) 334.
11. N. Arkani-Hamed and S. Dimopoulos, arXiv:hep-th/0405159; G. F. Giudice and A. Romanino, arXiv:hep-ph/0406088; N. Arkani-Hamed, S. Dimopoulos, G. F. Giudice and A. Romanino, arXiv:hep-ph/0409232.
12. S. Dimopoulos, S. Raby, F. Wilczek, Phys. Rev. D **24** (1981) 1681; L.E. Ibáñez, G.G. Ross, Phys. Lett. B **105** (1981) 439; M.B. Einhorn,

D.R. Jones, Nucl. Phys. B **196** (1982) 475; W. Marciano, G. Senjanović, Phys.Rev.D **25** (1982) 3092.

13. M. S. Chanowitz, J. R. Ellis and M. K. Gaillard, Nucl. Phys. B **128**, 506 (1977); A. J. Buras, J. R. Ellis, M. K. Gaillard and D. V. Nanopoulos, Nucl. Phys. B **135** (1978) 66.

14. B. Bajc, P. Fileviez Pérez and G. Senjanović, Phys. Rev. D **66** (2002) 075005; B. Bajc, P. Fileviez Pérez and G. Senjanović, arXiv:hep-ph/0210374; D. Emmanuel-Costa and S. Wiesenfeldt, Nucl. Phys. B **661** (2003) 62.

15. H. Georgi and C. Jarlskog, Phys. Lett. B **86**, 297 (1979).

16. R. Barbieri, J. R. Ellis and M. K. Gaillard, Phys. Lett. B **90** (1980) 249; E. K. Akhmedov, Z. G. Berezhiani and G. Senjanović, Phys. Rev. Lett. **69** (1992) 3013.

17. E. K. Akhmedov, Z. G. Berezhiani, G. Senjanovic and Z. j. Tao, Phys. Rev. D **47**, 3245 (1993) [arXiv:hep-ph/9208230].

18. R. N. Mohapatra, *Phys. Rev.* **D 34**, 3457 (1986); A. Font, L. E. Ibáñez and F. Quevedo, *Phys. Lett.* **B228**, 79 (1989); S. P. Martin, *Phys. Rev.* **D46**, 2769 (1992).

19. C.S. Aulakh, K. Benakli, G. Senjanović, Phys. Rev. Lett. **79** (1997) 2188; C. S. Aulakh, A. Melfo and G. Senjanović, Phys. Rev. D **57**, 4174 (1998); C. S. Aulakh, A. Melfo, A. Rašin and G. Senjanović, Phys. Lett. B **459** (1999) 557.

20. C. S. Aulakh, B. Bajc, A. Melfo, A. Rašin and G. Senjanović, Nucl. Phys. B **597** (2001) 89.

21. B. Bajc, G. Senjanović and F. Vissani, Phys. Rev. Lett. **90** (2003) 051802.

22. For useful reviews on spinors in SO(2N) see R. N. Mohapatra and B. Sakita, Phys. Rev. D **21** (1980) 1062 and F. Wilczek and A. Zee, Phys. Rev. D **25** (1982) 553; see also P. Nath and R. M. Syed, Nucl. Phys. B **618** (2001) 138 and C. S. Aulakh and A. Girdhar, arXiv:hep-ph/0204097.

23. K. S. Babu and R. N. Mohapatra, Phys. Rev. Lett. **70**, 2845 (1993).

24. E. Witten, Phys. Lett. B **91** (1980) 81.

25. K.S. Babu, S.M. Barr, Phys. Rev. D **51** (1995) 2463; K.S. Babu, R.N. Mohapatra, Phys. Rev. Lett. **74** (1995) 2418; G.R. Dvali, S. Pokorski, Phys. Lett. B **379** (1996) 126; S.M. Barr, S. Raby, Phys. Rev. Lett. **79** (1997) 4748; Z. Chacko, R.N. Mohapatra, Phys. Rev. D **59** (1999) 011702; K. S. Babu, J. C. Pati and F. Wilczek, Nucl. Phys. B **566** (2000) 33; C. H. Albright and S. M. Barr, Phys. Rev. Lett. **85** (2000) 244; M. C. Chen and K. T. Mahanthappa, Phys. Rev. D **62** (2000) 113007; Z. Berezhiani and A. Rossi, Nucl. Phys. B **594** (2001) 113; R. Kitano and Y. Mimura, Phys. Rev. D **63** (2001) 016008.

26. R. N. Mohapatra, Phys. Rev. Lett. **56**, 561 (1986).

27. S. M. Barr, Phys. Rev. Lett. **92** (2004) 101601.

28. C.S. Aulakh, R.N. Mohapatra, Phys. Rev. D **28** (1983) 217.

29. T. E. Clark, T. K. Kuo and N. Nakagawa, Phys. Lett. B **115** (1982) 26.

30. D. Chang, R. N. Mohapatra and M. K. Parida, Phys. Rev. D **30** (1984) 1052; X. G. He and S. Meljanac, Phys. Rev. D **41** (1990) 1620; D. G. Lee, Phys. Rev. D **49** (1994) 1417; D. G. Lee and R. N. Mohapatra, Phys. Rev. D **51**

(1995) 1353.

31. C. S. Aulakh, B. Bajc, A. Melfo, G. Senjanović and F. Vissani, Phys. Lett. B **588**, 196 (2004).

32. B. Bajc, G. Senjanović and F. Vissani, arXiv:hep-ph/0402140.

33. K. Matsuda, Y. Koide and T. Fukuyama, Phys. Rev. D **64** (2001) 053015; T. Fukuyama and N. Okada, JHEP **0211** (2002) 011.

34. L. Lavoura, Phys. Rev. D **48** (1993) 5440; B. Brahmachari and R. N. Mohapatra, Phys. Rev. D **58** (1998) 015001; K. y. Oda, E. Takasugi, M. Tanaka and M. Yoshimura, Phys. Rev. D **59** (1999) 055001.

35. H. S. Goh, R. N. Mohapatra and S. P. Ng, Phys. Lett. B **570** (2003) 215; H. S. Goh, R. N. Mohapatra and S. P. Ng, Phys. Rev. D **68** (2003) 115008.

36. S. Bertolini, M. Frigerio and M. Malinsky, arXiv:hep-ph/0406117; W. M. Yang and Z. G. Wang, arXiv:hep-ph/0406221.
B. Dutta, Y. Mimura and R. N. Mohapatra, arXiv:hep-ph/0406262;

37. B. Dutta, Y. Mimura and R. N. Mohapatra, Phys. Rev. D **69** (2004) 115014.

38. B. Bajc, A. Melfo, G. Senjanović and F. Vissani, Phys. Rev. D **70** (2004) 035007;

39. T. Fukuyama, A. Ilakovac, T. Kikuchi, S. Meljanac and N. Okada, arXiv:hep-ph/0401213; C. S. Aulakh and A. Girdhar, arXiv:hep-ph/0204097.

40. C. S. Aulakh and A. Girdhar, arXiv:hep-ph/0405074.

41. H. S. Goh, R. N. Mohapatra and S. Nasri, arXiv:hep-ph/0408139.

42. H. S. Goh, R. N. Mohapatra, S. Nasri and S. P. Ng, Phys. Lett. B **587** (2004) 105; T. Fukuyama, A. Ilakovac, T. Kikuchi, S. Meljanac and N. Okada, JHEP **0409** (2004) 052.

43. M. Fukugita and T. Yanagida, Phys. Lett. B **174** (1986) 45.

44. See the talk by T. Hambye at this conference, and references therein.

45. T. Hambye and G. Senjanović, Phys. Lett. B **582** (2004) 73; S. Antusch and S. F. King, Phys. Lett. B **597** (2004) 199; P. J. O'Donnell and U. Sarkar, Phys. Rev. D **49** (1994) 2118; E. Ma and U. Sarkar, Phys. Rev. Lett. **80** (1998) 5716; G. Lazarides and Q. Shafi, Phys. Rev. D **58** (1998) 071702.

46. P. Fileviez-Pérez, T. Hambye and G. Senjanović, work in preparation.

47. See e.g. D. Chang, R. N. Mohapatra, J. Gipson, R. E. Marshak and M. K. Parida, Phys. Rev. D **31** (1985) 1718.

48. G. Dvali and A. Vilenkin, arXiv:hep-th/0304043; G. Dvali, arXiv:hep-th/0410286.

49. L. F. Li, Phys. Rev. D **9** (1974) 1723.

50. B.Bajc, A. Melfo, G. Senjanović and F. Vissani, work in preparation.

51. See F. Acampora, G. Amelino-Camelia, F. Buccella, O. Pisanti, L. Rosa and T. Tuzi, Nuovo Cim. A **108**, 375 (1995); G. Amelino-Camelia, F. Buccella and L. Rosa, arXiv:hep-ph/9405334; M. Abud, F. Buccella, D. Falcone, G. Ricciardi and F. Tramontano, Mod. Phys. Lett. A **15** (2000) 15 [arXiv:hep-ph/9911238].

EVIDENCE FOR NEUTRINO MASS:
A DECADE OF DISCOVERY

KARSTEN M. HEEGER *

Lawrence Berkeley National Laboratory
Physics Division, MS50R5008
Berkeley, CA 94720, USA
E-mail: kmheeger@lbl.gov

Neutrino mass and mixing are amongst the major discoveries of recent years. From the observation of flavor change in solar and atmospheric neutrino experiments to the measurements of neutrino mixing with terrestrial neutrinos, recent experiments have provided consistent and compelling evidence for the mixing of massive neutrinos. The discoveries at Super-Kamiokande, SNO, and KamLAND have solved the long-standing solar neutrino problem and demand that we make the first significant revision of the Standard Model in decades. Searches for neutrinoless double-beta decay probe the particle nature of neutrinos and continue to place limits on the effective mass of the neutrino. Possible signs of neutrinoless double-beta decay will stimulate neutrino mass searches in the next decade and beyond. I review the recent discoveries in neutrino physics and the current evidence for massive neutrinos.

1. Neutrinos Within the Standard Model

In 1930 Pauli postulated the neutrino as a "desperate remedy" to the energy crisis of the time - the continuous energy spectrum of electrons emitted in nuclear β-decay. With the postulate of a new particle Pauli could account for the continuous spectrum. He assumed that nuclear β-decay emits a neutron together with an electron in such a way that the sum of the energies is constant. Sensitive measurements of the energy and momentum of β-decay electrons and the recoiling nuclei in cloud chambers indicated that substantial quantities of energy and momentum were missing. These experiments left little doubt that a third particle had to be involved. As early as 1932 Enrico Fermi provided a theoretical framework for β-decay which

---

*This work was supported by the Director, Office of Science, High Energy Physics, U.S. Department of Energy under contract no. DE-AC03-76SF00098.

included the neutrino but it took another 25 years before the neutrino was detected experimentally. In 1957 Frederick Reines and Clyde Cowan made the first observation of the free antineutrino through the inverse β-reaction $\overline{\nu}_e + \mathrm{p} \rightarrow \mathrm{e}^+ + \mathrm{n}$ utilizing the flux of $\overline{\nu}$ from the Savannah River nuclear reactor [1]. The muon neutrino was finally detected by Schwartz, Lederman, and Steinberger in 1961 [2]. Neutrinos from pion and kaon decays with energies of hundreds of MeV to several GeV were detected in a 10-ton spark chamber built from aluminum plates that provided distinct signals for the showering of electrons and the tracks of muons generated by neutrinos. The excess of muons produced in the chamber provided evidence that the neutrino produced in the decay of $\pi \rightarrow \mu + \nu$ was distinct from those produced in β-decays. The total number of light neutrino types, N_ν, has been deduced from the studies of Z production in e^+e^- collisions. Assuming that the invisible partial decay width is due to light neutrino species with partial width Γ_ν the number of active, light neutrino species is given by

$$N_\nu = \frac{\Gamma_{inv}}{\Gamma_l} \left(\frac{\Gamma_l}{\Gamma_\nu} \right) \tag{1}$$

The LEP and SLC lineshape measurement of Z bosons left little doubt about the existence of the ν_τ. In 2001 the existence of ν_τ was confirmed by the Fermilab DONUT experiment [3]. Seven decades after Pauli's postulate it was experimentally established that there are three neutrinos associated with the flavor of their leptonic interaction. In the absence of any other insight the neutrino was assumed to be massless, an ad-hoc assumption in the Standard Model of particle physics.

2. Birth of the Solar Neutrino Problem

Around the same time Pauli postulated the neutrino, Bethe and Critchfield proposed pp fusion, $\mathrm{p} + \mathrm{p} \rightarrow {}^2\mathrm{H} + \mathrm{e}^+ + \nu_e$, as the mechanism for solar energy generation. The solar nuclear reactions fuse protons into helium and release neutrinos with energies of up to 15 MeV. As early as 1946 and 1949, Bruno Pontecorvo and Luis Alvarez proposed independently neutrino detection via ν_e capture on chlorine through ${}^{37}\mathrm{Cl} + \nu_e \rightarrow {}^{37}\mathrm{Ar} + \mathrm{e}^-$. Using this idea Ray Davis built a chlorine detector in the 1960's to detect neutrinos from the Sun and *"to see into the interior of a star and thus verify directly the hypothesis of nuclear energy generation in stars"*. The efforts of Ray Davis and John Bahcall in the measurement of the solar neutrino flux and the development of solar models and the prediction of the solar neutrino flux resulted in the birth of neutrino astrophysics. In his experiment Ray Davis measured a significantly lower flux of solar neutrinos than predicted by current solar models. The results from this first solar neutrino experiment are shown in Figure 1.

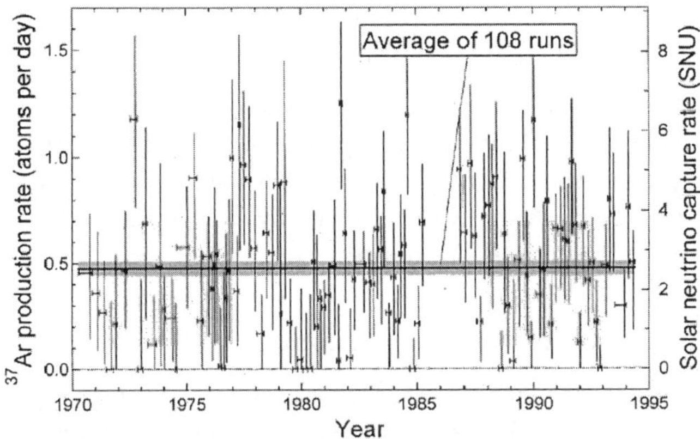

Figure 1. Measured solar neutrino capture rate in Ray Davis' chlorine experiment at Homestake between 1970 and 1995. The observed average solar neutrino flux was $2.56\pm0.16\pm0.16$ SNU, about a third of the current solar model prediction of $7.6^{+1.3}_{-1.1}$ SNU. The neutrino capture rate is given in Solar Neutrino Units (SNU). One SNU is equivalent to $10^{-36}\mathrm{s}^{-1}$ interactions per nucleus. Figure from [4].

For more than 30 years now, experiments have observed neutrinos produced in the thermonuclear fusion reactions which power the Sun. The solar neutrino flux has been measured through the charged-current $({}^{37}\mathrm{Cl} + \nu_e \rightarrow {}^{37}\mathrm{Ar} + e^-,\ {}^{71}\mathrm{Ga} + \nu_e \rightarrow {}^{71}\mathrm{Ge} + e^-)$ or elastic scattering $(\nu_x + e^- \rightarrow \nu_x + e^-)$ channels. Data from these solar neutrino experiments were found to be incompatible with the predictions of solar models. More precisely, the flux of neutrinos detected on Earth was less than expected, and the measured relative intensities of the neutrino sources in the Sun were incompatible with those predicted by solar models.

A variety of hypotheses including neutrino decay were postulated to explain the discrepancy between solar model expectations and the apparent deficit of solar neutrinos detected on Earth. As early as 1969, Bruno Pontecorvo proposed that neutrinos might oscillate between the electron and muon flavor, the only states known at the time [5]. Oscillations can occur if the physical neutrinos consist of a superposition of mass states. If neutrinos are massive an initially pure flavor state changes as the neutrino propagates. Neutrino mass and flavor mixing are not features of the Standard Model of particle physics and neutrino flavor change through oscillation requires the existence of massive neutrinos. For two neutrino flavors the survival probability of neutrinos in vacuum is given by

$$P(\nu_l \to \nu_k) = \sin^2 2\theta \sin^2 \left(1.27 \Delta m^2 \frac{L}{E_\nu} \right) \tag{2}$$

where $\Delta m^2 = \left| m_2^2 - m_1^2 \right|$ is given in eV^2, L in km and E_ν in GeV. The neutrino survival probability $P_{l \to k}$ depends on ratio of the distance traveled over the energy of the neutrino L/E and the mixing angle θ and the mass splitting Δm^2. L/E is usually determined by experiments while θ and Δm^2 are fundamental parameters of nature.

3. The Atmospheric Neutrino Anomaly

Strong indication for neutrino oscillation first came from the observation of atmospheric neutrinos. These neutrinos are the decay products of hadronic showers produced by cosmic ray interactions in the atmosphere. The pion production in the atmosphere determines the flux of atmospheric neutrinos incident on Earth. Around 1 GeV, where the product of flux and neutrino charged-current interactions cross-section reaches a maximum, the atmospheric neutrino flux is about 1 cm^{-2}s^{-1}. Atmospheric neutrinos span an energy range from $\sim$ 0.5-5 GeV. The path length of downgoing and upward going atmospheric neutrinos varies from 10-10,000 km and provides an opportunity for oscillation studies over a wide range of distances.

If all pions and muons decay we expect to observe about two μ-like neutrinos for each e-like neutrino. The ratio $\nu_e/\overline{\nu}_e$ is expected to be close to μ^+/μ^-, about 1.2 at 1 GeV. Several effects including the muon decay length at high energies above 2.5 GeV, the muon energy loss, the geomagnetic cut-off and the variations of the cosmic ray flux with the solar cycle affect this prediction. Figure 2 shows the results from various experiments that measured the total atmospheric neutrino flux. The "ratio of ratio" R_{atm} is used to compare the results of various experiments. A number of experiments found a statistically low value of R_{atm}.

The Super-Kamiokande experiment has measured the up-down asymmetry of the the atmospheric ν_μ flux as well as the zenith angle distribution [6]. Figure 3 shows the experimental result. Only a zenith-angle dependent transformation of neutrino flavors can explain these measurements. Up-

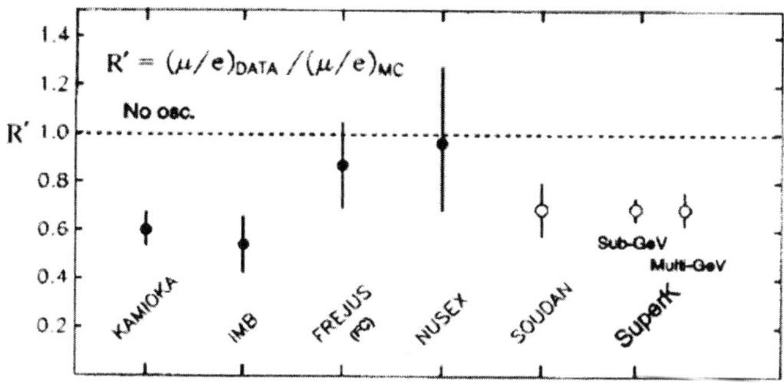

Figure 2. Measurements of the double ratio $R_{atm} = \dfrac{(\nu_\mu + \overline{\nu}_\mu)/(\nu_e + \overline{\nu}_e)_{Data}}{(\nu_\mu + \overline{\nu}_\mu)/(\nu_e + \overline{\nu}_e)_{MC}}$ in atmospheric neutrino experiments. The ratio denotes the ratio of the number of μ-like to e-like neutrino interactions. R_{atm} estimates the atmospheric neutrino flavor ratio and is expected to be 1.

ward going neutrinos traverse a much longer distance and have time to oscillate whereas downward going neutrinos do not. For electron neutrinos the event rate is independent of direction and the solid angle. The hypothesis of $\nu_\mu \rightarrow \nu_\tau$ oscillations fits well the angular distribution of the atmospheric neutrino flux. In the oscillation model, the mixing angle θ_{23} is found to be near maximal and the separation of mass states is about $\Delta m^2_{atm} \sim 2 \times 10^{-3}$ eV^2. A recent analysis of the L/E distribution of the data excludes neutrino decay and decoherence as possible explanations [6]. Besides oscillation, no other consistent particle physics explanation has been proposed to explain the atmospheric neutrino result.

4. Neutrino Flavor Change in Solar Neutrinos

Solar neutrinos are produced in the light element fusion reactions that power the Sun. Using input from astronomical observation, nuclear physics, and astrophysics, models have been developed that allow us to make detailed predictions of the life cycle of stars and their energy generation. Solar models trace the evolution of the Sun over the past 4.7 billion years of main sequence burning, thereby predicting the present-day temperature and composition profile of the solar core. It is believed that thermonuclear reaction chains generate the solar energy. Standard solar models (SSM) predict that over 98% of the solar energy is produced from the pp-chain conversion of four protons into ^{4}He, $4p \rightarrow {}^4\text{He} + 2e^+ + 2\nu_e$, while the proton burning through the CNO cycle contributes the remaining 2%. Ac-

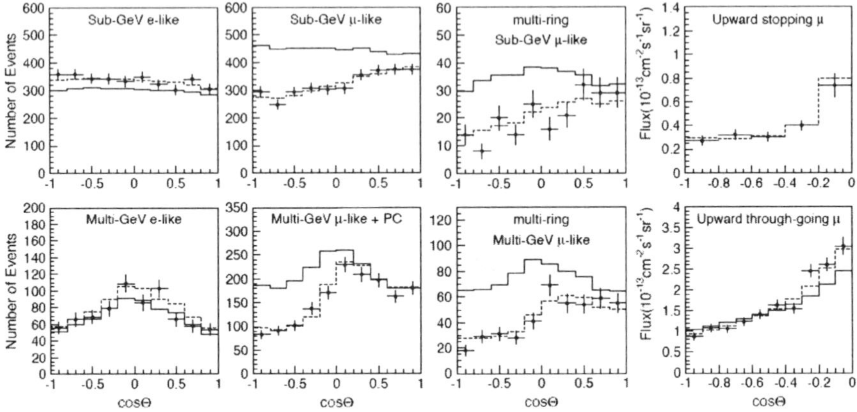

Figure 3. Zenith-angle distribution for fully-contained single-ring e-like and μ-like events, multi-ring μ-like events, partially contained events, and upward-going muons. The points are the data and the solid lines show the Monte Carlo events without neutrino oscillation. The dashed lines show the best-fit expectations for $\nu_\mu \to \nu_\tau$ oscillation. Figure from [6].

cording to standard solar models only 0.01% of the total solar neutrino emission is produced in the β-decay of $^8\mathrm{B} \to {}^8\mathrm{Be} + e^+ + \nu_e$. Solar models are constrained to produce today's solar radius, mass, and luminosity. The predictions of these models are in good agreement with recent observations from helioseismology and other observables.

Over the past 30 years the flux of electron neutrinos from the Sun has been detected and measured in a number of experiments using a variety of experimental techniques. All solar neutrino experiments found indications for a suppression of the solar neutrino flux. With different detection thresholds the experiments have provided information across the entire solar neutrino spectrum from sub-MeV to about 15 MeV. In all cases the solar neutrino flux measurements fall significantly below the predictions of the standard solar models. By the mid-1990's the data were beginning to suggest that one could not even in principle adjust solar models sufficiently to account for the effects. A model-independent analysis of the available data showed that no change in the solar models can completely account for the discrepancy between data and the energy-dependent solar neutrino flux predictions [7]. If the experimental uncertainties are correctly estimated and the Sun is generating energy by light-element fusion in quasistatic equilibrium, the probability of a solution to the solar neutrino problem within the minimal Standard Model of particle physics is less than 2%. Novel neutrino properties seemed to be called for. The long-standing Solar Neutrino Problem indicated that either solar models are incorrect and do not predict correctly the neutrino production and emission from the Sun or solar neutrinos undergo a flavor-changing transformation on their way from the Sun to Earth, and the electron solar neutrino flux detected in all first-generation solar neutrino experiments was only a component of the total solar neutrino flux.

With the recent measurements of the Sudbury Neutrino Observatory (SNO) [8] it has finally become possible to test solar model predictions and the particle properties of neutrinos independently. With D_2O as its target medium the SNO detector is uniquely suited to make a simultaneous measurement of the solar ν_e flux and the total flux of all active ^{8}B neutrinos. In SNO, solar ^{8}B neutrinos interact with deuterium in three different reactions: The charged-current interaction of electron neutrinos with deuterium (CC), the neutral-current dissociation of deuterium though the interaction with active neutrino flavors, and the elastic scattering off electrons. Only the charged-current reaction is exclusively sensitive to ν_e.

$$(CC) \quad \nu_e + d \rightarrow p + p + e^-$$
$$(NC) \quad \nu_x + d \rightarrow p + n + e^-$$
$$(ES) \quad \nu_x + e^- \rightarrow nu_x + e^-$$

The sensitivity of SNO to the neutral current channel, the total flux of active solar neutrinos, allows it to make several key measurements: Comparing the NC to CC interaction rate SNO can test directly for neutrino flavor change independent of any solar model predictions. The measurement of the total flux of solar ^{8}B neutrinos provides a good test of neutrino flux predictions in solar models. The diurnal time dependence and distortions in the neutrino spectrum are direct signatures of neutrino oscillation.

Using first pure D_2O and then heavy water with dissolved NaCl to increase the neutron capture energy and efficiency in the NC interaction channel, SNO has measured the total solar neutrino flux [9]

$$\phi_{total}^{SNO} = 5.21 \pm 0.27(\text{stat}) \pm 0.38(\text{syst}) \times 10^6 \text{cm}^{-2}\text{s}^{-1} \tag{3}$$

The interaction rates in the NC, CC, and ES channels are determined from a statistical separation of events using the angular distribution, the event isotropy, and characteristic detector distributions. This measurement of the solar ^{8}B ν flux is in excellent agreement with previous measurements and standard solar models. The ratio of the solar electron neutrino flux to the total flux of active ^{8}B neutrinos indicates a clear "deficit" of solar electron neutrinos:

$$\frac{\phi_{CC}^{SNO}}{\phi_{NC}^{SNO}} = 0.306 \pm 0.026(\text{stat}) \pm 0.024(\text{syst}) \tag{4}$$

Figure 4 shows a summary of the SNO solar neutrino flux measurements from the D$_2$O phase of the experiment [10].

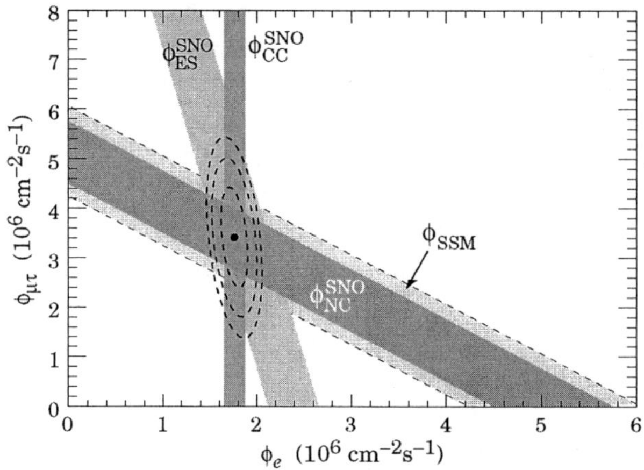

Figure 4. Measured ^{8}B solar ν_μ and ν_τ flux, $\phi_{\mu,\tau}$, versus the observed ν_e flux, ϕ_e, deduced from the NC, ES, and CC neutrino interaction rates in SNO. The dashed lines show the total ^{8}B neutrino flux as predicted by standard solar models. The bands intersect at the fit values for ϕ_e and $\phi_{\mu,\tau}$. This illustrates the $\nu_e \to \nu_{\mu,\tau}$ transformation of solar ^{8}B solar neutrinos. Figure from [20].

The ϕ_e and $\phi_{\mu,\tau}$ measurements of SNO alone provide direct evidence for the flavor change of solar neutrinos. Together with other solar neutrino experiments, the available experimental data probe different regions of the solar neutrino energy spectrum and test the energy-dependent oscillation effect of solar neutrinos. Solar neutrinos pass through dense solar matter before they escape from the surface of the Sun and travel to Earth. The interaction of neutrinos with matter in the Sun and Earth creates an additional effective potential for electron neutrinos and enhances the oscillation probability of ν_e by shifting the energy of the states through the so-called MSW effect [12]. The model of matter-enhanced neutrino oscillation provides an excellent description of the available solar neutrino data, as shown in

Figure 5.

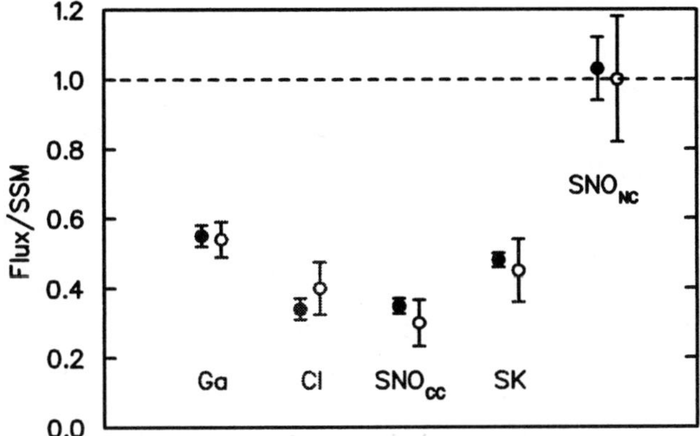

Figure 5.   Ratio of the measured solar neutrino flux to the predicted flux (in the absence of ν oscillations) for various experiments. The experimental data (filled circles) and the best-fit predictions from the ν oscillation hypothesis (open circles) are in good agreement. Figure from [11].

5. Signatures of Neutrino Oscillation in Reactor Experiments

Reactor neutrino experiments have played an important role in the history of neutrino physics. From the first direct detection of the antineutrino by Frederick Reines and Clyde Cowan [1] to the searches for neutrino oscillation and neutrino magnetic moment. For the past five decades nuclear power plants have been used as sources for low-energy studies with electron antineutrinos. Fission reactions in ^{238}U, ^{239}Pu, ^{241}Pu, ^{235}U, and other isotopes produce $\sim 6\ \overline{\nu}_e$ per fission with an energy release of about 200 MeV per fission. On average nuclear reactors produce $\sim 2 \times 10^{20}\ \overline{\nu}_e$ per GW$_{th}$-sec with energies up to ~ 8 MeV and an average energy of about 4 MeV.

The observation of neutrino flavor change with large mixing at a mass splitting of $\Delta m^2 \sim 7.1 \times 10^{-5}$ eV^2 in solar neutrinos suggests that neutrinos or antineutrinos undergo oscillations in vacuum with a baseline of $\mathcal{O}(100$ km). The Kamioka Liquid Scintillator Antineutrino Detector (KamLAND) located in the Kamioka underground laboratory in Japan is uniquely suited to measure the flux of reactor $\overline{\nu}_e$. With 1000 t of liquid scintillator detector KamLAND measures the interaction rate and energy spectrum of $\overline{\nu}_e$ using the coincidence signal of the e^+ annihilation and the neutron capture in the inverse β-reaction $\overline{\nu}_e + p \rightarrow e^+ + n$. About 95% of the $\overline{\nu}_e$ flux at KamLAND comes from commercial power plants in Japan. The flux-averaged mean baseline is about 180 km.

In 2003, KamLAND made the first direct observation of reactor $\overline{\nu}_e$ disappearance. Based on an exposure of 162 kt-yr KamLAND observed 54 events above 2.6 MeV compared to an expected number of 86.8 ± 5.6 events. The expected number of $\overline{\nu}_e$ interactions derives from the calculated flux of antineutrinos from the nuclear power plants. Under the assumption of CPT invariance the observed deficit in the reactor $\overline{\nu}_e$ flux and the observed flavor change of solar ν's point to neutrino oscillation as a consistent explanation of all experimental data. An oscillation analysis of the available data yields good agreement between the oscillation parameters for ν and $\overline{\nu}$ [17].

With a livetime of 766.3 ton-yr KamLAND has recently published a more accurate measurement of the reactor $\overline{\nu}_e$ flux and spectrum providing unambiguous evidence from KamLAND alone for the oscillation of reactor antineutrinos. The shape of the energy spectrum measured by KamLAND is inconsistent with the energy spectrum of reactor $\overline{\nu}_e$ in the absence of oscillations at the 99.6% C.L. For a constant baseline the neutrino survival probability $P_{ee} = 1 - \sin^2 2\theta \sin^2 \left(\Delta m^2 \frac{L}{E_\nu} \right)$ depends on the neutrino energy E_ν, and spectral distortions are a characteristic signature of the oscillation effect. The current limits on neutrino oscillation parameters from reactor and solar experiments are summarized in Figure 9.

6. Evidence for Neutrino Mass in Oscillation Experiments

Experimental studies of terrestrial, atmospheric, and solar neutrinos have established the flavor change and mixing of massive neutrinos. Measurements of atmospheric and accelerator experiments and solar and reactor neutrino observatories yield two different mass scales for the oscillation: $\Delta m_{atm}^2 \sim 2.0 \times 10^{-3}$ eV^2 and $\Delta m_{sol}^2 \sim 7.1 \times 10^{-5}$ eV^2. These measurements define the relative mass scale and allow two possible mass spectra, as shown in Figure 8. The absolute scale of the mass spectra is yet unknown but the minimum scale is given by the larger mass splitting $m \geq \sqrt{\Delta m_{atm}^2} \simeq 50$ meV. The mixing angles associated with the atmospheric and solar transitions are nearly maximal and large, respectively. Combining the current results from all oscillation experiments we obtain the allowed Δm^2-$\tan^2 \theta$ oscillation parameter regions shown in Figure 9.

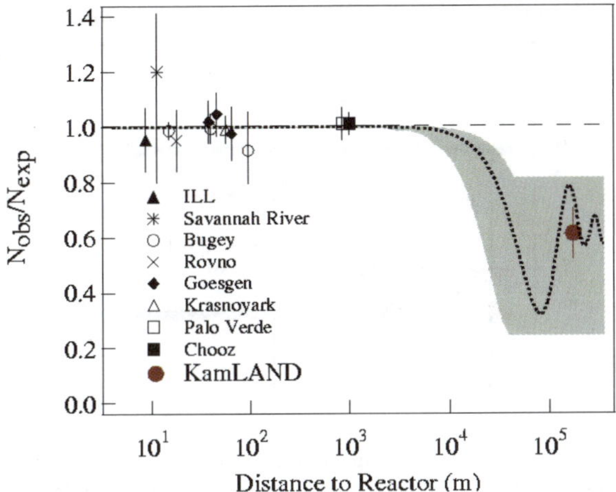

Figure 6. Ratio of the $\overline{\nu}_e$ flux measured in reactor experiments to the expected $\overline{\nu}_e$ flux in the absence of neutrino oscillation as a function of baseline. The shaded-region indicates the range of flux predictions corresponding to the 95% C.L. large-mixing-angle region found in a global analysis of solar ν data. KamLAND made the first observation of the disappearance of $\overline{\nu}_e$ and confirmed the oscillation predictions from solar neutrino experiments. Figure from [15].

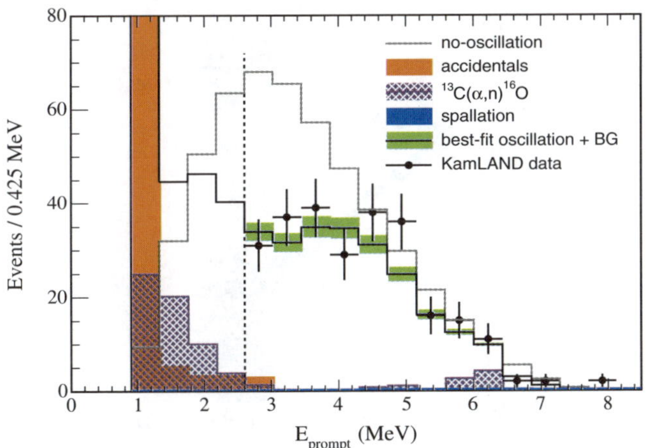

Figure 7. Prompt energy spectrum of $\overline{\nu}_e$ candidate events and associated background spectra. The shaded region indicates the systematic error in the best-fit reactor spectrum above 2.6 MeV. The observed $\overline{\nu}_e$ spectrum is not only suppressed but incompatible with the expected spectrum at 99.6% C.L. Figure from [16].

76

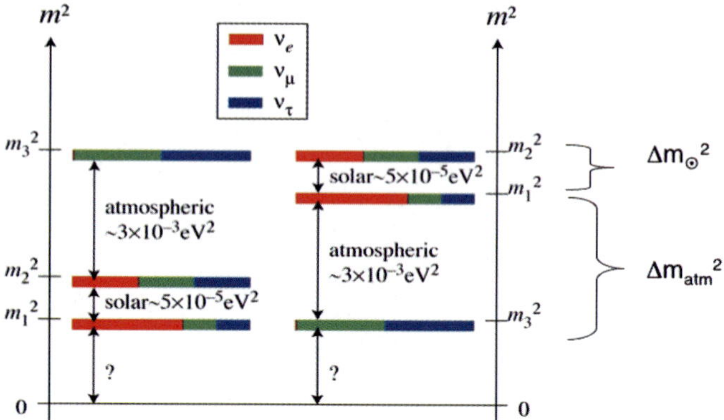

Figure 8. Normal and inverted mass spectrum for three neutrino states. The mass differences have been measured precisely in oscillation experiments with various baselines using neutrino and antineutrino sources with different energies.

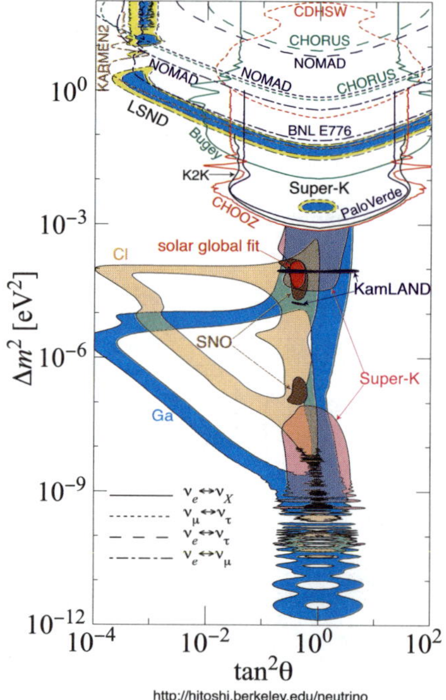

Figure 9. Neutrino oscillation parameters as measured by atmospheric and accelerator experiments ("atmospheric region") and solar neutrinos and reactor antineutrinos ("solar region"). The global fit of all solar experiments is consistent with the oscillation parameters of reactor antineutrinos under CPT invariance. Figure from [17].

7. Direct Neutrino Mass Measurements

Over the past decades there has been steady progress in probing neutrino masses through direct measurements of decay kinematics. Direct kinematical measurements of neutrino masses give values consistent with zero. Techniques for measuring the mass of the electron neutrino involve the search for a distortion in the shape of the β-spectrum in the endpoint region. Tritium β-decay is commonly used for this measurement because of its low endpoint energy and simple nuclear and atomic structure. Tritium β-decay experiments use electromagnetic or magnetic spectrometers to analyze the momentum of the electrons and to infer the endpoint energy of the spectrum. The current best limits of $m_{\nu_e} \leq 2.2$ eV/c^2 at 90% C.L. comes from the Mainz and Troitsk neutrino mass experiments [18]. A new experiment, the Karlsruhe Tritium Neutrino Experiment (KATRIN), with an expected sensitivity of 0.2 eV at 90% C.L. is under construction [19].

Direct limits on both the muon and tau neutrino masses are based on kinematic measurements using semileptonic, weak particle decays. The observables in these measurements are either invariant mass or the decay particle momentum. As these measurements rely on knowing the particle mass and momentum the sensitivity of these measurements to the neutrino mass is limited. Cosmology and nucleonsynthesis as well as the supernova

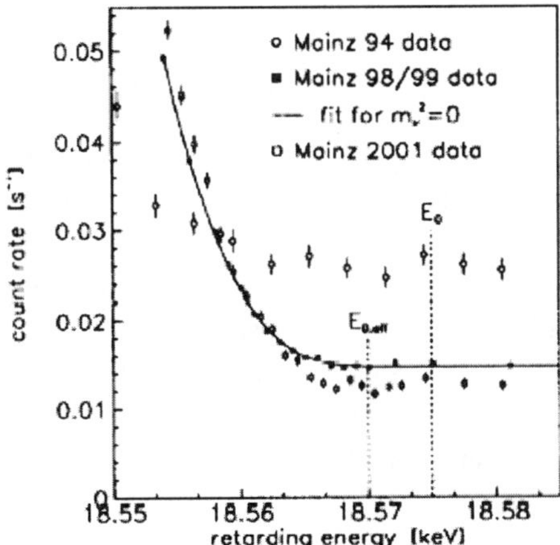

Figure 10.   Average count rate of tritium β-decays near the endpoint of 18.6 keV in the Mainz neutrino mass experiment. The count rate is shown as a function of the retarding energy of the spectrometer. An analysis of this data yields an upper limit on the neutrino mass of $m_{\nu_e} \leq 2.2$ eV/c^2 at 90% C.L. Figure from [18].

1987A set limits far lower than those placed by direct mass measurements. The direct experimental limits on neutrino mass as reported by the Particle Data Group [20] are summarized in Table 1.

Direct kinematic methods have not yet measured a non-zero neutrino mass. At present there is no direct indication from these experiments for new physics beyond the Standard Model and other searches for the signature of massive neutrinos are needed.

Neutrino Mass	Mass Limit	Decay Mode	Experiment
m_{ν_e}	< 2.2 eV	$^3\text{H} \to {}^3\text{He} + e^- + \nu_e$	Mainz [18]
m_{ν_μ}	< 190 keV	$\pi^+ \to \mu^+ + \nu_\mu$	PSI [21]
m_{ν_τ}	< 18.2 MeV	$\tau^- \to 2\pi^-\pi + \nu_\tau$	ALEPH [22]
		$\tau^- \to 3\pi^-2\pi + \nu_\tau$	

8. Neutrino Constraints from Cosmology

Stable neutrinos with masses as large as the limits from direct kinematic measurements would certainly overclose the Universe, i.e. contribute such a large cosmological density that the Universe could have never attained its present age. Cosmology implies a much lower upper limit on these neutrino masses. Considering the freezeout of neutrinos in the early Universe it can be shown that the mass density and the sum of the neutrino masses are related as

$$\sum m_{\nu_x} = 93\Omega_m h^2 \text{ eV} \tag{5}$$

where Ω_m is the mass contribution to the cosmological constant. Analysis of the cosmic microwave background anisotropy combined with the galaxy redshift surveys and other data yield a constraint on the the sum of the neutrino masses of $\sum m_{\nu_i} \leq 0.7$ eV [23]. The model dependence of this result is presently under discussion. Big Bang nucleonsynthesis constrains the parameters of possible sterile neutrinos which do not interact and are produced only by mixing. The current limit on the total number of neutrinos from Big Bang nucleosynthesis is $1.7 \leq N_\nu \leq 4.3$ at 95% C.L.

9. Probing the Nature of Neutrinos and ν Mass in $0\nu\beta\beta$

Another unique signature of massive neutrinos is neutrinoless double β-decay, a lepton-number-violating process also known as $0\nu\beta\beta$. The process $(A, Z) \to (A, Z + 2) + 2e^-$ can be mediated by an exchange of a

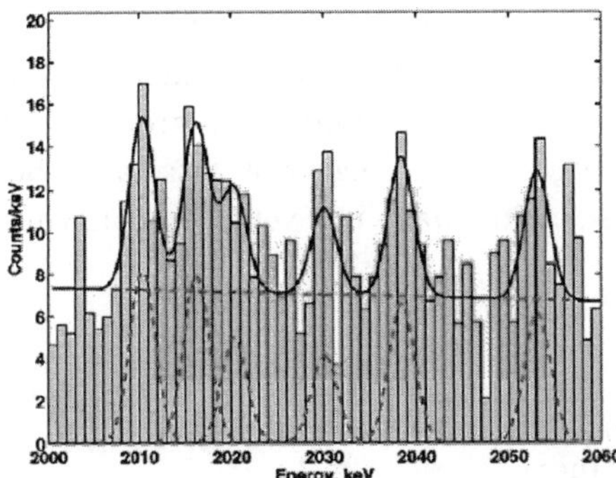

Figure 11.   Energy spectrum of 10.96 kg enriched ^{76}Ge in the range of 2000-2060 keV. The line indicates the identified peaks including Bi backgrounds at 2010.7, 2016.7, 2021.8, 2052.9 keV and an additional signal at ~2039 keV. This corresponds to the Q value of the $0\nu\beta\beta$ process. Figure from [25].

light Majorana particle, or an exchange of other articles. The existence of $0\nu\beta\beta$ requires the existence of Majorana neutrino mass independent of the mechanism of the process. Neutrinoless double-beta decay is the only experimental approach known to date that distinguishes between Majorana and Dirac masses. The experimental signature of $0\nu\beta\beta$ is a peak in the combined electron spectrum at the $Q_{\beta\beta}$-value of the reaction. The observable $0\nu\beta\beta$-decay rate $1/T_{1/2}^{0\nu}$ is proportional to the effective Majorana mass squared $|\langle m_{\beta\beta}\rangle|^2$

$$1/T_{1/2}^{0\nu} = G^0\nu \left|M^{0\nu}\right|^2 \left|\langle m_{\beta\beta}\rangle\right|^2 \tag{6}$$

with $\langle m_{\beta\beta}\rangle = \sum_i U_{ei}^2 m_{\nu i}$. The lifetime measurement is translated into an effective Majorana mass using nuclear structure calculations which in turn can be used to set upper limits on the neutrino mass. The phase factor $G^{0\nu}$ can be calculated reliably but there is significant uncertainty in the calculations of the matrix elements $M^{0\nu}$.

The best current limits on $T^{0\nu}$ and $\langle m_{\beta\beta}\rangle$ come from the Heidelberg-Moscow experiment which used 11 kg of enriched ^{76}Ge with an isotopic abundance of 86% [24]. Until recently, the Heidelberg-Moscow collaboration reported a lower limit on the half-life and upper limit on the effective neutrino mass:

$$T_{1/2}^{0\nu} \geq 1.9 \times 10^{25} \text{ yr } (90\% \text{ C.L.})$$
$$m_{\beta\beta} \leq 0.35 \text{ eV } (90\% \text{ C.L.})$$

A recent analysis of data from this experiment by Klapdor-Kleingrothaus *et al.* led to the announcement of the discovery of neutrinoless double-beta decay. Klapdor-Kleingrothaus *et al.* report a 4.2 σ evidence for $0\nu\beta\beta$ based on 71.7 kg-yr of data taken between August 1990-May 2003 [25]. These claims have not yet been confirmed.

References

1. F. Reines and C.L. Cowan, Phys. Rev. **117**, 159 (1960); F. Reines and C.L. Cowan, Phys. Rev. **92**, 830 (1953); C.L. Cowan and F. Reines, Phys. Rev. **106**, 825 (1957).
2. G. Danby *et al.*, Phys. Rev. Lett. **9**, 36 (1962).
3. K. Kodama *et al.* (DONUT Collaboration), Phys. Lett. **B504**, 218 (2001); T. Patzak, Europhys. News **32**, 56 (2001).
4. R. Davis, Int. J. Mod. Phys. **A18**, 3089 (2003).
5. B. Pontecorvo, Zh, Eksp. Teor. Fiz. 33 549 (1957) and 34, 247 (1958). **C34**, 1729 (1980).
6. Y. Fukuda *et al.* (Super-Kamiokande Collaboration), Phys. Rev. Lett. **81**, 1562 (1998); Y. Ashie *et al.* (Super-Kamiokande Collaboration), Phys. Rev. Lett. **93**,101801 (2004).
7. K.M. Heeger and R.G.H. Robertson, Phys. Rev. Lett. **77**, 3720 (1996).
8. J. Boger *et al.* (SNO Collaboration), Nucl. Instrum. Meth. **A449**, 172 (2000).
9. S.N. Ahmed *et al.* (SNO Collaboration), Phys. Rev. Lett. **92**, 181301 (2004).
10. Q.R. Ahmad *et al.* (SNO Collaboration), Phys. Rev. Lett. **89**, 011301 (2002).
11. R.D. McKeown and P. Vogel, Phys. Rept **394**, 315 (2004).
12. L. Wolfenstein, Phys. Rev. **D17**, 2369 (1978); S.P. Mikheyev and A. Yu. Smirnov, Sov. J. Nucl. Phys. **42**, 913 (1985).
13. F. Böhm *et al.* (Palo Verde Collaboration), Phys. Rev. **D64**,112001 (2001).
14. M. Apollonio *et al.* (CHOOZ Collaboration), Eur. Phys. J. **C27**, 331 (2003).
15. K. Eguchi *et al.* (KamLAND Collaboration), Phys. Rev. Lett. **90**, 021802 (2003).
16. T. Araki *et al.* (KamLAND Collaboration), hep-ex/0406035 (2004).
17. H. Murayama, http://hitoshi.berkeley.edu/neutrino/, (2004).
18. Ch. Weinheimer *et al.*, Phys. Lett. **B460**, 219 (1999); Ch. Weinheimer *et al.*, Phys. Lett. **B464**, 352 (1999); Ch. Weinheimer, Nucl. Phys. **B118**, 279 (2003).
19. A. Osipowicz *et al*, arXive:hep-ex/010903 (2001); http://www-ik1.fzk.de/tritium/.
20. S. Eidelman *et al.* (Particle Data Group), Phys. Lett. **B592**, 1 (2004).
21. K. Assamagan *et al.*, Phys Rev. **D53**, 6065 (1996).
22. R. Barate *et al.* Eur. Phys. J. **C2**, 395 (1998).
23. N. Spergel *et al.* Astrophys. J. Suppl. **148**, 175 (2003).
24. H.V. Klapdor-Kleingrothaus *et al.*, Eur. Phys. J. **A12**, 147 (2001).
25. H.V. Klapdor-Kleingrothaus *et al.*, Phys. Lett. **B586**, 198 (2004).

TEXTURES AND FLAVOUR MODELS

G. G. ROSS

The Rudolf Peierls Centre for Theoretical Physics
1 Keble Road
Oxford
OX1 3NP, UK

We discuss the implications of the measurements of neutrino masses and mixing angles for an underlying theory of flavour. The near bi- tri- maximal mixing observed is significantly different from the small mixing angles observed in the quark sector. If this structure is due to an underlying family symmetry it suggests the symmetry be non-Abelian in nature in order to relate the different family Yukawa couplings responsible for generating the mixing angles. We discuss how the combination of a non-Abelian family symmetry together with the see-saw mechanism can explain why there are large mixing angles in the neutrino sector while the mixing angles in the quark sector are small. Phenomenological implications of such a family symmetry are discussed.

1. Introduction

In the Standard Model or its minimal supersymmetric extension there is no explanation for the pattern of fermion masses and mixings of the quarks and leptons. This has stimulated extensive work trying to explain the observed structure. Perhaps the most conservative suggestion for their origin is that the symmetry of the Standard Model is extended to include a (spontaneously broken) family symmetry. If one restricts the discussion to the quark sector it is possible to build quite elegant examples which generate the observed hierarchical structure of masses and mixing angles. However attempts to extend this to the leptons has proved very difficult, mainly because of the large mixing angles needed to explain neutrino oscillation is quite different from the small mixing angles observed in the quark sector. However this discrepancy may not be as bad as it seems at first sight because the neutrino mass matrix cannot be directly related to the quark and charged lepton mass matrices. This follows because the quark and lepton masses are pure Dirac masses while the light neutrinos masses can arise as a result of a combination of Dirac and Majorana masses through

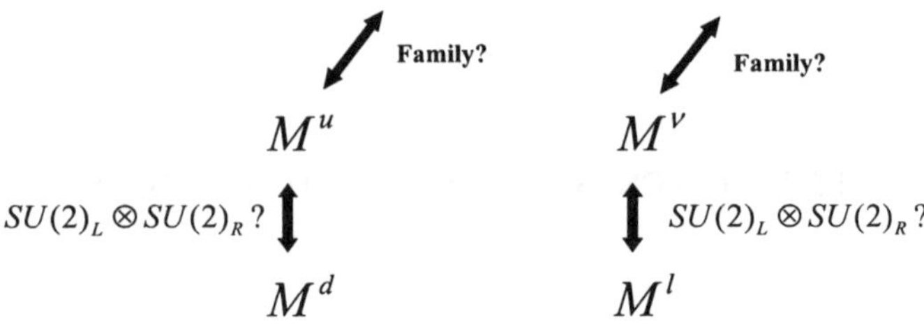

Figure 1.   The possible symmetries relating the quark and lepton mass matrices

the see-saw mechanism. As we shall discuss this can explain the observed differences between quark and lepton mass matrix structure and may allow a complete description of fermion masses and mixings to come from an underlying extension of the Standard Model symmetry.

What is the most symmetric possibility for an extended symmetry group [1]? The possibilities are illustrated in Figure 1. The fact that, after including radiative corrections, the bottom quark and the tau lepton are closely degenerate suggests a GUT type relation between down quark and charged lepton masses. Although the light down quarks are not degenerate as we show below their mass too can be accommodated in a simple GUT such as $SU(5)$ and the full form of the down quark and charged lepton mass matrices is consistent with an underlying GUT. This suggests that there may be a similar GUT mass relation connecting the up quark mass matrix to the Dirac matrix of neutrino masses. Of course the GUT gauge group will contain the Standard Model gauge group and possibly an $SU(2)_R$ factor as arises in an underlying $SO(10)$ GUT. As shown in Figure 1 this can relate the up to the down mass matrices and the Dirac matrix of neutrino masses to the charged lepton mass matrix.

The extended symmetry group may also include a family symmetry. The recent data, c.f. Figure 2, suggests that neutrino mixing is consistent with near (bi-) maximal mixing between the tau and the muon neutrinos for the atmospheric neutrino and nearly equal (tri-maximal) mixing between the electron, muon and tau neutrinos for the solar neutrino.

If an underlying family symmetry is to predict such mixing it must relate the associated Yukawa couplings involving different families. Only a non-Abelian family symmetry has this property and for this reason I will concentrate in this talk on the possibility that there is a non-Abelian family symmetry [2] such as $SU(3)$ [3], [4]. Indeed if the underlying GUT symmetry

$$V_{MNS} = \begin{pmatrix} 0.72\text{--}0.89 & 0.45\text{--}0.69 & <0.2 \\ 0.24\text{--}0.58 & 0.39\text{--}0.76 & 0.52\text{--}0.84 \\ 0.24\text{--}0.58 & 0.39\text{--}0.76 & 0.53\text{--}0.84 \end{pmatrix}$$

$$\approx \begin{pmatrix} \sqrt{\frac{2}{3}} & \frac{1}{\sqrt{3}} & \approx 0 \\ -\frac{1}{\sqrt{6}} & \frac{1}{\sqrt{3}} & -\frac{1}{\sqrt{2}} \\ -\frac{1}{\sqrt{6}} & \frac{1}{\sqrt{3}} & \frac{1}{\sqrt{2}} \end{pmatrix}$$

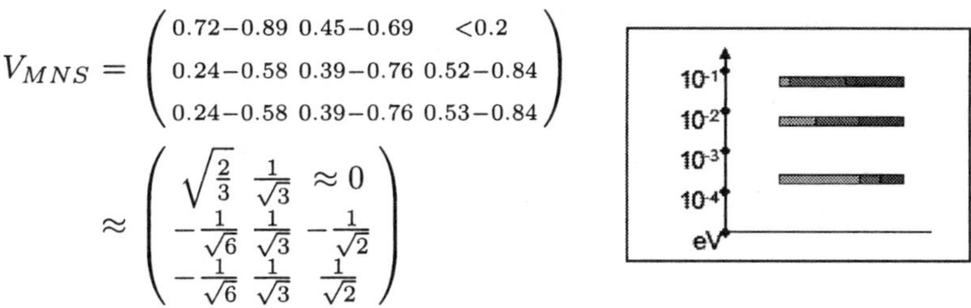

Figure 2. The lepton mixing matrix showing a near bi- tri- maximal mixing pattern. This is consistent with the normal hierarchy for neutrino masses shown where the composition of the mass eigenstates is indicated by the shading.

is $SO(10)$ or larger the maximal non-Abelian family symmetry group is $SU(3)$. The representations of the fermions under the family symmetry must be determined but is severely restricted if the full symmetry involves a product of a GUT group and the family symmetry group. In this case the representation of all states in a single GUT representation must be the same. For the case of $SO(10)$ all the states of a single family fit in a 16 of $SO(10)$ and so all these states must transform the same way under the family symmetry. For the case of $SU(3)$ this means all the states of the Standard Model must transform as family triplets. An immediate consequence of this is that, before $SO(10)$ breaking, the mass matrix must be symmetric. As we shall discuss such a structure is phenomenologically favoured.

Finally, in order to obtain a phenomenologically viable theory, the symmetry group must be extended further to include further symmetries, S, needed to restrict the allowed Yukawa couplings. As a result the final symmetry has the form $GUT \otimes SU(3)_{family} \otimes S$. In this talk I will discuss how the observed fermion masses and mixings can emerge from such a structure. First, however, I start with a brief review of the attempts at an experimental determination of the quark mass matrices and the indications for texture zero structure which may indicate an underlying family symmetry.

2. Quark Textures

A fundamental difficulty in identifying a family symmetry is that the measured masses and mixing angles provide incomplete information about the full mass matrices and the related matrix of Yukawa couplings. However the mild assumption that the smallness of the mixing angles observed in

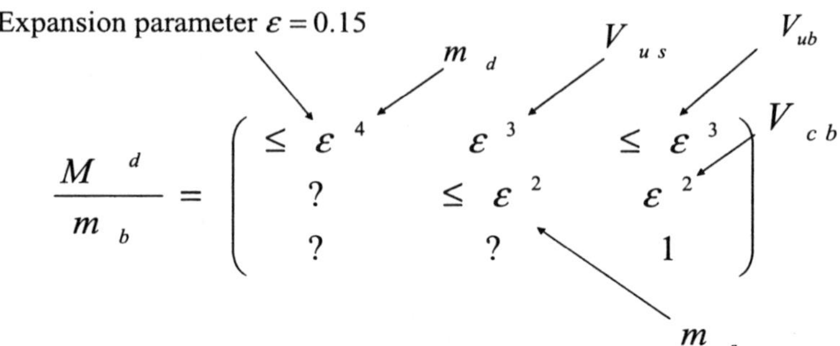

Figure 3. The constraints on the down quark mass matrix elements on and above the diagonal coming from mixing angle and mass measurements.

the quark sector is due to the smallness of the off-diagonal mixing matrix elements of the up and the down quark mass matrices allows us to make considerable progress. Making an expansion in the small off diagonal terms one obtains the relation of the CKM matrix elements V_{ij} to the mass matrix elements given by

$$V_{23} \cong \frac{m_{23}}{m_{33}} + \frac{m_{32}}{m_{33}}\frac{m_s}{m_b}$$
$$V_{12} \cong \frac{m_{12}}{m_{22}} + \frac{m_{21}}{m_{22}}\frac{m_d}{m_s} \tag{1}$$
$$V_{13} \cong \frac{m_{13}}{m_{33}} + \frac{m_{31}}{m_{33}}\frac{m_d}{m_b}$$

From this it is clear that in leading order in the ratio of light to heavier quark masses the data determines the components of the mass matrix on the diagonal and above. To determine the precise value of the matrix element requires a knowledge of the order of magnitude of the elements below the diagonal. If one constrains the elements below the diagonal to be no larger than those above the diagonal elements, the elements above the diagonal in the down quark mass matrix are quite well determined with magnitudes indicated in Figure 3, together with the data responsible for determining it.

A detailed fit to the down quark mass matrix [5] assuming a symmetric form for the mass matrices gives the structure of eq(2). In this we have also shown the up quark mass matrix. This fit shows the up quark mass matrix can have a similar form to that of the down quark mass matrix provided the expansion parameter is smaller as is needed to describe the more hierarchical structure of the up quark masses. Note that the off diagonal elements are very poorly constrained because, due to the smaller expansion parameter, the contribution of these elements to the observed mixing angles is much smaller than that of the down quark sector. In this fit a texture zero has been assumed in the $(1,1)$ element and this is discussed further below.

$$\frac{M^d}{m_b} = \begin{pmatrix} 0 & 1.5\varepsilon^3 & 0.4e^{i20}\varepsilon^3 \\ 1.5\varepsilon^3 & \varepsilon^2 & 1.3\varepsilon^2 \\ 0.4e^{i20}\varepsilon^3 & 1.3\varepsilon^2 & 1 \end{pmatrix}, \ \varepsilon = 0.15 \tag{2}$$

$$\frac{M^u}{m_t} = \begin{pmatrix} 0 & \varepsilon'^3 & ?\varepsilon'^3 \\ \varepsilon'^3 & \varepsilon'^2 & ?\varepsilon'^2 \\ ?\varepsilon'^3 & ?\varepsilon'^2 & 1 \end{pmatrix}, \ \varepsilon' = 0.05$$

For comparison we present in eq(3) an asymmetrical fit in which the elements below the diagonal saturate the bound of Figure 2. The fit requires texture zeros in the $(1,1)$ and $(3,1),(1,3)$ positions but only requires a symmetric form for the mass matrices in the $(1,2)$ sector:

$$\frac{M^d}{m_b} = \begin{pmatrix} 0 & 1.7\varepsilon^3 & 0 \\ 1.7\varepsilon^3 & 0 & 5\varepsilon^2 \\ 0 & 0.3 & 1 \end{pmatrix}$$

$$\frac{M^u}{m_t} = \begin{pmatrix} 0 & \varepsilon'^3 & ?\varepsilon'^3 \\ \varepsilon'^3 & \varepsilon'^2 & ?\varepsilon'^2 \\ ?\varepsilon'^3 & ?\varepsilon'^2 & 1 \end{pmatrix} \tag{3}$$

The difference between these fits clearly demonstrates the difficulty in extracting the mass matrix elements. It is partly due to this uncertainty that there are so many competing ideas for the origin of the fermion mass structure. One particularly important aspect of this is the possibility that the $(3,2)$ and $(3,3)$ elements are comparable in magnitude. In $SU(5)$ the lepton doublet is in the same multiplet as the charge conjugate down quarks meaning that, if $SU(5)$ relates the down quark mass matrix and the lepton mass matrix, comparable $(3,2)$ and $(3,3)$ charged lepton mass matrix elements can generate near maximal mixing in the $(2,3)$ lepton sector.

The immediate question is therefore to determine whether one structure is favoured. An obvious difference between these two schemes is the large family mixing implied by the non-symmetric fit coming from the elements below the diagonal. In the Standard Model physics is not sensitive to these elements because the weak interactions involve the LH doublets and, c.f. eq(1), are insensitive to elements below the diagonal. However in a supersymmetric theory the flavour changing effects coming from the squark sector will be sensitive to these elements and there will be constraints on the full mass matrix elements.

What these constraints are depends on what drives the hierarchical structure of the mass matrices. There have been various suggestions for generating this structure. One possibility, which I shall concentrate on here, is that there is an underlying family symmetry which, when unbroken, allows only the third family to acquire a mass. The symmetry is then spontaneously broken by one, or more, familon fields θ which carry family quantum numbers and generate the Yukawa couplings and associated mass matrix structure through higher dimension operators of the form coming from the superpotential W given by

$$W = \left(\frac{\theta}{M}\right)^{\alpha_{ij}} H_a Q_{Li} q^c_{Rj} \qquad (4)$$

This structure of Yukawa couplings can give rise to significant flavour changing neutral currents (FCNC) and (flavour conserving) CP violating effects which are known to be strongly suppressed. This comes about because the θ field(s) acquire non-vanishing F−terms, $F_\theta = \beta m_{3/2}\langle\theta\rangle$. For the case of SUGRA mediated supersymmetry breaking the resultant effects are important. A study of various models shows that the expectation is that $\beta = O(1)$ [6,7] although in models with numerous intermediate scales of symmetry breaking a suppression is possible. The F−term then induces trilinear soft supersymmetry breaking terms

$$A_{ij}\hat{Y}^{ij} = F^\eta \hat{K}_\eta Y^{ij} + \alpha_{ij}\,\frac{e^{K/2}}{M}\left(\frac{\theta}{M}\right)^{\alpha_{ij}-1}\beta m_{3/2}\theta \qquad (5)$$

with $\hat{K}_\eta = \partial\hat{K}/\partial\eta$ where K is the Kähler potential and $Y^{ij} = e^{K/2}(\theta/M)^{\alpha_{ij}}$ are the Yukawa couplings. From Eq. (5) we see that, in any model which explains the hierarchy in the Yukawa textures through non-renormalisable operators, the trilinear couplings are necessarily non-universal. Moreover, due to the factor α_{ij} these trilinear terms are not diagonalised at the same time as the Yukawa couplings and hence the fermion masses are diagonalised. They thus give rise to FCNC. Moreover, even in the most conservative case where all soft SUSY breaking parameters and μ are real, we know that the Yukawa matrices contain phases $\mathcal{O}(1)$. If the trilinear terms are non-universal, these phases are not completely removed from the diagonal elements of Y^A in the SCKM basis and hence can give rise to large EDMs [6,18].

The overall limits [7] on the mass matrix elements coming from this analysis is shown in Figure 4. The most significant bound on the Yukawa cou-

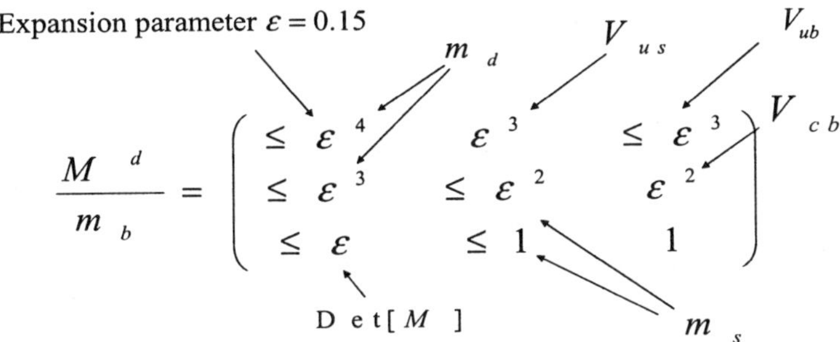

Figure 4. Bounds on the mass matrix elements below the diagonal coming from FCNC and CP violating signals in a SUGRA model of supersymmetry breaking.

plings is provided by the mercury Electric Dipole Moment (EDM) bound. In particular it imposes a significant constraint on the Yukawa coupling matrix elements below the diagonal in the $(3,1)$ and $(3,2)$ positions. This is particularly interesting because these are the terms responsible for right handed mixing which are very poorly constrained in non-supersymmetric theories due to the absence of right-handed weak currents. Moreover in $SU(5)$ based models the (left handed) neutrino mixing angles are related to the down quark right handed mixing angles.. Demanding that the terms of eq(5) do not violate the experimental bounds on the mercury EDM constrains the $(3,1)$ and $(3,2)$ elements to be $\leq O((\theta/M)^3)$ and $\leq O(((\theta/M)^2)$ respectively and means they should be no larger than the $(1,3)$ and $(2,3)$ elements It is also in conflict with $SU(5)$ based models in which the large neutrino mixing angles are related to large down quark right-handed mixing angles.

The trilinear couplings implied by eq(5) also contribute significantly to quark and lepton flavour violation. With the Yukawa coupling necessary to generate the fermion mass structure the $b \to s\gamma$ rate should be close to the present bound. Even more interesting is the lepton flavour violation process $\mu \to e\gamma$. If there is an underlying relation between charged lepton and down quark mass matrices we expect off diagonal elements in the charged lepton Yukawa couplings which will lead to lepton flavour violation. As we discuss below to generate the correct muon and electron mass we follow Georgi and Jarlskog's suggestion and put a relative factor of 3 in the $(2,2)$ entry. We also put a factor of 3 in the $(2,3)$ and $(3,2)$ entries as is required by non-Abelian models which seek to explain the near equality in the down quark mass matrix of the $(2,2)$ and $(2,3)$ elements. With this to keep

$\mu \to e\gamma$ at the level of current experimental bounds requires a slepton mass greater than $320\ GeV$. At this level $\mu \to e\gamma$ should be seen by the proposed experiments in the near future.

To summarize this section, in the context of SUGRA mediated SUSY breaking the symmetric form of eq(2) is favoured over that of eq(3) if the origin of fermion masses is through the spontaneous breaking of a family symmetry giving the Yukawa matrix element structure the form of eq(4).

2.1. *Texture zeros*

The fit of eq(2) has a simultaneous zero in the up and down quark mass matrix in the $(1,1)$ position while the fit of eq(3) has simultaneous zeros in the $(1,1), (1,3)$ and $(3,1)$ positions. Such simultaneous zeros are called "texture zeros" and are important because they may signal some underlying dynamics, such as a family symmetry, that guarantees these entries are anomalously small [9]. The most significant evidence for a texture zero applies to the $(1,1)$ texture zero [10]. In this case, for a symmetric mass matrix, the $(1,2)$ sub-matrix for either the up or the down quarks has the form

$$\begin{pmatrix} 0 & m_{12} \\ m_{12} & m_{22} \end{pmatrix}$$

This has now only two parameters (the phase can be absorbed in a redefinition of the fermion fields). Hence there is a relation between its two mass eigenvalues and the rotation diagonalising it. One may readily see from eq(1) that this implies [10,5]

$$|V_{us}| = |\sqrt{\frac{M_d}{M_s}} - \sqrt{\frac{M_u}{M_c}}\ e^{i\sigma}| \tag{6}$$

$$c.f.(0.216 - 0.214) = |(0.16 - 0.33) - (0.047 - 0.076)e^{i\sigma}|$$

$$= 0.213 - 0.223, \quad \delta = 90^0$$

Here σ is a phase that enters when combining the up and the down rotations and is in fact the CP violating phase of the Standard Model. The experimental comparison is also shown and one may see this prediction is in good agreement with experiment (due to the smallness of the up quark contribution the σ dependence is quite small but favours a large CP violating phase in agreement with the observations of CP violation). Note that the success of this result also supports the postulate that the mass matrix should have symmetric magnitude for the mass matrix elements.

3. Extension to leptons

We have seen that the quark mass matrices are consistent with the choice that all the states of the (left-handed) components of a given family have the same transformation properties under the family symmetry. This is suggestive of a larger underlying GUT symmetry. The GUT $SO(10)$ is particularly promising as all the (left-handed) states of a family, plus the charge conjugate of the right handed neutrino, fit into a single 16 dimensional representation. If $SO(10)$ is an underlying symmetry of the theory any family symmetry must commute with it implying all the charges of a given family must be the same, consistent with the form of the mass matrix discussed above. An associated advantage of such a family symmetry is that the mixed anomalies of the Standard Model gauge group with the family symmetry will automatically cancel because of the structure of the underlying $SO(10)$. It also contains $SU(2)_R$ and can relate the up to the down sectors as discussed above. In addition a GUT symmetry can relate quark and lepton mass matrices and this is the issue we wish to study here.

3.1. *Charged leptons*

We first consider the charged leptons. Given the same family properties, the form of the charged lepton mass matrix will be the same as that of the down quark matrices in eq(2) up to coefficients of $O(1)$. If there is an underlying GUT the coefficients too may be related. After including radiative corrections to fermion masses, the relation $m_b = m_\tau$ at the unification scale is in good agreement with the measured masses. Such an equality applies in $SU(5)$ if the Higgs responsible for the third generation masses transforms as a $\bar{5}$ of $SU(5)$. In $SO(10)$ equality applies if the Higgs belongs to a 10 representation, this also gives equality between the top quark and the third generation Dirac neutrino mass, something we explore below.

What about the two lighter generations? In this case we must address the question whether the expansion parameters are related. From a phenomenological point of view, note that, after taking radiative corrections into account, the relation $Det[M^d] = Det[M^l]$ at the unification scale is also in good agreement with the experimental measurements. However it is not possible to have *identical* charged lepton and down quark mass matrices because, after taking account of the radiative correction on going from high to low scale, gives approximately a factor of 3 increase in the quark masses, the relations $m_s = m_\mu$ and $m_d = m_e$ are in gross disagreement with experiment. As pointed out by Georgi and Jarlskog [8] this discrepancy

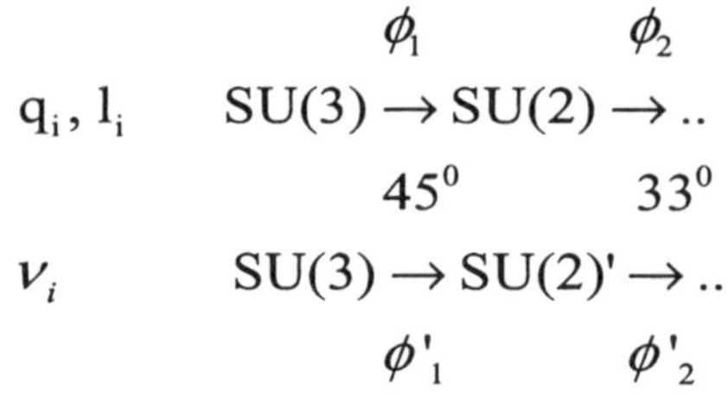

$$
\begin{array}{ccc}
 & \phi_1 & \phi_2 \\
\mathrm{q_i}, \mathrm{l_i} & SU(3) \to SU(2) \to .. & \\
 & 45^0 \quad\quad 33^0 & \\
\nu_i & SU(3) \to SU(2)' \to .. & \\
 & \phi'_1 \quad\quad \phi'_2 &
\end{array}
$$

Figure 5. The pattern of family symmetry breaking needed in the quark and charged lepton sector compared to that needed in the neutrino sector. The vevs of the fields ϕ_1 and ϕ'_1 must be misaligned as described in the text.

is readily explained if there is an underlying GUT through the appearance of Clebsch Gordon factors in the matrix element coefficients. In particular if the Higgs responsible for the $(2,2)$ matrix element should belong to a $\overline{45}$ of $SU(5)$ (or 126 of $SO(10)$) the lepton coupling is a factor -3 times the down quark coupling. In this case, taking account of the equality of the determinants, the relations for the light generations are modified to give $m_s = m_\mu/3$ and $m_d = 3m_e$. Including the radiative corrections needed to determine the masses at laboratory scales, these relations are in excellent agreement with the measured masses. Remarkably this structure for the lepton masses is consistent with a charged lepton mass matrix equal, up to the Georgi Jarlskog factor, to the mass matrix of the down quarks with the following structure:

$$
\frac{M_l}{m_\tau} = \begin{pmatrix} 0 & \varepsilon^3 & ?\varepsilon^3 \\ \varepsilon^3 & 3\varepsilon^2 & ?\varepsilon^2 \\ ?\varepsilon^3 & ?\varepsilon^2 & 1 \end{pmatrix}
\tag{7}
$$

3.2. *Inclusion of neutrinos*

As stressed above, if the near bi-tri-maximal mixing is to have a quantitative explanation coming from a broken family symmetry it is necessary to have a non-Abelian family structure such as $SU(3)_{family}$. However, by itself this does not explain why the mixing angles are small in the quark sector while they are large in the lepton sector. If this is to be consistent with an underlying spontaneously broken family symmetry there must be a mismatch between the symmetry breaking pattern in the quark and charged lepton sectors and the symmetry breaking pattern in the neutrino sector.

This is illustrated in Figure 5. In this the dominant spontaneous break-ing of the $SU(3)_{family}$ symmetry in the quark sector and charged lep-ton sector is generated by the scalar field ϕ_3 with vev $\langle\phi_3\rangle = v_1(0,0,1)$. The heaviest quark and charged lepton states then come from the Yukawa couplings proportional to $q_i\phi_3^i u_j^c \phi_3^j$, $q_i\phi_3^i d_j^c \phi_3^j$ and $l_i\phi_3^i e_j^c \phi_3^j$ times the ap-propriate Higgs doublet field. Here q_i, l_i are the left handed (LH) quark and lepton doublets and u_i^c, d_i^c e_i^c are the (LH) charge conjugate of the up quark, down quark and charged lepton $SU(2)$ singlet fields. All trans-form as $SU(3)_{family}$ triplets, consistent with there being an underlying $SO(10)$ family symmetry. The second generation of quarks and charged leptons acquire their mass through the further breaking of the family sym-metry via the field ϕ_2 corresponding to the second breaking stage of Fig-ure 5 To reproduce the symmetric structure of the $2-3$ block of eq(2) we need[a] $\langle\phi_2\rangle = v_2(0,1,1)$ with the Yukawa couplings proportional to $a^u q_i\phi_2^i u_j^c \phi_2^j$, $a^d q_i\phi_2^i d_j^c \phi_2^j$ and $a^e l_i\phi_2^i e_j^c \phi_2^j$ times the appropriate Higgs doublet field. Here a^i is the Georgi Jarlskog factor needed to generate the difference between quark and lepton masses. Finally the lightest generation of quarks and charged leptons acquire their mass through higher order terms of the form $q_i\phi_2^i u_j^c \phi_1^{j\,i} u_j^c \phi_1^j$, $q_i\phi_2^i d_j^c \phi_1^j$ and $l_i\phi_2^i e_j^c \phi_1^j$ plus $(2 \leftrightarrow 1)$ terms, where[b] $\langle\phi_1\rangle \propto (1,0,0)$. Together these give the charged quark and lepton mass matrices of the form

$$\frac{M_D}{m_3} = \begin{pmatrix} 0 & \varepsilon^3 & \varepsilon^3 \\ \varepsilon^3 & a\varepsilon^2 & a\varepsilon^2 \\ \varepsilon^3 & a\varepsilon^2 & 1 + a\varepsilon^2 \end{pmatrix}, \qquad \begin{array}{l} \varepsilon^d = 0.15, \ \ a^d = 1 \\ \varepsilon^e = 0.15, \ a^e = -3 \\ \varepsilon^u = 0.05, \ \ a^u = 2 \end{array} \qquad (8)$$

Here the expansion parameters are given by $\varepsilon^d = v_2/M^d$, $\varepsilon^u = v_2/M^u$ and $\varepsilon^e = v_2/M^e$ where M^d, M^u and M^d are the dominant messenger masses in the Froggatt Nielsen mechanism [11] and we expect $M^d \simeq M^e$ if the breaking of the underlying $SO(10)$ leaves $SU(5)$ unbroken. Once one includes the overall coefficients of O(1) which are not fixed by the non-Abelian symmetries this form is in good agreement with the fit of eq(2).

On the other hand to reproduce near maximal mixing in the atmospheric neutrino sector the effective neutrino mass matrix should have heaviest mass driven by a term proportional to $\nu_i\phi_2^i \nu_j \phi_2^j$ times a combination of fields which transform as an $SU(2)$ triplet. The result is that the heaviest neutrino is in the $2+3$ direction, corresponding to maximal mixing. It

---

[a]The origin of the vacuum alignment needed to obtain this vev is discussed below.

[b]In specific models $\phi_1^i = \epsilon_{ijk}\overline{\phi}_2^j\overline{\phi}_1^k$ where $\overline{\phi}_{1,2}$ are the vector-like partners of $\phi_{1,2}$ needed to maintain unbroken supersymmetry when the family symmetry is broken.

is important that the term proportional to $\nu_i\phi_3^i\nu_j\phi_3^j$ should be small in order to explain the difference between neutrinos and charged fermions. As is illustrated in Figure 5 this corresponds to a different pattern of family symmetry breaking from that of the quarks.

Of course the question is how can this come about? At first sight it appears quite unnatural. However if neutrino masses are generated by the see-saw mechanism in fact it can readily arise even if all quark and lepton Dirac masses, including those of the neutrinos, having similar forms up to Georgi Jarlskog type factors. To see this consider the general form of the see-saw mechanism [12]

$$M_\nu = M_D^\nu \, M_M^{-1} \, M_D^{\nu T}$$

where M_D^ν is the Dirac mass matrix coupling ν to ν^c and M_M is the Majorana mass matrix coupling ν^c to ν^c. Assuming the *same* structure for neutrino Dirac masses as is given by eq(8) with $\varepsilon^\nu = 0.05$ and a factor[c] $a^\nu \simeq 0$, we include the Yukawa couplings $l_i\phi_3^i\nu_j^c\phi_3^j$, $a^e l_i\phi_{23}^i\nu_j^c\phi_{23}^j$ and $(l_i\phi_{23}^i\nu_j^c\phi_1^j + (23 \leftrightarrow 1))$. We consider the case the Majorana mass matrix also has an hierarchical structure of the form

$$M_M \approx \begin{pmatrix} M_1 & & \\ & M_2 & \\ & & M_3 \end{pmatrix} \quad M_1 \ll M_2 \ll M_3$$

and we allow for the Yukawa couplings of doublet neutrinos to the singlet neutrinos responsible for generating the Dirac mass M_{Dirac}^ν. For a sufficiently strong hierarchy this gives rise to sequential domination [13] in which the heaviest of the three light eigenstates gets its mass from the exchange of the lightest right-handed singlet neutrino with mass M_1 through a product of the Yukawa coupling terms $l_i\phi_{23}^i\nu_j^c\phi_1^j$ giving a light effective neutrino mass term proportional to $\nu_i\phi_{23}^i\nu_j\phi_{23}^j/M_1$. This corresponds to the bi-maximally mixed neutrino mass eigenstate $\propto (\nu_\tau + \nu_\mu)$. In this case the contribution of the ϕ_3 field, given by the see-saw mechanism through a product of the Yukawa coupling terms $l_i\phi_3^i\nu_j^c\phi_3^j$ is suppressed by the relative factor M_1/M_3 as the resultant light effective neutrino mass is proportional to $\nu_i\phi_3^i\nu_j\phi_3^j/M_3$. Provided $v_3^2/M_3 \ll v_2^2/M_2$ the required term dominates and one obtains near maximal mixing for the atmospheric neutrino sector.

The message from this is that any underlying quark lepton symmetry is necessarily broken in the neutrino sector due to the Majorana masses of

---

[c]This Georgi Jarlskog factor naturally arises in the breaking of $SO(10)$ along the $(SU(5)$ conserving) direction $B - L + kT_{3R}$ with $k = 2$ [4].

the RH neutrino states and, through the see-saw mechanism, this feeds into the neutrino masses and the lepton mixing angles. This example illustrates how this effect can hide an underlying quark lepton symmetry in the Dirac mass sector.

3.3. *Bi-Tri-maximal mixing*

The $SU(3)$ family model presented above naturally generates a near maximal mixing in the atmospheric neutrino sector and can lead to large mixing between the three neutrino current eigenstates for the solar neutrino. However this relies on unknown Yukawa couplings of $O(1)$ which are not predicted by the family symmetry. If one is to generate tri-maximal mixing for the solar neutrino through the underlying $SU(3)$ symmetry alone a modification of the symmetry breaking structure is necessary. In particular it is necessary that the field ϕ_1 has a vev given by $\langle\phi_1\rangle = v_1(1,1,-1)$. This does not change the leading order structure of eq(8):

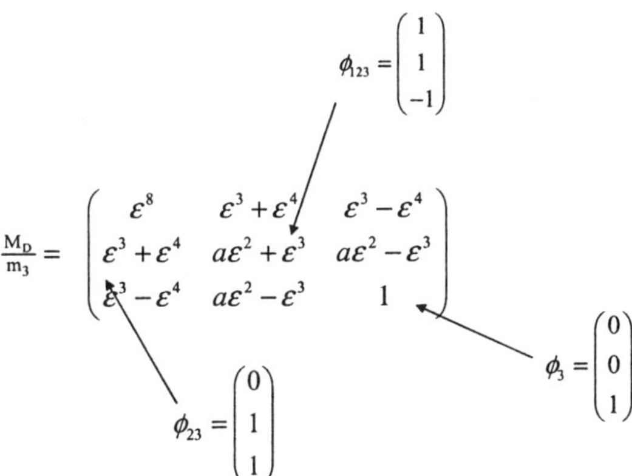

$$\phi_{23} = \begin{pmatrix} 1 \\ 1 \\ -1 \end{pmatrix}$$

$$\frac{M_D}{m_3} = \begin{pmatrix} \varepsilon^8 & \varepsilon^3 + \varepsilon^4 & \varepsilon^3 - \varepsilon^4 \\ \varepsilon^3 + \varepsilon^4 & a\varepsilon^2 + \varepsilon^3 & a\varepsilon^2 - \varepsilon^3 \\ \varepsilon^3 - \varepsilon^4 & a\varepsilon^2 - \varepsilon^3 & 1 \end{pmatrix}$$

$$\phi_{23} = \begin{pmatrix} 0 \\ 1 \\ 1 \end{pmatrix} \qquad \phi_3 = \begin{pmatrix} 0 \\ 0 \\ 1 \end{pmatrix}$$

However in this case the exchange of the right handed neutrino with mass M_2 leads to the light effective neutrino mass term proportional to $\nu_i\phi_1^i\nu_j\phi_1^j/M_1$. This corresponds to the tri-maximally mixed neutrino mass eigenstate $\propto (-\nu_\tau + \nu_\mu + \nu_e)$. Its mass is proportional to v_1^2/M_2 and can readily be lighter than the bi-maximal mixed state coming from the term $\nu_i\phi_{23}^i\nu_j\phi_{23}^j/M_1$ with mass proportional to v_2^2/M_1.

Of course one has to explain why such a pattern of vevs for the fields $\phi_{3,23,1}$ arises when the $SU(3)$ family symmetry is broken. In fact it may be shown that this can arise in quite an elegant way, the symmetry breaking vevs being related by the underlying symmetry [14]. Given this structure, the above discussion shows that the properties of the neutrino system can be related to those of the quarks and charged leptons by an extended symmetry made up of a combination of a Grand unified group and a non-Abelian family group. Due to the see-saw mechanism and the symmetry breaking effects in the Majorana sector, the bi-tri-maximal mixing results for the leptons while the quarks have small mixing angles.

3.4. *Texture zeros in the neutrino sector*

As we have discussed above, the quark and charged lepton mass matrices are in excellent agreement with a texture zero in the $(1,1)$ element together with a symmetric structure. It is possible that this zero structure extends too to the Dirac mass matrix structure of the leptons as the example given above demonstrates. It is interest to ask whether there is a model independent test of such a texture zero in the neutrino sector analogous to that for the quarks given in eq(6). This is made difficult due to the see saw mechanism and the uncertainties introduced by the right handed neutrino sector. However it is possible to determine the implications for a variety of simplifying cases [15],[16]. Perhaps the most interesting is the case that one of the right handed neutrinos is very heavy $M_{1,2} \ll M_3$, a weaker ordering than used in the model discussed above. Using the general parameterization introduced by Casas and Ibarra [17] it is straightforward to explore various cases. For the particularly interesting case $Y_D^\nu(1,1) = Y_D^e(1,1) = 0$ and $Y_D^{\nu,e}(1,2) = Y_D^{\nu,e}(2,1)$ which duplicates the input for the GST relation for the quarks we find

$$W_{13}^2 = -\frac{m_2}{m_1}\, W_{12}^2 \,+\, \sqrt{\frac{M_1}{M_2}} \cdot \sqrt{\frac{m_2}{m_1}} \cdot \frac{W_{31}^*}{\det W^*}$$

where W is the MNS mixing matrix not including the contribution of the charged lepton rotation from current to mass basis[d]. Here $m_{1,2}$ are the measured neutrino masses. From this one may obtain predictions for the Chooz angle and for the CP violating phase relevant to leptogenesis and double beta decay

$$\sin\theta_{13} \approx \left| \sqrt{\frac{m_2}{m_3}} \sin\theta_{12} \pm \sqrt{\frac{m_e}{2m_\mu}} \right|$$

$$\delta - \phi/2 \approx (2n+1)\frac{\pi}{2}$$

results which apply for $M_1/M_2 < 10^{-2}$. For other values of M_1/M_2 the results are shown in Figure 6 where we also show how close to maximal CP

---

[d]This is very small if the charged leptons have the mass matrix shown in eq(7)

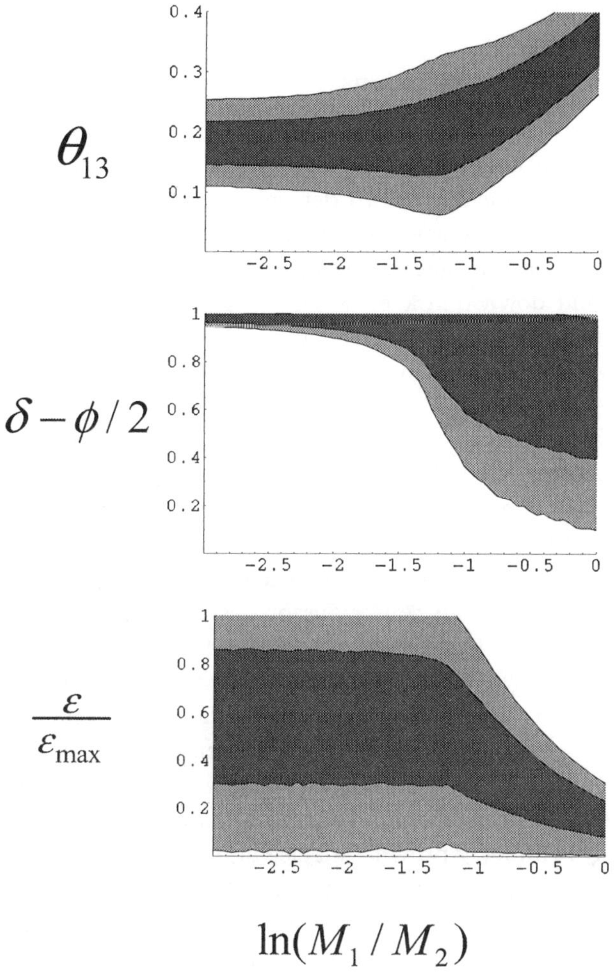

Figure 6. The predictions for θ_{13}, the CP violating angle relevant to double beta decay and the fraction of maximal CP violation in neutrino oscillation that result from a $(1,1)$ texture zero in a symmetric Dirac neutrino matrix.

violation in neutrino oscillation experiments one may get. One may see that there is now a band of possibilities due to unknown phases appearing in the seesaw mechanism but one still gets significant predictive power. It is possible to use similar techniques to study the phenomenological implications of texture zeros located in other positions in the mass matrix [15],[16] and these signals will be useful in discriminating between the various possible structures.

4. Summary and conclusions

The extraction of Yukawa couplings is crucial to our understanding the origin of fermion masses and mixings. However the measured masses and mixing angles are insufficient to determine these couplings and some additional information is needed. In this talk I have concentrated on the possibility that the off diagonal mass matrix elements are small separately for the up and down quark mass matrices and further that the matrices are (anti)symmetric or Hermitian. With these assumptions the mass matrices are well constrained by the data and consistent with the same form for the up and down matrices but with a different expansion parameter. The texture and texture zeros suggested by this fit hints at an underlying structure, perhaps a combination of a Grand Unified and a family symmetry. In the case of a supersymmetric theory with supersymmetry breaking communicated by supergravity effects such a structure seems necessary to keep flavour changing processes small and to explain the smallness of CP violating effects in flavour conserving processes. At first sight it seems that the structure of the leptons cannot be similar to that of quarks because of the gross differences in the mixing angle structure of the two sectors. However, due to the see-saw mechanism, the quark, charged lepton and neutrino masses and mixing angles is consistent with a similar structure for their Dirac mass matrices. Such a structure hints at an extended underlying (spontaneously broken) symmetry perhaps as large as $SO(10) \times SU(3)$. A $SU(3)$ family symmetry provides an elegant explanation for the suppression of flavour changing neutral currents. If CP is violated by the spontaneous breaking of the family symmetry the dominant CP violating effects occurs in the the flavour changing sector and is strongly suppressed in the flavour conserving sector thus avoiding the SUSY CP problem. Perhaps most compelling is the fact that the near bi-tri-maximal mixing observed in the lepton sector suggests that the Yukawa couplings between different families are related in a manner that only a non-Abelian family symmetry group can explain. For this to be consistent with a unified description of quark and lepton masses requires the see saw mechanism to explain the differences between the quark and charged lepton sector and the neutrino sector. Thus it may be that the see-saw mechanism for neutrino masses and mixing angles has provided us with the first definite indication of a non-Abelian family symmetry.

References

1. For a review and further references see:
   G.G.Ross, "Models of Fermion masses", published in TASI 2000 ed. J.L.Rosner (World Scientific, New Jersey, 2001)
2. For an early discussion of non-abelian family symmetries see P. Ramond, "The Family Group In Grand Unified Theories," in *C79-02-25.3* hep-ph/9809459;
   for sU(2) models see :R. Barbieri, P. Creminelli and A. Romanino, Nucl. Phys. B **559** (1999) 17 [hep-ph/9903460]; R. Barbieri, L. J. Hall and A. Romanino, Phys. Lett. B **401** (1997) 47 [hep-ph/9702315];
   R. Barbieri, L. J. Hall, S. Raby and A. Romanino, Nucl. Phys. B **493** (1997) 3 [hep-ph/9610449];
   T. Blazek, S. Raby and K. Tobe, Phys. Rev. D **62** (2000) 055001 [hep-ph/9912482];
   Z. Berezhiani and A. Rossi, JHEP **9903** (1999) 002 [hep-ph/9811447];
   A. Aranda, C. D. Carone and R. F. Lebed, Phys. Rev. D **62** (2000) 016009 [hep-ph/0002044];
   A. Aranda, C. D. Carone and R. F. Lebed, Phys. Lett. B **474** (2000) 170 [hep-ph/9910392].
3. The literature is vast, including :
   J.Chkareuli, JETP Lett.32 (1980)684;
   Z. G. Berezhiani and J. L. Chkareuli, JETP Lett. **35** (1982) 612 [Pisma Zh. Eksp. Teor. Fiz. **35** (1982) 494].
   Z. G. Berezhiani, Phys. Lett. B **129** (1983) 99;
   Z. G. Berezhiani, Phys. Lett. B **150** (1985) 177;
   A. A. Anselm and Z. G. Berezhiani, Phys. Lett. B **162** (1985) 349;
   Z. Berezhiani, Phys. Lett. B **417** (1998) 287 [hep-ph/9609342];
   Z. Berezhiani and A. Rossi, Nucl. Phys. B **594** (2001) 113 [hep-ph/0003084];
   M. Soldate, M. H. Reno and C. T. Hill, Phys. Lett. B **179** (1986) 95;
   Y. Koide and S. Oneda, Phys. Rev. D **36** (1987) 2867;
   R. Kitano and Y. Mimura, Phys. Rev. D **63** (2001) 016008 [hep-ph/0008269].
4. S. F. King and G. G. Ross, Phys. Lett. B **520**, 243 (2001) [hep-ph/0108112] G. G. Ross and L. Velasco-Sevilla, Nucl. Phys. B **653**, 3 (2003) [hep-ph/0208218]. S. F. King and G. G. Ross, "Fermion masses and mixing angles from SU(3) family symmetry and Phys. Lett. B **574**, 239 (2003) [hep-ph/0307190].
   G. G. Ross, L. Velasco-Sevilla and O. Vives, Nucl. Phys. B **692** (2004) 50 [hep-ph/0401064];
   O. Vives, G. G. Ross and L. Velasco-Sevilla, *Prepared for International Workshop on Astroparticle and High-Energy Physics (AHEP-2003), Valencia, Spain, 14-18 Oct 2003*
5. R. G. Roberts, A. Romanino, G. G. Ross and L. Velasco-Sevilla, Nucl. Phys. B **615**, 358 (2001) [hep-ph/0104088].
6. S. Abel, S. Khalil and O. Lebedev, Phys. Rev. Lett. **89**, 121601 (2002) [hep-

ph/0112260].

7. G. G. Ross and O. Vives, Phys. Rev. D **67**, 095013 (2003) [hep-ph/0211279].

8. H.Georgi and C.Jarlskog, Phys.Lett.B86 (1979) 297; H.Georgi and D.V.Nanopoulos, Nucl.Phys. B159 (1979) 16; J.Harvey, P.Ramond and D.B.Reiss, Phys.Lett. B92 (1980) 309; Nucl.Phys. B199 (1982) 223.

9. P. Ramond, R. G. Roberts and G. G. Ross, Nucl. Phys. B **406**, 19 (1993) [hep-ph/9303320].

10. R. Gatto, G. Sartori and M. Tonin, Phys. Lett. B **28**, 128 (1968).

11. C. D. Froggatt and H. B. Nielsen, Nucl. Phys. B **147**, 277 (1979).

12. P. Minkowski, Phys Lett 67B 421 (1977);
M. Gell-Mann, P. Ramond and R. Slansky in Sanibel Talk, CALT-68-709, Feb 1979, and in *Supergravity* (North Holland, Amsterdam 1979);
T. Yanagida in *Proc. of the Workshop on Unified Theory and Baryon Number of the Universe*, KEK, Japan, 1979; S.L.Glashow, Cargese Lectures (1979);
R. N. Mohapatra and G. Senjanovic, Phys. Rev. Lett. **44** (1980) 912;
J. Schechter and J. W. Valle, Phys. Rev. D **25** (1982) 774.

13. S. F. King, Phys. Lett. B **439** (1998) 350 [hep-ph/9806440]. S. F. King, Nucl. Phys. B **562** (1999) 57 [hep-ph/9904210]; S. F. King, Nucl. Phys. B **576** (2000) 85 [hep-ph/9912492]. G. Altarelli, F. Feruglio and I. Masina, Phys. Lett. B **472** (2000) 382 [hep-ph/9907532]; A. Y. Smirnov, Phys. Rev. D **48** (1993) 3264 [hep-ph/9304205].

14. G.G.Ross, I. de Medeiros Varzielas, in preparation.

15. A. Ibarra and G. G. Ross, Phys. Lett. B **591** (2004) 285 [hep-ph/0312138].

16. A. Ibarra and G. G. Ross, Phys. Lett. B **575** (2003) 279 [hep-ph/0307051].

17. J. A. Casas and A. Ibarra, Nucl. Phys. B **618**, 171 (2001) [hep-ph/0103065].

18. Y. Nir and R. Rattazzi, Phys. Lett. B **382**, 363 (1996) [hep-ph/9603233] ; J. L. Chkareuli, C. D. Froggatt and H. B. Nielsen, Nucl. Phys. B **626**, 307 (2002) [hep-ph/0109156].

THE SEESAW MECHANISM AND RENORMALIZATION GROUP EFFECTS

M. LINDNER

Physik Department, Technische Universität München
James-Franck-Str., D-85748 Garching/München, Germany
E-mail: lindner@ph.tum.de

Neutrino mass models predict masses and mixings typically at very high scales, while the measured values are determined at low energies. The renormalization group running which connects models with measurements is discussed in this paper. Analytic formulae for the running which include both Dirac- and Majorana CP phases are provided and they allow a systematic understanding of all effects. Some applications and numerical examples are shown.

1. Introduction

The determination of neutrino masses and mixings has made enormous progress in recent years. Furthermore it is expected that precision neutrino physics will become possible in the future such that the lepton sector may ultimately provide the most precise information on flavour structures. Already now exists enough information to try to understand the patterns of masses and mixings in different models of flavour, but this will become much more interesting in the future with growing precision. One class of models is, for example, given by discrete flavour symmetries which might emerge as unbroken subgroups of broken flavour gauge symmetries. There are different reasons why the scale where an understanding of flavour becomes possible is very high. This has the consequence, that like in the quark sector [1,2] renormalization group (RGE) effects must potentially be taken into account when high energy predictions are compared with low energy measurements. We will show that such RGE effects can be important and that the pattern which needs to be explained at high energies may differ substantially from that at low energies.

2. Running below the seesaw scale

We will discuss neutrino masses which can be described by the lowest-dimensional neutrino mass operator compatible with the gauge symmetries of the Standard Model (SM). This dimension 5 operator reads in the SM

$$\mathscr{L}_\kappa = \frac{1}{4}\kappa_{gf}\,\overline{\ell^C_{L\,c}}^{\,g}\varepsilon^{cd}\phi_d\,\ell^f_{Lb}\varepsilon^{ba}\phi_a + \text{h.c.}\,, \tag{1}$$

and in its minimal supersymmetric extension, the MSSM,

$$\mathscr{L}_\kappa^{\text{MSSM}} = \mathscr{W}_\kappa\big|_{\theta\theta} + \text{h.c.} = -\tfrac{1}{4}\kappa_{gf}\,\mathbb{l}^g_c\varepsilon^{cd}\mathbb{h}^{(2)}_d\,\mathbb{l}^f_b\varepsilon^{ba}\mathbb{h}^{(2)}_a\big|_{\theta\theta} + \text{h.c.}\,. \tag{2}$$

κ_{gf} has mass dimension -1 and is symmetric under interchange of the generation indices f and g, ε is the totally antisymmetric tensor in 2 dimensions, and ℓ^C_L is the charge conjugate of a lepton doublet. $a, b, c, d \in \{1,2\}$ are $SU(2)_L$ indices. $\mathbb{l}$ and $\mathbb{h}$ denote lepton doublets and the up-type Higgs superfield in the MSSM. After electroweak (EW) symmetry breaking, a Majorana neutrino mass matrix proportional to κ emerges as shown in Fig. 1. The d=5 mass operator describes neutrino masses in a rather model-

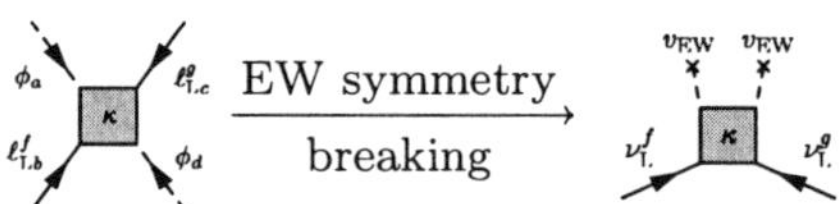

Figure 1.   The Majorana mass matrix for light neutrinos from the dimension 5 operator.

independent way to (see e.g. [3]). Integrating out heavy singlet fermions and/or Higgs triplets as for instance in left-right-symmetric extensions of the SM or MSSM leads to tree-level realizations which are usually referred to as type I and type II see-saw mechanisms. The energy dependence of the effective neutrino mass matrix below the scale M_1 where the operator is generated is given by its RGE. At the one-loop level we have [4,5,6,7]

$$16\pi^2\frac{\mathrm{d}\kappa}{\mathrm{d}t} = C\,(Y_e^\dagger Y_e)^T\,\kappa + C\,\kappa\,(Y_e^\dagger Y_e) + \alpha\,\kappa\,, \tag{3}$$

where $t = \ln(\mu/\mu_0)$ and μ is the renormalization scale[a] and where

$$C = 1 \qquad \text{in the MSSM and} \quad C = -\frac{3}{2} \quad \text{in the SM}\,. \tag{4}$$

---

[a]In the MSSM, the RGE is known at two-loop [8]. In this study, we will, however, focus on the one-loop equation.

In the SM and in the MSSM, α reads

$$\alpha_{\mathrm{SM}} = -3g_2^2 + 2(y_\tau^2 + y_\mu^2 + y_e^2) + 6\left(y_t^2 + y_b^2 + y_c^2 + y_s^2 + y_d^2 + y_u^2\right) + \lambda, \tag{5a}$$

$$\alpha_{\mathrm{MSSM}} = -\frac{6}{5}g_1^2 - 6g_2^2 + 6\left(y_t^2 + y_c^2 + y_u^2\right). \tag{5b}$$

Here Y_f ($f \in \{e, d, u\}$) represent the Yukawa coupling matrices of the charged leptons, down- and up-quarks, respectively, g_i denote the gauge couplings[b] and λ the Higgs self-coupling. We work in the basis where Y_e is diagonal. The masses are proportional to the eigenvalues of κ and the mixing angles and physical phases are given by the leptonic mixing matrix [9]

$$U = V(\theta_{12}, \theta_{13}, \theta_{23}, \delta)\,\mathrm{diag}(e^{-i\varphi_1/2}, e^{-i\varphi_2/2}, 1)\,, \tag{6}$$

which diagonalizes κ in this basis. V is the leptonic analogon to the CKM matrix in the quark sector and we use the standard parameterization [10].

3. Analytical Formulae

In [11] explicit RGEs for the all physical parameters (including CP phases) are given. y_e and y_μ are neglected against y_τ and the expansion parameter

$$\zeta := \frac{\Delta m_{\mathrm{sol}}^2}{\Delta m_{\mathrm{atm}}^2}\,, \tag{7}$$

is introduced, whose LMA best-fit value is about 0.03. We furthermore define $m_i(t) := v^2\,\kappa_i(t)/4$ with $v = 246\,\mathrm{GeV}$ in the SM or $v = 246\,\mathrm{GeV}\cdot\cos\beta$ in the MSSM and, as usual, $\Delta m_{\mathrm{sol}}^2 := m_2^2 - m_1^2$ and $\Delta m_{\mathrm{atm}}^2 := m_3^2 - m_2^2$. With these conventions, we obtain for the mixing angles:

$$\dot{\theta}_{12} = -\frac{Cy_\tau^2}{32\pi^2}\sin 2\theta_{12}\,s_{23}^2\,\frac{|m_1\,e^{i\varphi_1} + m_2\,e^{i\varphi_2}|^2}{\Delta m_{\mathrm{sol}}^2} + \mathcal{O}(\theta_{13})\,, \tag{8}$$

$$\dot{\theta}_{13} = \frac{Cy_\tau^2}{32\pi^2}\sin 2\theta_{12}\sin 2\theta_{23}\,\frac{m_3}{\Delta m_{\mathrm{atm}}^2\,(1+\zeta)}\times [m_1\cos(\varphi_1 - \delta)$$
$$- (1+\zeta)\,m_2\cos(\varphi_2 - \delta) - \zeta m_3\cos\delta] + \mathcal{O}(\theta_{13})\,, \tag{9}$$

$$\dot{\theta}_{23} = -\frac{Cy_\tau^2}{32\pi^2}\sin 2\theta_{23}\,\frac{1}{\Delta m_{\mathrm{atm}}^2}\left[c_{12}^2\,|m_2\,e^{i\varphi_2} + m_3|^2\right.$$
$$\left. + s_{12}^2\,\frac{|m_1\,e^{i\varphi_1} + m_3|^2}{1+\zeta}\right] + \mathcal{O}(\theta_{13})\,. \tag{10}$$

---

[b]We are using GUT charge normalization for g_1.

The $\mathcal{O}(\theta_{13})$ terms in the above RGEs can become important if θ_{13} is not too small and if cancellations appear in the leading terms. This is, for example, the case for $|\varphi_1 - \varphi_2| = \pi$ in (8). The RGE for the Dirac phase is given by

$$\dot{\delta} = \frac{Cy_\tau^2}{32\pi^2} \frac{\delta^{(-1)}}{\theta_{13}} + \frac{Cy_\tau^2}{8\pi^2} \delta^{(0)} + \mathcal{O}(\theta_{13}) , \tag{11a}$$

$$\delta^{(-1)} = \sin 2\theta_{12} \sin 2\theta_{23} \frac{m_3}{\Delta m_{\text{atm}}^2 (1+\zeta)} \times [m_1 \sin(\varphi_1 - \delta)$$
$$- (1+\zeta) m_2 \sin(\varphi_2 - \delta) + \zeta m_3 \sin\delta] , \tag{11b}$$

$$\delta^{(0)} = \frac{m_1 m_2 s_{23}^2 \sin(\varphi_1 - \varphi_2)}{\Delta m_{\text{sol}}^2}$$
$$+ m_3 s_{12}^2 \left[\frac{m_1 \cos 2\theta_{23} \sin\varphi_1}{\Delta m_{\text{atm}}^2 (1+\zeta)} + \frac{m_2 c_{23}^2 \sin(2\delta - \varphi_2)}{\Delta m_{\text{atm}}^2} \right]$$
$$+ m_3 c_{12}^2 \left[\frac{m_1 c_{23}^2 \sin(2\delta - \varphi_1)}{\Delta m_{\text{atm}}^2 (1+\zeta)} + \frac{m_2 \cos 2\theta_{23} \sin\varphi_2}{\Delta m_{\text{atm}}^2} \right] . \tag{11c}$$

The physical Majorana phases are given by

$$\dot{\varphi}_1 = \frac{Cy_\tau^2}{4\pi^2} \left\{ m_3 \cos 2\theta_{23} \frac{m_1 s_{12}^2 \sin\varphi_1 + (1+\zeta) m_2 c_{12}^2 \sin\varphi_2}{\Delta m_{\text{atm}}^2 (1+\zeta)} \right.$$
$$\left. + \frac{m_1 m_2 c_{12}^2 s_{23}^2 \sin(\varphi_1 - \varphi_2)}{\Delta m_{\text{sol}}^2} \right\} + \mathcal{O}(\theta_{13}) , \tag{12}$$

$$\dot{\varphi}_2 = \frac{Cy_\tau^2}{4\pi^2} \left\{ m_3 \cos 2\theta_{23} \frac{m_1 s_{12}^2 \sin\varphi_1 + (1+\zeta) m_2 c_{12}^2 \sin\varphi_2}{\Delta m_{\text{atm}}^2 (1+\zeta)} \right.$$
$$\left. + \frac{m_1 m_2 s_{12}^2 s_{23}^2 \sin(\varphi_1 - \varphi_2)}{\Delta m_{\text{sol}}^2} \right\} + \mathcal{O}(\theta_{13}) . \tag{13}$$

The above expressions can be further simplified by neglecting ζ against 1. Note that singularities can appear in the $\mathcal{O}(\theta_{13})$-terms at points in parameter space, where the phases are not well-defined. For the masses, the results for $y_e = y_\mu = 0$ but arbitrary θ_{13} are

$$16\pi^2 \dot{m}_1 = \left[\alpha + Cy_\tau^2 \left(2s_{12}^2 s_{23}^2 + F_1 \right) \right] m_1 , \tag{14a}$$
$$16\pi^2 \dot{m}_2 = \left[\alpha + Cy_\tau^2 \left(2c_{12}^2 s_{23}^2 + F_2 \right) \right] m_2 , \tag{14b}$$
$$16\pi^2 \dot{m}_3 = \left[\alpha + 2Cy_\tau^2 c_{13}^2 c_{23}^2 \right] m_3 , \tag{14c}$$

where F_1 and F_2 contain terms proportional to $\sin\theta_{13}$,

$$F_1 = -s_{13} \sin 2\theta_{12} \sin 2\theta_{23} \cos\delta + 2s_{13}^2 c_{12}^2 c_{23}^2 , \tag{15a}$$
$$F_2 = s_{13} \sin 2\theta_{12} \sin 2\theta_{23} \cos\delta + 2s_{13}^2 s_{12}^2 c_{23}^2 . \tag{15b}$$

These formulae lead to RGEs for the mass squared differences,

$$8\pi^2 \frac{d}{dt}\Delta m^2_{\text{sol}} = \alpha\,\Delta m^2_{\text{sol}} + Cy_\tau^2\left[2s_{23}^2\left(m_2^2\,c_{12}^2 - m_1^2\,s_{12}^2\right) + F_{\text{sol}}\right], \quad (16\text{a})$$

$$8\pi^2 \frac{d}{dt}\Delta m^2_{\text{atm}} = \alpha\,\Delta m^2_{\text{atm}} + Cy_\tau^2\left[2m_3^2\,c_{13}^2\,c_{23}^2 - 2m_2^2\,c_{12}^2\,s_{23}^2 + F_{\text{atm}}\right] \quad (16\text{b})$$

where

$$\begin{aligned}
F_{\text{sol}} &= \left(m_1^2 + m_2^2\right)s_{13}\,\sin 2\theta_{12}\,\sin 2\theta_{23}\,\cos\delta \\
&\quad + 2s_{13}^2\,c_{23}^2\left(m_2^2\,s_{12}^2 - m_1^2\,c_{12}^2\right),
\end{aligned} \quad (17\text{a})$$

$$F_{\text{atm}} = -m_2^2\,s_{13}\,\sin 2\theta_{12}\,\sin 2\theta_{23}\,\cos\delta - 2m_2^2\,s_{13}^2\,s_{12}^2\,c_{23}^2. \quad (17\text{b})$$

4. RG Evolution of θ_{13}, θ_{23}, θ_{12}

An interesting question is if deviations from $\theta_{13} = 0$ and $\theta_{23} = \pi/4$ at low energies could be the consequence of radiative corrections. Therefore we study RG corrections to θ_{13} and θ_{23} from the running of the effective neutrino mass operator between the see-saw scale and the electroweak scale[c].

The corrections to θ_{13} are in a good approximation described by the leading term which does not depend on θ_{13}. Then $\dot{\theta}_{13} \simeq$ const. in Eq. (9), i.e. a constant slope depending on the Dirac CP phase δ and the Majorana phases φ_1 and φ_2. Assuming at some high scale M_1 where neutrino masses are generated $\theta_{13} = 0$, Eq. (9) allows to determine the RG corrections at $10^2\,\text{GeV}$. For the examples we take $M_1 = 10^{12}\,\text{GeV}$ and the approximate size of the RG corrections to $\sin^2 2\theta_{13}$ in the MSSM is shown in Fig. 2. In the upper diagram it is plotted as a function of $\tan\beta$ and the lightest neutrino mass m_1 for constant Majorana phases $\varphi_1 = 0$ and $\varphi_2 = \pi$. The lower diagram shows the dependence of the corrections on φ_1 and φ_2 for $\tan\beta = 50$ and $m_1 = 0.08\,\text{eV}$ in the case of a normal mass hierarchy. The diagrams look rather similar for an inverted hierarchy. Analytically, the pattern of the upper plot is easy to understand, and for the lower one there is a simple explanation as well. Consider partially or nearly degenerate neutrino masses. Then Eq. (9) yields to a reasonably good approximation

$$\begin{aligned}
\dot{\theta}_{13} &\approx \frac{Cy_\tau^2}{32\pi^2}\,\sin 2\theta_{12}\,\sin 2\theta_{23}\,\frac{m^2}{\Delta m^2_{\text{atm}}}\left[\cos(\varphi_1 - \delta) - \cos(\varphi_2 - \delta)\right] \\
&\propto \sin\frac{\varphi_1 + \varphi_2 - 2\delta}{2}\,\sin\frac{\varphi_1 - \varphi_2}{2}.
\end{aligned} \quad (18)$$

---

[c]The potential of future long-baseline neutrino oscillation experiments to determine deviations from maximal $\nu_\mu - \nu_\tau$ mixing was discusses in [12].

This gives an understanding of the diagonal bands in Fig. 2, in particular the white one corresponding to $\varphi_1 - \varphi_2 = 0$.

Planned reactor experiments [13] and next generation superbeam experiments [14,15] are expected to have an approximate sensitivity on $\sin^2 2\theta_{13}$ of 10^{-2}. From Fig. 2 we find that the radiative corrections exceed this value for large regions of the currently allowed parameter space, unless there are cancellations due to Majorana phases, i.e. $\varphi_1 = \varphi_2$. If so, the effects are generically smaller than 10^{-2} as can be seen from the lower diagram. Future upgraded superbeam experiments like JHF-HyperKamiokande have the potential to further push the sensitivity to about 10^{-3} and with a neutrino factory even about 10^{-4} might be reached.

From the theoretical point of view, one would expect that even if some model predicted $\theta_{13} = 0$ at the energy scale of neutrino mass generation, RG effects would at least produce a non-zero value of the order shown in Fig. 2. Consequently, experiments with such a sensitivity have a large discovery potential for θ_{13}. We should point out that this is a conservative estimate, since if neutrino masses are e.g. determined by GUT scale physics, model-dependent radiative corrections in the region between M_1 and M_{GUT} contribute as well [16,17,18,19] and there can be additional corrections from physics above the GUT scale [20]. On the other hand, if experiments do not measure θ_{13}, this will improve the upper bound on θ_{13}. Parameter space regions where the corrections are larger than this bound will then appear unnatural from the theoretical side.

Next we consider the RG corrections to θ_{23}, where very interesting questions arise if θ_{23} turns out to be close to $\pi/4$. The deviation of θ_{23} from $\pi/4$ might again be an effect of the RG running which induces a deviation of θ_{23} from $\pi/4$. In order to understand the corrections we use the analytical formula (10) with a constant right-hand side in order to calculate the running in the MSSM between M_Z and the see-saw scale, where we take again $M_1 = 10^{12}\,\mathrm{GeV}$ for our examples. As initial conditions we assume small θ_{13} at M_1 and low-energy best-fit values for the remaining lepton mixings and the neutrino mass squared differences. In leading order in θ_{13}, the evolution is of course independent of the Dirac phase δ.

The size of the RG corrections in the MSSM is shown in Fig. 3. From the upper diagram it can be read off for desired values of $\tan\beta$ and the lightest mass eigenvalue m_1 in an example with vanishing Majorana phases. The lower diagram shows its dependence on the Majorana phases φ_1 and φ_2 for $\tan\beta = 50$, $m_1 = 0.1\,\mathrm{eV}$ and a normal mass hierarchy. The diagrams look rather similar in the case of an inverted hierarchy. The effects of the

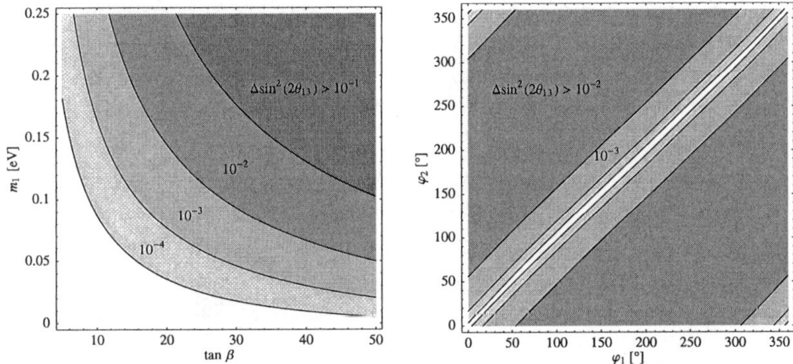

Figure 2. Corrections to θ_{13} from the RG evolution between 10^2 and 10^{12} GeV in the MSSM, calculated using the analytical approximations with initial conditions $\theta_{13} = 0$ and LMA best-fit values for the remaining parameters. The upper diagram shows the dependence on $\tan\beta$ and on the mass of the lightest neutrino for the case of a normal mass hierarchy and phases $\varphi_1 = 0$ and $\varphi_2 = \pi$. In the lower diagram the dependence on the Majorana phases φ_1 and φ_2 is shown for $\tan\beta = 50$ and $m_1 = 0.08$ eV. The contour lines are defined as in the upper diagram.

Majorana phases can easily be understood from Eq. (10). In the region with $\varphi_1 \approx \varphi_2 \approx \pi$, both $|m_2\, e^{i\varphi_2} + m_3|^2$ and $|m_1\, e^{i\varphi_1} + m_3|^2$ are small for quasi-degenerate neutrinos, which gives the ellipse with small radiative corrections in the center of the lower diagram. Such cancellations are not possible with hierarchical masses, but the RG effects are generally not very large in this case, as shown by the upper plot.

Even if a model predicted $\theta_{23} = \pi/4$ at some high energy scale, we would thus expect radiative corrections to produce at least a deviation from this value of the size shown in Fig. 3, so that experiments with such a sensitivity are expected to measure a deviation of θ_{23} from $\pi/4$. The sensitivity to $\sin^2 2\theta_{23}$ of future superbeam experiments like T2K is expected to be approximately 1% (see e.g. [21]). This can now be compared with Fig. 3. We find that the radiative corrections exceed this value for large regions of the currently allowed parameter space, where no significant cancellations due to Majorana phases occur. This means that φ_1 and φ_2 must not be to close to π. Otherwise, the effects are generically smaller as can be seen from the lower diagram. Upgraded superbeam experiments or a neutrino factory might even reach a sensitivity of about 0.5%. As argued for the case of θ_{13}, if experiments measure θ_{23} rather close to $\pi/4$, parameter combinations implying larger radiative corrections than the measured deviation will appear unnatural from the theoretical point of view.

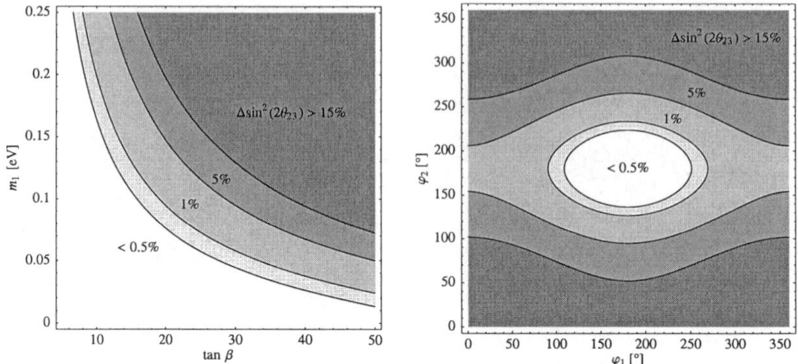

Figure 3.   Corrections to θ_{23} from the RG evolution between 10^2 GeV and 10^{12} GeV in the MSSM, calculated from the analytical approximation Eq. (10) with initial conditions $\theta_{23} = \pi/4$, small $\theta_{13} = 0$ and LMA best-fit values for the remaining parameters. The upper diagram shows the dependence on $\tan\beta$ and on the mass m_1 of the lightest neutrino for the case of a normal mass hierarchy and phases $\varphi_1 = \varphi_2 = 0$. In the lower diagram the dependence on the Majorana phases φ_1 and φ_1 is shown for the example $\tan\beta = 50$ and $m_1 = 0.1$. Note that for small θ_{13} the results are independent of the Dirac phase to a good approximation.

Next we discuss the RG Evolution of θ_{12}. From Eq. 8 we see that the running of the solar angle θ_{12} is proportional to $(\Delta m_{sol}^2)^{-1}$, while the running of the other angles is proportional to $(\Delta m_{atm}^2)^{-1}$. Therefore, $\Delta m_{sol}^2) \ll \Delta m_{atm}^2)$ explains why θ_{12} has generically the strongest RG effects among the mixing angles, especially for for quasi-degenerate neutrinos and for the case of an inverted mass hierarchy. Furthermore, it is known that in the MSSM the solar angle always increases when running down from M_1 for $\theta_{13} = 0$ [22]. This is confirmed by our formula (8). From the term $|m_1 e^{i\varphi_1} + m_2 e^{i\varphi_2}|^2$ in Eq. (8), we see that a non-zero value of the difference $|\varphi_1 - \varphi_2|$ of the Majorana phases damps the RG evolution. The damping becomes maximal if this difference equals π, which corresponds to an opposite CP parity of the mass eigenstates m_1 and m_2. This is in agreement with earlier studies, e.g. [23,24,25].

Let us now compare the analytical approximation for $\dot{\theta}_{12}$ of Eq. (8) with the numerical solution for the running in the case of nearly degenerate masses, which is shown in Fig. 4. The dark-gray region shows the evolution with LMA best-fit values for the neutrino parameters, θ_{13} varying in the interval $[0°, 9°]$ and all CP phases equal to zero. The medium-gray regions show the evolution for $|\varphi_1 - \varphi_2| \in \{0°, 90°, 180°, 270°\}$, $\theta_{13} \in [0°, 9°]$ and $\delta \in \{0°, 90°, 180°, 270°\}$, confirming the expectation of the damping influence of φ_1 and φ_2. The flat line at low energy stems from the SM running below M_{SUSY}, which is negligible.

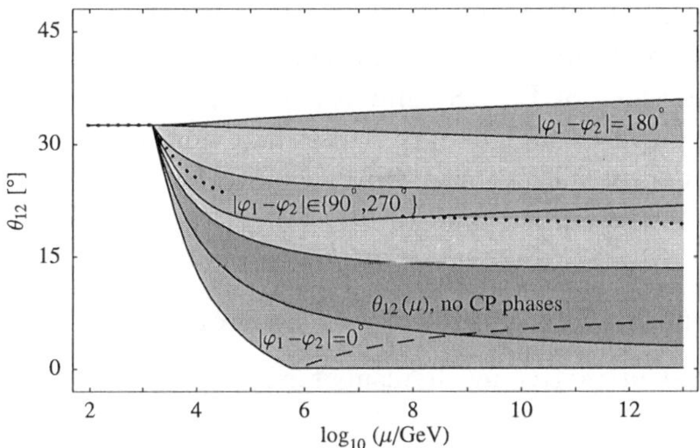

Figure 4. RG evolution of θ_{12} in the MSSM with $\tan\beta = 50$, a normal mass hierarchy and $m_1 = 0.1\,\text{eV}$. The dark-gray region shows the evolution with best-fit values for the neutrino parameters, $\theta_{13} \in [0°, 9°]$ and all CP phases equal to zero. The medium-gray regions show the evolution for $|\varphi_1 - \varphi_2| = 0°$, $|\varphi_1 - \varphi_2| \in \{90°, 270°\}$ and $|\varphi_1 - \varphi_2| = 180°$. They emerge from varying $\theta_{13} \in [0°, 9°]$ and $\delta \in \{0°, 90°, 180°, 270°\}$. The light-gray regions can be reached by choosing specific values for the CP phases different from the ones listed above. The dashed line shows the RG evolution with $|\varphi_1 - \varphi_2| = 0$, $\theta_{13} = 9°$ and $\delta = 180°$. Note that for the numerics we use the convention where θ_{12} is restricted to the interval $[0°, 45°]$, so that the angle increases again after reaching 0. The dotted line shows the evolution with $|\varphi_1 - \varphi_2| = 90°$ and $\theta_{13} = 0°$.

In the case of large cancellations by phases, the $\mathcal{O}(\theta_{13})$-term in the RGE turns out to be important. The dominant contribution to the next-to-leading term is given by Υ where

$$\Upsilon = \frac{C y_\tau^2}{32\pi^2} \frac{m_2 + m_1}{m_2 - m_1} \cos\left(\frac{\varphi_1 - \varphi_2}{2}\right) \times$$
$$\times \left[\cos(2\theta_{12}) \cos\delta \cos\left(\frac{\varphi_1 - \varphi_2}{2}\right) + \sin\delta \sin\left(\frac{\varphi_1 - \varphi_2}{2}\right)\right] \cdot \theta_{13} \,. \quad (19)$$

Clearly, the RG evolution of θ_{12} is independent of the Dirac phase δ only in the approximation $\theta_{13} = 0$. The largest running, where θ_{12} can even become zero, occurs for θ_{13} as large as possible ($9°$), $\delta = \pi$ and $\varphi_1 - \varphi_2 = 0$. In this case the leading and the next-to-leading term add up constructively. It is also interesting to observe that due to $\mathcal{O}(\theta_{13})$ effects θ_{12} can run to slightly larger values. The damping due to the Majorana phases is maximal in this case, which almost eliminates the leading term. Then, all the running comes from the next-to-leading term (19).

In the inverted scheme, $m_1 \gg m_2 - m_1$ always holds, so that large RG effects are generic, i.e. always present except for the case of cancellations due to Majorana phases. For a normal mass hierarchy with a small m_1, the running of the solar mixing is of course rather insignificant.

Finally, we would like to emphasize that it is not appropriate to assume the right-hand sides of Eq. (8) and Eq. (19) to be constant in order to interpolate θ_{12} up to a high energy scale, since non-linear effects especially from the running of $\sin 2\theta_{12}$ and $\Delta m_{\mathrm{sol}}^2$ cannot be neglected here. This is easily seen from the curved lines in Fig. 4.

5. RG Corrections to Leptogenesis Parameters

One of the most attractive mechanisms for explaining the observed baryon asymmetry of the universe, $\eta_B = (6.5^{+0.4}_{-0.8}) \cdot 10^{-10}$ [26], is leptogenesis [27]. In this scenario, η_B is generated by the out-of-equilibrium decay of the same heavy singlet neutrinos which are responsible for the suppression of light neutrino masses in the see-saw mechanism. The masses of the heavy neutrinos are typically assumed to be some orders of magnitude below the GUT scale. Though the parameters entering the leptogenesis mechanism cannot be completely expressed in terms of low-energy neutrino mass parameters, it is possible to derive bounds on the neutrino mass scale from requiring successful leptogenesis [28]. However, since leptogenesis occurs at high temperatures and correspondingly at high scales, one cannot directly use the low energy parameters and the RGE evolution has to be taken into account. The neutrino masses experience corrections of about 20-25% in the MSSM or more than 60% in the SM and we expect therefore sizable corrections to the leptogenesis bounds. The maximal baryon asymmetry generated in the thermal version of this scenario is given by [29,30,28]

$$\eta_B^{\mathrm{max}} \simeq 0.96 \cdot 10^{-2} \, \varepsilon_1^{\mathrm{max}} \, \kappa_{\mathrm{f}} \,. \tag{20}$$

κ_{f} is a dilution factor which can be computed from a set of coupled Boltzmann equations (see, e.g. [31]). In [28], an analytic expression for the maximal relevant CP asymmetry was derived,

$$\varepsilon_1^{\mathrm{max}}(m_1, m_3, \widetilde{m}_1) = \frac{3}{16\pi} \frac{M_1 \, m_3}{(v/\sqrt{2})^2} \left[1 - \frac{m_1}{m_3} \left(1 + \frac{m_3^2 - m_1^2}{\widetilde{m}_1^2} \right)^{1/2} \right] \tag{21}$$

which refines the older bound

$$\varepsilon_1^{\mathrm{max}}(m_1, m_3) = \frac{3}{16\pi} \frac{M_1}{(v/\sqrt{2})^2} \frac{\Delta m_{\mathrm{atm}}^2 + \Delta m_{\mathrm{sol}}^2}{m_3} \tag{22}$$

and is valid for a normal mass hierarchy in the SM as well as in the MSSM.[d] $\widetilde{m}_1$ is defined by

$$\widetilde{m}_1 = \frac{(m_{\mathrm{D}}^\dagger m_{\mathrm{D}})_{11}}{M_1} \tag{23}$$

with $m_{\mathrm{D}} \sim Y_\nu$ being the neutrino Dirac mass and typically lies between m_1 and m_3. It can be constrained by the requirement of successful leptogenesis because it controls the dilution of the generated asymmetry. The authors of [28] introduced the 'neutrino mass window for baryogenesis' which corresponds to the region in the $\widetilde{m}_1$-M_1 plane allowing for successful thermal leptogenesis. The shape and size of the 'mass window' depends on $\overline{m} = \sqrt{m_1^2 + m_2^2 + m_3^2}$, i.e. it becomes smaller for increasing $\overline{m}$, and $\overline{m} \geq 0.2\,\mathrm{eV}$ is not compatible with thermal leptogenesis.

The calculations relevant for leptogenesis, however, refer to processes at very high energies, and therefore the RG evolution of the input parameters has to be taken into account [32]. The size of the error arising if RG effects are neglected has been estimated in [11]. It was found that there are two effects in opposite directions: The CP asymmetry is enhanced because the mass squared differences increase, and the dilution of the baryon asymmetry is more effective since the overall mass scale rises due to RG effects. As the dependence of the dilution factor on the mass scale is stronger than that of the CP asymmetry, it is expected that the mass window for baryogenesis shrinks when RG effects are included in the analysis. An exception is the case of large $\tan\beta$, where the situation is more complicated.

There exist also different other, non-thermal baryogenesis mechanisms [35] in which the masses of the light neutrinos may be almost degenerate [36]. In these kinds of scenarios, RG effects increase the baryon asymmetry, since ε_1 increases, while the effects from the expected decrease of the dilution factor do not occur.

6. Conclusions

We discussed the RG running of neutrino parameters. Analytical solutions of the running mixings and masses were presented and phenomenologiacl consequences were discussed for leptogenesis. Further phenomenological issues and possibilities as, for example, the radiative generation of CP Phases,

---

[d]To use these formulae in our conventions for the inverted scheme, one would have to replace $(m_1, m_2, m_3) \rightarrow (m_3, m_1, m_2)$.

are discussed in [11]. Running above the see-saw threshold leads also to important effects and this will be discussed in a forthcoming paper.

References

1. B. Grzadkowski, M. Lindner, Phys. Lett. **B193** (1987) 71.
2. B. Grzadkowski, M. Lindner, S. Theisen, Phys. Lett. **B198** (1987) 64.
3. E. Ma, Phys. Rev. Lett. **81** (1998), 1171.
4. P. H. Chankowski, Z. Pluciennik, Phys. Lett. **B316** (1993), 312.
5. K. S. Babu, C. N. Leung, J. Pantaleone, Phys. Lett. **B319** (1993), 191.
6. S. Antusch, M. Drees, J. Kersten, M. Lindner, M. Ratz, Phys. Lett. **B519** (2001), 238.
7. S. Antusch, M. Drees, J. Kersten, M. Lindner, M. Ratz, Phys. Lett. **B525** (2002), 130.
8. S. Antusch, M. Ratz, JHEP **07** (2002), 059.
9. Z. Maki, M. Nakagawa, S. Sakata, Prog. Theor. Phys. **28** (1962), 870.
10. standard parametrization
11. S. Antusch, J. Kersten, M. Lindner, M. Ratz, Nucl. Phys. **B674** (2003) 401.
12. S. Antusch, P. Huber, J. Kersten, T. Schwetz, W. Winter, **hep-ph/0404268**.
13. P. Huber, M. Lindner, T. Schwetz, W. Winter, Nucl.Phys. **B665** (2003) 487.
14. P. Huber, M. Lindner, W. Winter, Nucl. Phys. **B654** (2003), 3.
15. H. Minakata, H. Nunokawa, S. Parke, Phys.Rev. **D68** (2003) 013010.
16. S. F. King, N. N. Singh, Nucl. Phys. **B591** (2000), 3.
17. S. Antusch, J. Kersten, M. Lindner, M. Ratz, Phys. Lett. **B538** (2002) 87.
18. S. Antusch, J. Kersten, M. Lindner, M. Ratz, Phys. Lett. **B544** (2002) 1.
19. S. Antusch, M. Ratz, JHEP **11** (2002), 010.
20. F. Vissani, M. Narayan, V. Berezinsky, Phys.Lett. **B571** (2003) 209.
21. Y. Itow et al., *The JHF-Kamioka neutrino project*, (2001), **hep-ex/0106019**.
22. T. Miura, T. Shindou, E. Takasugi, Phys. Rev. **D66** (2002), 093002.
23. K. R. S. Balaji, A. S. Dighe, R. N. Mohapatra, M. K. Parida, Phys. Rev. Lett. **84** (2000), 5034.
24. N. Haba, Y. Matsui, N. Okamura, Eur. Phys. J. **C17** (2000), 513.
25. P. H. Chankowski, S. Pokorski, Int. J. Mod. Phys. **A17** (2002), 575.
26. D. N. Spergel et al., (2003), **astro-ph/0302209**.
27. M. Fukugita, T. Yanagida, Phys. Lett. **174B** (1986), 45.
28. W. Buchmüller, P. D. Bari, M. Plümacher, (2003), **hep-ph/0302092**.
29. K. Hamaguchi, H. Murayama, T. Yanagida, Phys. Rev. **D65** (2002), 043512.
30. S. Davidson, A. Ibarra, Phys. Lett. **B535** (2002), 25.
31. W. Buchmüller, M. Plümacher, Int. J. Mod. Phys. **A15** (2000), 5047.
32. R.Barbieri, P.Creminelli, A.Strumia, N.Tetradis, Nucl.Phys. **B575** (2000),61.
33. H. B. Nielsen, Y. Takanishi, Nucl. Phys. **B636** (2002), 305.
34. P. Di Bari, AIP Conf.Proc. **655** (2003) 208.
35. K. Kumekawa, T. Moroi, T. Yanagida, Prog. Theor. Phys. **92** (1994), 437.
36. M. Fujii, K. Hamaguchi, T. Yanagida, Phys. Rev. **D65** (2002), 115012.

SEESAW MECHANISM AND SUPERSYMMETRY

ANTONIO MASIERO, SUDHIR K. VEMPATI

Dip. di Fisica 'G. Galilei', Univ. di Padova
and INFN, Sezione di Padova, via F. Marzolo 8,
I-35131, Padua, Italy.
E-mail: masiero@pd.infn.it

OSCAR VIVES

Theory Group,
Physics Department, CERN, CH-1211,
Geneva 23, Switzerland.
Email:vives@mail.cern.ch

After almost three decades of intense search for new physics beyond the Standard Model (SM), two ideas stand out to naturally cope with (i) small neutrino masses and (ii) a light higgs boson : Seesaw and SUSY. The combination of these two ideas, i.e. SUSY seesaw exhibits a potentially striking signature: a strong (or even very strong) enhancement of lepton flavour violation (LFV), which on the contrary remains unobservable in the SM seesaw. Indeed, even when supersymmetry breaking is completely flavour blind, Renormalisation Group running effects are expected to generate large lepton flavour violating entries at the weak scale. We explicitly show that in a class of SUSY SO(10) GUTs there exist cases where LFV and CP violation can constitute a major road in simultaneously confirming the ideas of Seesaw and low-energy SUSY.

1. Introduction

In the Standard Model (SM) with massless neutrinos the three Lepton Flavour (LF) numbers are exactly conserved (at any order in perturbation theory). The introduction of a mass for the neutrinos leads to LF violation (LFV) analogously to the violation of flavour numbers (strangeness, charm, etc.) in the quark sector. However, given that LFV in the SM has to be proportional to the neutrino masses, we conclude that within the SM we expect any LFV process other than neutrino oscillations to be affected by suppression factors proportional to some power of the ratio of neutrino mass to the W mass. Hence, although the discovery that neutrinos are massive entails that LF numbers are no longer conserved in the SM, we can safely state that, as long as the SM represents the correct physical description, no LFV process like $\mu \to e + \gamma$ should ever be observed.

The situation radically changes when we move from the SM to its supersymmetric (SUSY) extensions. The main difference lies in the fact that now we have also the scalar partners of the leptons (sleptons) which carry LF numbers and hence, a priori, one may expect that there are contributions to LFV processes where ratios of flavour off-diagonal slepton masses to some average SUSY mass appear which may easily be orders of magnitude larger than the ratio m_ν / M_W.

This is indeed the case. The reason for a conspicuous value of the above mentioned LFV entries in the slepton mass matrices is twofold. In SUSY extensions of the SM where the terms which break SUSY softly are not flavour universal, one could even imagine the off-diagonal LFV entries to be of the same order as the flavour conserving diagonal entries. This would be disastrous just because LFV would become too large (and the same would happen also in the hadronic sector if flavour non-universality in the squarks is maximal). However, even assuming the opposite case, namely exact flavour universality of the soft breaking terms (at the superlarge scale at which they appear in a supergravity framework) , there exists a remarkable property of the RG running of the slepton masses which may yield seizable off-diagonal slepton masses at the scale of interest for our experiments, i.e. the electroweak scale. This occurs whenever the lepton superfields have new large Yukawa couplings. The seesaw [1,2] mechanism represents a typical context where this can be implemented. This was first pointed out in the SUSY seesaw model in the work of Borzumati and Masiero in 1986 [3] (as I said in the talk, this work was prompted by some discussion that I previously had with Marciano and Sanda on the issue of LFV in SUSY). At that time we individuated the two quantities which crucially determine the size of the RG-induced FV off-diagonal slepton mass matrix entries: the Yukawa couplings responsible for the Dirac entries of the neutrino mass matrices and the rotating matrix establishing the mismatch in the diagonalisation of the lepton and slepton mass matrices. Unfortunately in 1986, still little was known about neutrino masses and mixings (indeed, to be sure, not even the fact that neutrinos were massive and mixed was established!).

The enormous amount of literature dealing with LFV in SUSY seesaw after the discovery of neutrino oscillations can be easily understood. Although, honestly, from the experimental data we find neither the Dirac neutrino Yukawa couplings nor the mentioned mixing matrix, it is true that all the information we have collected in recent years on neutrino masses and mixings provides important clues on the above quantities relevant in SUSY seesaw. Complementarily, various experiments have improved the limits on the rare LFV decay processes over the years and in the near future, they are expected to do furthermore. To have an idea where we stand, here we provide a list of present and upcoming experimental limits:

Present limits:

$$\begin{aligned}
BR(\mu \to e\gamma) &\leq 1.2 \times 10^{-11} & (^4)\\
BR(\tau \to \mu\gamma) &\leq 3.1 \times 10^{-7} & (^5)\\
BR(\tau \to e\gamma) &\leq 3.7 \times 10^{-7} & (^6)
\end{aligned}$$

Upcoming limits:

$$\begin{aligned}
BR(\mu \to e\gamma) &\leq 10^{-13} \div 10^{-14} & (^7)\\
BR(\tau \to \mu\gamma) &\leq 10^{-8} & (^6)\\
BR(\tau \to e\gamma) &\leq 10^{-8} & (^6)
\end{aligned}$$

In this talk we are going to provide an example of how the interplay between new experimental data on these decays and theoretical progress may help in shedding light on the quantitative predictions on LFV in SUSY seesaw in general, and, more specifically, in the context of the SUSY $SO(10)$ scheme. Other LFV processes like $\mu \to e$ conversion in Nuclei [8], (Higgs mediated) $\tau \to 3\mu$ [9], flavour violating Z-decays [10], Higgs decays[11] and other collider processes [12] are also being investigated in the literature.

2. Supersymmetric Seesaw and Leptonic Flavour Violation

The seesaw mechanism can be incorporated in the Minimal Supersymmetric Standard Model in the similar manner as in the Standard Model, by adding right handed neutrino superfields to the MSSM superpotential:

$$W = W_{Y_Q} + h^e_{ij} L_i e^c_j H_1 + h^\nu_{ij} L_i \nu^c_j H_2 + M_{R_{ij}} \nu^c_i \nu^c_j, \tag{1}$$

where the leptonic part has been detailed, while the quark Yukawa couplings and the μ parameter are contained in W_{Y_Q}. i, j are generation indices. M_R represents the (heavy) Majorana mass matrix for the right-handed neutrinos. Eq.(1) leads to the standard seesaw formula for the (light) neutrino mass matrix

$$\mathcal{M}_\nu = -h^\nu M_R^{-1} h^{\nu\,T} v_2^2, \tag{2}$$

where v_2 is the vacuum expectation value (VEV) of the up-type Higgs field, H_2. Under suitable conditions on h^ν and M_R, the correct mass splittings

114

and mixing angles in $\mathcal{M}_\nu$ can be obtained. Detailed analyses deriving these conditions are already present in the literature [13].

The above lagrangian has to be supplemented by a part containing supersymmetry breaking soft terms. The flavour structure of these terms would depend on the mechanism which breaks supersymmetry and conveys it to the observable sector. However, the accumulating concordance between the Standard Model (SM) expectations and the vast range of FCNC and CP violation [14] point out to a SUSY breaking mechanism which is flavour blind, as in mSUGRA, Gauge-Mediation (GMSB) and Anomaly Mediation (AMSB) and its variants. The main observation of [3] was that in spite of possible flavour-blindness of SUSY breaking, the supersymmetrization of the seesaw leads to new sources of LFV[a]. This occurs because the flavour-blindness of the slepton mass matrices is no longer invariant under RG evolution from the large SUSY breaking scale down to the electroweak (seesaw) scale in the presence of the new (seesaw) couplings [17].

The amount of lepton flavour violation generated by the SUSY seesaw at the weak scale crucially depends on the flavour structure of h^ν and M_R, shown in the eq.(1), the 'new' sources of flavour violation not present in the MSSM. To see this, one has to solve the Renormalisation Group Equations (RGE) for the slepton mass matrices from the high scale to the scale of the right handed neutrinos. Below this scale, the running of the FV slepton mass terms is RG-invariant as the right handed neutrinos decouple from the theory. For the purpose of our illustration, a leading log estimate can easily be obtained for these equations. Assuming the flavour blind mSUGRA specified by the high-scale parameters: m_0, the common scalar mass, A_0, the common trilinear coupling and $M_{1/2}$, the universal gaugino mass, the flavour violating entries in these mass matrices at the weak scale are given as:

$$(m_{\tilde{L}}^2)_{ij(i \neq j)} \approx -\frac{3m_0^2 + A_0^2}{8\pi^2} \sum_k (h_{ik}^\nu h_{jk}^{\nu *}) \ln \frac{M_{GUT}}{M_{R_k}} \tag{3}$$

where M_R is the scale of the right handed neutrinos. Given this, the branching ratios for LFV rare decays, $l_j \to l_i, \gamma$ are roughly estimated as[3,18,19,20]:

$$\mathrm{BR}(l_j \to l_i \gamma) \approx \frac{\alpha^3 \ ([m_{\tilde{L}}^2]_{ij})^2}{G_F^2 \ m_{SUSY}^8} \tan^2 \beta, \tag{4}$$

---

[a]Of the above mentioned SUSY breaking mechanisms, this is always true in a gravity mediated supersymmetry breaking model, but, applies also to other mechanisms under some specific conditions [15,16].

where m_{SUSY} represents the typical soft supersymmetric breaking mass, determined by $m_0, M_{1/2}$, etc., at the weak scale.

From above it is obvious that if either the neutrino Yukawa couplings or the flavour mixings present in h^ν are very tiny, the strength of LFV will be significantly reduced. Further, if the right handed neutrino masses are heavier than the supersymmetry breaking scale (as in GMSB models), these effects would vanish.

However, to make a more quantitative analysis of LFV in susy seesaw models, say, in terms of the supersymmetry breaking parameters, one needs to make further assumptions on the seesaw couplings of the model. This is because despite the huge successes we had in the neutrino physics, information from neutrino masses is nonetheless not sufficient to determine all the seesaw parameters [21] in eq.(2), which are crucial to compute the relevant LFV rates[b]. To remedy this, either a top-down approach with specific SUSY-GUT models and/or flavour symmetries [22,23,24,25] or a bottom-up approach with specific parameterisations of low energy unknowns have been adopted in the literature [26,27,28].

3. SO(10) and SUSY Seesaw

In the $SO(10)$ gauge theory, all the known fermions and the right handed neutrinos are unified in a single representation of the gauge group, the **16**. The product of two **16** matter representations can only couple to **10**, **120** or **126** representations which can be formed either by a single Higgs field representation or a non-renormalisable product of representations of several Higgs fields. In either case, the Yukawa matrices resulting from the couplings to **10** and **126** are complex symmetric whereas they are anti-symmetric when the couplings are to the **120**. Thus, the most general $SO(10)$ superpotential relevant for fermion masses can be written as

$$W_{SO(10)} = h_{ij}^{10} 16_i \, 16_j \, 10 + h_{ij}^{126} 16_i \, 16_j \, 126 + h_{ij}^{120} 16_i \, 16_j \, 120, \quad (5)$$

where i, j refer to the generation indices. In terms of the SM fields, the Yukawa couplings relevant for fermion masses are given by [29]:

$$16 \; 16 \; 10 \; \supset 5 \; (uu^c + \nu\nu^c) + \bar{5} \; (dd^c + ee^c),$$

$$16 \; 16 \; 126 \; \supset 1 \; \nu^c\nu^c + 15 \; \nu\nu + 5 \; (uu^c - 3 \; \nu\nu^c) + \bar{45} \; (dd^c - 3 \; ee^c),$$

$$16 \; 16 \; 120 \; \supset 5 \; \nu\nu^c + 45 \; uu^c + \bar{5} \; (dd^c + ee^c) + \bar{45} \; (dd^c - 3 \; ee^c), \quad (6)$$

---

[b]This can be seen from a simple parameter counting on either sides of the seesaw equation, eq.(2). h^ν contains 9 complex parameters, M_R, three real whereas we only have information about two mass squared differences and three mixing angles in $\mathcal{M}_\nu$.

116

where we have specified the corresponding $SU(5)$ Higgs representations for each of the couplings and all the fermions are left handed fields. The resulting mass matrices can be written as

$$M^u = M_{10}^5 + M_{126}^5 + M_{120}^{45}, \tag{7}$$

$$M_{LR}^\nu = M_{10}^5 - 3\,M_{126}^5 + M_{120}^5, \tag{8}$$

$$M^d = M_{10}^{\bar 5} + M_{126}^{\overline{45}} + M_{120}^{\bar 5} + M_{120}^{\overline{45}}, \tag{9}$$

$$M^e = M_{10}^{\bar 5} - 3M_{126}^{\overline{45}} + M_{120}^{\bar 5} - 3M_{120}^{\overline{45}}, \tag{10}$$

$$M_{LL}^\nu = M_{126}^{15}, \tag{11}$$

$$M_R^\nu = M_{126}^1. \tag{12}$$

A simple analysis of the above mass matrices leads us to the following result: *At least one of the Yukawa couplings in* $h^\nu = v_u^{-1}\,M_{LR}^\nu$ *has to be as large as the top Yukawa coupling* [25]. This result holds true in general independently from the choice of the Higgses responsible for the masses in Eqs. (7, 8) provided that no accidental fine tuned cancellations of the different contributions in Eq. (8) are present. If contributions from the **10**'s solely dominate, h^ν and h^u would be equal. If this occurs for the **126**'s, then $h^\nu = -3\,h^u$. In case both of them have dominant entries, barring a rather precisely fine tuned cancellation between M_{10}^5 and M_{126}^5 in Eq. (8), we expect at least one large entry to be present in h^ν. A dominant antisymmetric contribution to top quark mass due to the **120** Higgs is phenomenologically excluded since it would lead to at least a pair of heavy degenerate up quarks.

Apart from sharing the property that at least one eigenvalue of both M^u and M_{LR}^ν has to be large, for the rest it is clear from (7) and (8) that these two matrices are not aligned in general, and hence we may expect different mixing angles appearing from their diagonalisation. We find two cases which would serve as 'benchmark' scenarios for seesaw induced lepton flavour violation in SUSY $SO(10)$. The first one corresponds to a case where the mixing present in h^ν is small and CKM-like. This is typical of the models where fermions attain their masses through 10-plets. We will call this case, 'the minimal case'. As a second case, we consider scenarios where the mixing in h^ν is no longer small, but large like the observed PMNS mixing. We will call this case the 'the maximal case'.

3.1. *The minimal Case: CKM mixings in h^ν*

The minimal Higgs spectrum to obtain obtain the required CKM mixing includes two **10**-plets [30] :

$$W_{SO(10)} = \frac{1}{2}\ h_{ij}^{u,\nu}16_i\ 16_j\ 10_u + \frac{1}{2}\ h_{ij}^{d,e}16_i\ 16_j\ 10_d + \frac{1}{2}\ h_{ij}^{R}\ 16_i\ 16_j\ 126.$$

We further assume the **126** dimensional Higgs field gives Majorana mass *only* to the right handed neutrinos. From the above, it is clear that the following mass relations hold between the quark and leptonic mass matrices at the GUT scale[c]:

$$h^u = h^\nu\ \ ;\ \ h^d = h^e. \tag{13}$$

In the above basis, the symmetric h^u is diagonalised by:

$$V_{CKM}\ h^u\ V_{CKM}^T = h^u_{diag}. \tag{14}$$

Hence from (13):

$$h^\nu = V_{CKM}^T\ h^u_{diag}\ V_{CKM}. \tag{15}$$

Given this, the induced off-diagonal entries relevant for $l_j \to l_i, \gamma$ are of the order,

$$(m_{\tilde{L}}^2)_{21} \approx -\frac{3m_0^2 + A_0^2}{8\pi^2}\ h_t^2 V_{td} V_{ts} \ln \frac{M_{GUT}}{M_{R_3}} + \mathcal{O}(h_c)^2, \tag{16}$$

$$(m_{\tilde{L}}^2)_{32} \approx -\frac{3m_0^2 + A_0^2}{8\pi^2}\ h_t^2 V_{tb} V_{ts} \ln \frac{M_{GUT}}{M_{R_3}} + \mathcal{O}(h_c)^2, \tag{17}$$

$$(m_{\tilde{L}}^2)_{31} \approx -\frac{3m_0^2 + A_0^2}{8\pi^2}\ h_t^2 V_{tb} V_{td} \ln \frac{M_{GUT}}{M_{R_3}} + \mathcal{O}(h_c)^2. \tag{18}$$

The required right handed neutrino Majorana mass matrix consistent with both the observed low energy neutrino masses and mixings as well as with CKM like mixings in h^ν is determined easily from the seesaw formula defined at the scale of right handed neutrinos. Neglecting the small CKM mixing in h^ν we have

$$M_R \approx v_u^2 \begin{pmatrix} h_u^2[m_\nu^{-1}]_{11} & \star & \star \\ h_u h_c[m_\nu^{-1}]_{12} & h_c^2[m_\nu^{-1}]_{22} & \star \\ h_u h_t[m_\nu^{-1}]_{13} & h_c h_t[m_\nu^{-1}]_{23} & h_t^2[m_\nu^{-1}]_{33} \end{pmatrix}. \tag{19}$$

---

[c]Clearly this relation cannot hold for the first two generations of down quarks and charged leptons. One expects, small corrections due to non-renormalisable operators or suppressed renormalisable operators [31] can be invoked.

118

It is clear from above that the hierarchy in the M_R mass matrix goes as the square of the hierarchy in the up-type quark mass matrix. it straightforward to check that all the right handed neutrino mass eigenvalues are controlled by the smallest left-handed neutrino mass.

$$M_{R_3} \approx \frac{m_t^2}{4\, m_{\nu_1}} \; ; \; M_{R_2} \approx \frac{m_c^2}{4\, m_{\nu_1}} \; ; \; M_{R_1} \approx \frac{m_u^2}{2\, m_{\nu_1}} \; . \tag{20}$$

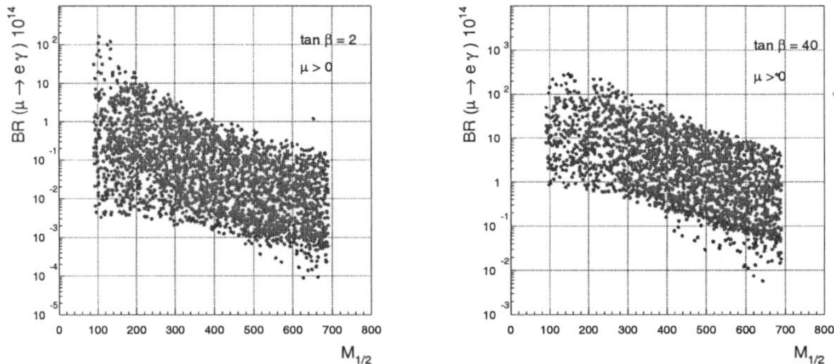

Figure 1.   The scatter plots of branching ratios of $\mu \to e, \gamma$ decays vs. $M_{1/2}$ are shown for the (minimal) CKM case for tan $\beta = 2$ and 40. Results do not alter significantly with the change of sign(μ).

We now present numerical results for this situation in the framework of minimal Supergravity (mSUGRA). In Fig. 1 we show the scatter plots for BR($\mu \to e, \gamma$) for the CKM case and tan $\beta = 2$ and tan $\beta = 40$ respectively. As expected from eq.(4), the BR scales with the second power in tanβ. For tan $\beta = 40$ reaching a sensitivity of 10^{-14} for BR($\mu \to e\gamma$) would allow us to probe 'completely' the SUSY spectrum up to $M_{1/2} = 300$ GeV (notice that this corresponds to gluino and squark masses of order 750 GeV) and would still probe a large regions in parameter space up to $M_{1/2} = 700$ GeV.

Thus in summary, though the present limits on BR($\mu \to e, \gamma$) would not induce any significant constraints on the supersymmetry breaking parameter space, an improvement in the limit to$\sim \mathcal{O}(10^{-14})$, as being foreseen, would start imposing non-trivial constraints especially for the large tan β region.

3.2. *The maximal case: PMNS mixing angles in h^ν*

The minimal $SO(10)$ model presented in the previous sub-section would inevitably lead to small mixing in h^ν. To generate mixing angles larger than CKM angles, asymmetric mass matrices have to be considered. In the present case, we assume that the down-sector couples to a combination of Higgs representations (symmetric and anti-symmetric) [d] Φ, leading to an asymmetric mass matrix in the basis where the up-sector is diagonal. As we will see below this would also require that the right handed Majorana mass matrix to be diagonal in this basis. We have :

$$W_{SO(10)} = \frac{1}{2}\, h_{ii}^{u,\nu}\, 16_i\, 16_i 10^u + \frac{1}{2}\, h_{ij}^{d,e}\, 16_i\, 16_j \Phi + \frac{1}{2}\, h_{ii}^{R}\, 16_i\, 16_i 126\ , \quad (21)$$

where the **126**, as before, generates only the right handed neutrino mass matrix. To study the consequences of these assumptions, we see that at the level of $SU(5)$, we have

$$W_{SU(5)} = \frac{1}{2}\, h_{ii}^{u}\, 10_i\, 10_i\, 5_u + h_{ii}^{\nu}\, \bar{5}_i\, 1_i\, 5_u + h_{ij}^{d}\, 10_i\, \bar{5}_j\, \bar{5}_d + \frac{1}{2}\, M_{ii}^{R}\, 1_i 1_i, \quad (22)$$

where we have decomposed the 16 into $10 + \bar{5} + 1$ and 5_u and $\bar{5}_d$ are components of 10_u and Φ respectively. To have large mixing $\sim U_{PMNS}$ in h^ν we see that the asymmetric matrix h^d should now be able to generate both the CKM mixing as well as PMNS mixing. This is possible if

$$V_{CKM}^{T}\, h^d\, U_{PMNS}^{T} = h_{diag}^{d}. \quad (23)$$

This would mean that the 10 which contains the left handed down-quarks would be rotated by the CKM matrix whereas the $\bar{5}$ which contains the left handed charged leptons would be rotated by the U_{PMNS} matrix to go into their respective mass bases [35]. Thus we have, in analogy with the previous sub-section, the following relations hold true in the basis where charged leptons and down quarks are diagonal:

$$h^u = V_{CKM}\, h_{diag}^{u}\, V_{CKM}^{T}\ , \quad (24)$$

$$h^\nu = U_{PMNS}\, h_{diag}^{u}. \quad (25)$$

Using the seesaw formula of Eq. (2) and Eq. (25) we have

$$M_R = \mathrm{Diag}\{\frac{m_u^2}{m_{\nu_1}},\ \frac{m_c^2}{m_{\nu_2}},\ \frac{m_t^2}{m_{\nu_3}}\}. \quad (26)$$

---

[d]The couplings of Φ in the superpotential can be either renormalisable or non-renormalisable. See [35] for a non-renormalisable example.

We now turn our attention to lepton flavour violation in the scenario. The branching ratio, $BR(\mu \to e, \gamma)$ would now be dependent on:

$$[h^\nu h^{\nu\,T}]_{21} = h_t^2 \, U_{\mu 3} \, U_{e3} + h_c^2 \, U_{\mu 2} \, U_{e2} + \mathcal{O}(h_u^2). \tag{27}$$

It is clear from the above that in contrast to the CKM case, the dominant contribution to the off-diagonal entries depends on the unknown magnitude of the element U_{e3} [23]. If U_{e3} is very close to its present limit ~ 0.2 [36], the first term on the RHS of the Eq. (27) would dominate. Moreover, this would lead to large contributions to the off-diagonal entries in the slepton masses with $U_{\mu 3}$ of $\mathcal{O}(1)$. We have :

$$(m_{\tilde{L}}^2)_{21} \approx -\frac{3m_0^2 + A_0^2}{8\pi^2} \, h_t^2 U_{e3} U_{\mu 3} \ln \frac{M_{GUT}}{M_{R_3}} + \mathcal{O}(h_c)^2. \tag{28}$$

The above contribution is large by a factor $(U_{\mu 3} U_{e3})/(V_{td} V_{ts}) \sim 140$ compared to the CKM case. From Eq. (4) we see that it would mean about a factor 10^4 times larger than the CKM case in $BR(\mu \to e, \gamma)$. In case U_{e3} is very small, *i.e*, either zero or $\lesssim (h_c^2/h_t^2) \, U_{e2} \sim 4 \times 10^{-5}$, the second term $\propto h_c^2$ in Eq. (27) would dominate. However the off-diagonal contribution in slepton masses, now being proportional to charm Yukawa could be much smaller, in fact, even smaller than the CKM contribution by a factor

$$\frac{h_c^2 \, U_{\mu 2} \, U_{e2}}{h_t^2 \, V_{td} \, V_{ts}} \sim 7 \times 10^{-2}. \tag{29}$$

If U_{e3} is close to it's present limit, the current bound on $BR(\mu \to e, \gamma)$ would already be sufficient to produce stringent limits on the SUSY mass spectrum. Similar U_{e3} dependence can be expected in the $\tau \to e$ transitions where the off-diagonal entries are given by :

$$(m_{\tilde{L}}^2)_{31} \approx -\frac{3m_0^2 + A_0^2}{8\pi^2} \, h_t^2 U_{e3} U_{\tau 3} \ln \frac{M_{GUT}}{M_{R_3}} + \mathcal{O}(h_c)^2. \tag{30}$$

The $\tau \to \mu$ transitions are instead U_{e3}-independent probes of SUSY, whose importance was first pointed out in Ref. [37]. As in the rest of the cases, the off-diagonal entry in this case is given by :

$$(m_{\tilde{L}}^2)_{32} \approx -\frac{3m_0^2 + A_0^2}{8\pi^2} \, h_t^2 U_{\mu 3} U_{\tau 3} \ln \frac{M_{GUT}}{M_{R_3}} + \mathcal{O}(h_c)^2. \tag{31}$$

In the PMNS scenario Fig. 2 (first panel) shows the plot for $BR(\mu \to e, \gamma)$ for $\tan \beta = 40$. In this plot, the value of U_{e3} chosen is very close to the present experimental upper limit [36]. As long as $U_{e3} \gtrsim 4 \times 10^{-5}$, the plots scale as U_{e3}^2, while for $U_{e3} \lesssim 4 \times 10^{-5}$ the term proportional to m_c^2 in Eq. (28) starts dominating and then, the result is insensitive to the

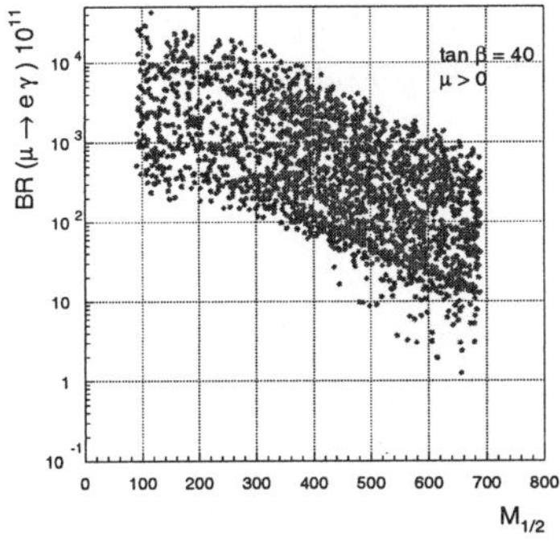
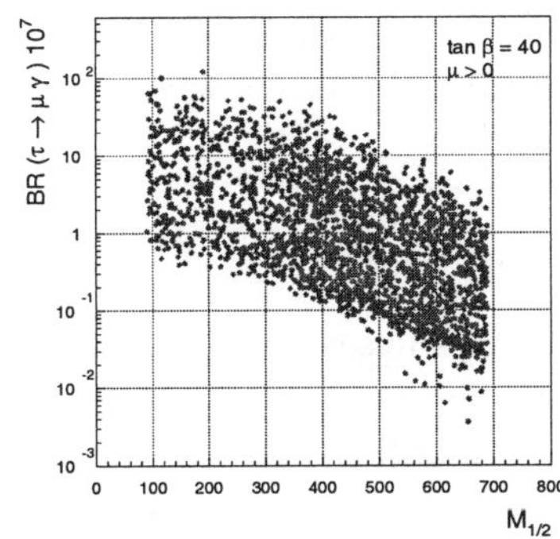

Figure 2. The scatter plots of branching ratios of $\mu \to e, \gamma$ and $\tau \to \mu\gamma$ decays vs. $M_{1/2}$ are shown for the (maximal) PMNS case for $\tan \beta = 40$. Results do not alter significantly with the change of sign(μ).

choice of U_{e3}. For instance, a value of $U_{e3} = 0.01$ would reduce the BR by a factor of 225 and still a significant amount of the parameter space for $\tan \beta = 40$ would be excluded. We further find that with the present limit on BR($\mu \to e, \gamma$), all the parameter space would be completely excluded up to $M_{1/2} = 300$ GeV for $U_{e3} = 0.15$, for any vale of $\tan \beta$.

In the $\tau \to \mu\gamma$ decay the situation is similarly constrained. For $\tan \beta = 2$, the present bound of 3×10^{-7} starts probing the parameter space up to $M_{1/2} \leq 150$ GeV. The main difference is that this does not depend on the value of U_{e3}, and therefore it is already a very important constraint on the parameter space of the model. In fact, for large $\tan \beta = 40$, as shown in Fig. 2 (second panel), reaching the expected limit of 10^{-8} would be able to rule out completely this scenario up to gaugino masses of 400 GeV and only a small portion of the parameter space with heavier gauginos would survive. In the limit $U_{e3} = 0$, this decay mode would provide a stronger constraint on the model, than $\mu \to e, \gamma$ which would now be suppressed as it contains only contributions proportional to h_c^2, as shown in eq.(28).

In summary, in the PMNS/maximal mixing case, even the present limits from BR($\mu \to e, \gamma$) can rule out large portions of the supersymmetric breaking parameter space, if U_{e3} is either close to its present limit or within an order of magnitude of it (as soon, the planned experiments might find out [38]). These are more severe for the large $\tan \beta$ case. In the extreme situation of U_{e3} being zero or very small $\sim \mathcal{O}(10^{-4} - 10^{-5})$, BR($\tau \to \mu, \gamma$) will start playing an important role, with its present constraints already disallowing large regions of the parameter space at large $\tan\beta$. A more detailed study of the importance of U_{e3} in SUSY search strategies can be found in [39,40].

4. Conclusions

Undoubtedly, the seesaw mechanism represents (one of) the best proposals to generate naturally small neutrino masses. But, how can we make sure that this is indeed the Nature's choice ? Even establishing the Majorana nature of the neutrinos through a positive evidence of neutrinoless double beta decay, it will be difficult to assess that such Majorana masses come from a seesaw. Indeed, as we said at the beginning, in the SM seesaw we expect very tiny charged LFV effects, probably without any chance to ever observe them. When moving to SUSY seesaw we add an important handle to our effort to establish the presence of a seesaw. In fact, as we tried to show in this talk, SUSY extensions of the SM with a seesaw have a general "tendency" to enhance (or even strongly enhance) rare LFV processes. Hence the combination of the observation of neutrinoless double beta decay and of some charged LFV phenomenon would constitute an important clue for the assessment of SUSY seesaw in Nature. In addition to this, seesaw incorporated in Grand Unified theories has the potentiality to lead to spectacular effects in hadronic physics [41,35,42].

There is no doubt that after the discovery of the neutrino masses, among the indirect tests of SUSY through FCNC and CP violating phenomena, LFV processes have acquired a position of utmost relevance. It would be spectacular, if by the time LHC observes the first SUSY particle, we could see also a muon decaying to an electron and a photon ! After thirty years, we could have the simultaneous confirmation of two of the most challenging physics ideas: seesaw and low energy SUSY.

Acknowledgments: We would like to thank all the organisers and the participants for providing such a lively and enjoyable atmosphere at Paris.

References

1. P. Minkowski, Phys. Lett. B **67**, 421 (1977); T. Yanagida in *Proc. Workshop on Unified Theories &c.,* eds. O. Sawada and A. Sugamoto (Tsukuba, 1979); M. Gell-Mann, P. Ramond, and R. Slansky, in *Supergravity,* eds. D. Freedman *et al.,* (North Holland 1980 Amsterdam).
2. S.L. Glashow, in *Quarks and Leptons, Cargèse 1979,* eds. M. Lévy, *et al.,* (Plenum 1980 New York), p. 707; R. N. Mohapatra and G. Senjanovic, Phys. Rev. Lett. **44**, 912 (1980).

3. F. Borzumati and A. Masiero, Phys. Rev. Lett. **57**, 961 (1986).

4. M. L. Brooks *et al.* [MEGA Collaboration], Phys. Rev. Lett. **83** (1999) 1521 [arXiv:hep-ex/9905013].

5. K. Abe *et al.* [Belle Collaboration], arXiv:hep-ex/0310029.

6. K. Inami, for the Belle Colloboration, Talk presented at the 19th International Workshop on Weak Interactions and Neutrinos (WIN-03) October 6th to 11th, 2003, Lake Geneva, Wisconsin, USA.

7. Web page: `http://meg.psi.ch`

8. See for example, R. Kitano, M. Koike and Y. Okada, Phys. Rev. D **66**, 096002 (2002) [arXiv:hep-ph/0203110]; R. Kitano, M. Koike, S. Komine and Y. Okada, Phys. Lett. B **575**, 300 (2003) [arXiv:hep-ph/0308021] and references there in.

9. K. S. Babu and C. Kolda, Phys. Rev. Lett. **89**, 241802 (2002) [arXiv:hep-ph/0206310]; A. Dedes, J. R. Ellis and M. Raidal, Phys. Lett. B **549**, 159 (2002) [arXiv:hep-ph/0209207].

10. See for example, J. I. Illana and T. Riemann, Phys. Rev. D **63**, 053004 (2001) [arXiv:hep-ph/0010193]; J. I. Illana and M. Masip, Phys. Rev. D **67**, 035004 (2003) [arXiv:hep-ph/0207328]; J. Cao, Z. Xiong and J. M. Yang, Eur. Phys. J. C **32** (2004) 245 [arXiv:hep-ph/0307126], and references there in.

11. A. Brignole and A. Rossi, Phys. Lett. B **566**, 217 (2003) [arXiv:hep-ph/0304081]; A. Brignole and A. Rossi, arXiv:hep-ph/0404211.

12. See for example, J. Hisano, M. M. Nojiri, Y. Shimizu and M. Tanaka, Phys. Rev. D **60**, 055008 (1999) [arXiv:hep-ph/9808410]; F. Deppisch, H. Pas, A. Redelbach, R. Ruckl and Y. Shimizu, Phys. Rev. D **69**, 054014 (2004), and references there in.

13. Some recent reviews are: G. Altarelli and F. Feruglio, arXiv:hep-ph/0206077; arXiv:hep-ph/0306265; I. Masina, Int. J. Mod. Phys. A **16**, 5101 (2001) [arXiv:hep-ph/0107220]; R. N. Mohapatra, arXiv:hep-ph/0211252; arXiv:hep-ph/0306016; S. F. King, arXiv:hep-ph/0310204; A. Y. Smirnov, arXiv:hep-ph/0311259.

14. For a recent review, see A. Masiero and O. Vives, Ann. Rev. Nucl. Part. Sci. **51** (2001) 161 [arXiv:hep-ph/0104027]; A. Masiero and O. Vives, New Jour. Phys. **4** (2002) 4.

15. K. Tobe, J. D. Wells and T. Yanagida, Phys. Rev. D **69**, 035010 (2004) [arXiv:hep-ph/0310148].

16. M. Ibe, R. Kitano, H. Murayama and T. Yanagida, arXiv:hep-ph/0403198.

17. L. J. Hall, V. A. Kostelecky and S. Raby, Nucl. Phys. B **267**, 415 (1986).

18. J. Hisano, T. Moroi, K. Tobe and M. Yamaguchi, Phys. Lett. B **391**, 341 (1997) [Erratum-ibid. B **397**, 357 (1997)] [arXiv:hep-ph/9605296].

19. F. Gabbiani, E. Gabrielli, A. Masiero and L. Silvestrini, Nucl. Phys. B **477**, 321 (1996) [arXiv:hep-ph/9604387].

20. J. Hisano, T. Moroi, K. Tobe, M. Yamaguchi and T. Yanagida, Phys. Lett. B **357** (1995) 579 [arXiv:hep-ph/9501407]; J. Hisano, T. Moroi, K. Tobe and M. Yamaguchi, Phys. Rev. D **53** (1996) 2442 [arXiv:hep-ph/9510309] ; J. Hisano and D. Nomura, Phys. Rev. D **59**, 116005 (1999) [arXiv:hep-ph/9810479]. See also: I. Masina and C. A. Savoy, Nucl. Phys. B **661**, 365

(2003) [arXiv:hep-ph/0211283].

21. J. A. Casas and A. Ibarra, Nucl. Phys. B **618** (2001) 171 [arXiv:hep-ph/0103065].

22. J. Hisano, D. Nomura and T. Yanagida, Phys. Lett. B **437**, 351 (1998) [arXiv:hep-ph/9711348]; S. F. King and M. Oliveira, Phys. Rev. D **60**, 035003 (1999) [arXiv:hep-ph/9804283]; W. Buchmuller, D. Delepine and F. Vissani, Phys. Lett. B **459**, 171 (1999) [arXiv:hep-ph/9904219]; J. R. Ellis, M. E. Gomez, G. K. Leontaris, S. Lola and D. V. Nanopoulos, Eur. Phys. J. C **14**, 319 (2000) [arXiv:hep-ph/9911459]; S. Baek, T. Goto, Y. Okada and K. i. Okumura, Phys. Rev. D **63**, 051701 (2001) [arXiv:hep-ph/0002141]; J. Hisano and K. Tobe, Phys. Lett. B **510**, 197 (2001) [arXiv:hep-ph/0102315]; D. F. Carvalho, J. R. Ellis, M. E. Gomez and S. Lola, Phys. Lett. B **515**, 323 (2001) [arXiv:hep-ph/0103256]; S. Baek, T. Goto, Y. Okada and K. i. Okumura, Phys. Rev. D **64**, 095001 (2001) [arXiv:hep-ph/0104146]; R. Gonzalez Felipe and F. R. Joaquim, JHEP **0109**, 015 (2001) [arXiv:hep-ph/0106226]; A. Romanino and A. Strumia, Nucl. Phys. B **622**, 73 (2002) [arXiv:hep-ph/0108275]; A. Kageyama, S. Kaneko, N. Shimoyama and M. Tanimoto, Phys. Lett. B **527**, 206 (2002) [arXiv:hep-ph/0110283]; A. Kageyama, S. Kaneko, N. Shimoyama and M. Tanimoto, Phys. Rev. D **65**, 096010 (2002) [arXiv:hep-ph/0112359]; F. Deppisch, H. Paes, A. Redelbach, R. Ruckl and Y. Shimizu, Eur. Phys. J. C **28**, 365 (2003) [arXiv:hep-ph/0206122]; A. Rossi, Phys. Rev. D **66**, 075003 (2002) [arXiv:hep-ph/0207006]. D. Falcone, Mod. Phys. Lett. A **17**, 2467 (2002) [arXiv:hep-ph/0207308]; J. Hisano, arXiv:hep-ph/0209005; K. S. Babu, B. Dutta and R. N. Mohapatra, Phys. Rev. D **67**, 076006 (2003) [arXiv:hep-ph/0211068]; T. Blazek and S. F. King, Nucl. Phys. B **662**, 359 (2003) [arXiv:hep-ph/0211368]; K. Hamaguchi, M. Kakizaki and M. Yamaguchi, Phys. Rev. D **68**, 056007 (2003) [arXiv:hep-ph/0212172]; J. Hisano and Y. Shimizu, Phys. Lett. B **565**, 183 (2003) [arXiv:hep-ph/0303071]; C. S. Huang, T. j. Li and W. Liao, Nucl. Phys. B **673**, 331 (2003) [arXiv:hep-ph/0304130]; T. Fukuyama, T. Kikuchi and N. Okada, Phys. Rev. D **68**, 033012 (2003) [arXiv:hep-ph/0304190]; B. Dutta and R. N. Mohapatra, Phys. Rev. D **68**, 056006 (2003) [arXiv:hep-ph/0305059]; T. F. Feng, T. Huang, X. Q. Li, X. M. Zhang and S. M. Zhao, Phys. Rev. D **68**, 016004 (2003) [arXiv:hep-ph/0305290]; S. F. King and I. N. R. Peddie, Nucl. Phys. B **678**, 339 (2004) [arXiv:hep-ph/0307091]; J. I. Illana and M. Masip, arXiv:hep-ph/0307393; K. S. Babu, T. Enkhbat and I. Gogoladze, Nucl. Phys. B **678**, 233 (2004) [arXiv:hep-ph/0308093]; A. Ibarra and G. G. Ross, arXiv:hep-ph/0312138; G. G. Ross, L. Velasco-Sevilla and O. Vives, arXiv:hep-ph/0401064; B. Dutta, Y. Mimura and R. N. Mohapatra, arXiv:hep-ph/0402113. P. H. Chankowski, J. R. Ellis, S. Pokorski, M. Raidal and K. Turzynski, arXiv:hep-ph/0403180.

23. J. Sato, K. Tobe and T. Yanagida, Phys. Lett. B **498**, 189 (2001) [arXiv:hep-ph/0010348].

24. J. Sato and K. Tobe, Phys. Rev. D **63**, 116010 (2001) [arXiv:hep-ph/0012333]. X. J. Bi, Eur. Phys. J. C **27**, 399 (2003) [arXiv:hep-

ph/0211236]; J. R. Ellis, M. Raidal and T. Yanagida, arXiv:hep-ph/0303242; S. M. Barr, Phys. Lett. B **578**, 394 (2004) [arXiv:hep-ph/0307372].

25. A. Masiero, S. K. Vempati and O. Vives, Nucl. Phys. B **649**, 189 (2003) [arXiv:hep-ph/0209303].

26. S. Davidson and A. Ibarra, JHEP **0109**, 013 (2001) [arXiv:hep-ph/0104076]; S. Lavignac, I. Masina and C. A. Savoy, Phys. Lett. B **520**, 269 (2001) [arXiv:hep-ph/0106245]; J. R. Ellis, J. Hisano, M. Raidal and Y. Shimizu, Phys. Rev. D **66**, 115013 (2002) [arXiv:hep-ph/0206110]. S. Pascoli, S. T. Petcov and W. Rodejohann, Phys. Rev. D **68**, 093007 (2003) [arXiv:hep-ph/0302054]; S. Davidson, JHEP **0303**, 037 (2003) [arXiv:hep-ph/0302075].

27. S. T. Petcov, S. Profumo, Y. Takanishi and C. E. Yaguna, Nucl. Phys. B **676**, 453 (2004) [arXiv:hep-ph/0306195].

28. S. Pascoli, S. T. Petcov and C. E. Yaguna, Phys. Lett. B **564**, 241 (2003) [arXiv:hep-ph/0301095].

29. R. Barbieri, D. V. Nanopoulos, G. Morchio and F. Strocchi, Phys. Lett. B **90**, 91 (1980); P. Langacker, Phys. Rept. **72** (1981) 185.

30. S. Dimopoulos and L. J. Hall, Phys. Lett. B **344**, 185 (1995) [arXiv:hep-ph/9411273]; W. Buchmuller and D. Wyler, Phys. Lett. B **521** (2001) 291 [arXiv:hep-ph/0108216].

31. H. Georgi and C. Jarlskog, Phys. Lett. B **86** (1979) 297.

32. For a review, see P. H. Chankowski and S. Pokorski, Int. J. Mod. Phys. A **17**, 575 (2002) [arXiv:hep-ph/0110249].

33. J. R. Ellis and S. Lola, Phys. Lett. B **458**, 310 (1999) [arXiv:hep-ph/9904279] ; J. A. Casas, J. R. Espinosa, A. Ibarra and I. Navarro, Nucl. Phys. B **556**, 3 (1999) [arXiv:hep-ph/9904395] ; N. Haba and N. Okamura, Eur. Phys. J. C **14**, 347 (2000) [arXiv:hep-ph/9906481] .

34. S. Antusch, J. Kersten, M. Lindner and M. Ratz, Phys. Lett. B **544**, 1 (2002) [arXiv:hep-ph/0206078] ; S. Antusch and M. Ratz, JHEP **0211**, 010 (2002) [arXiv:hep-ph/0208136].

35. D. Chang, A. Masiero and H. Murayama, Phys. Rev. D **67**, 075013 (2003) [arxiv:hep-ph/0205111].

36. M. Apollonio *et al.* [CHOOZ Collaboration], Phys. Lett. B **466**, 415 (1999) [arXiv:hep-ex/9907037].

37. T. Blazek and S. F. King, Phys. Lett. B **518**, 109 (2001) [arXiv:hep-ph/0105005].

38. M. Goodman, arXiv:hep-ex/0404031.

39. A. Masiero, S. Profumo, S. K. Vempati and C. E. Yaguna, JHEP **0403**, 046 (2004) [arXiv:hep-ph/0401138].

40. A. Masiero, S. K. Vempati and O. Vives, arXiv:hep-ph/0407325.

41. T. Moroi, JHEP **0003**, 019 (2000) [arXiv:hep-ph/0002208]; T. Moroi, Phys. Lett. B **493**, 366 (2000) [arXiv:hep-ph/0007328]; N. Akama, Y. Kiyo, S. Komine and T. Moroi, Phys. Rev. D **64**, 095012 (2001) [arXiv:hep-ph/0104263].

42. M. Ciuchini, A. Masiero, L. Silvestrini, S. K. Vempati and O. Vives, Phys. Rev. Lett. **92**, 071801 (2004) [arXiv:hep-ph/0307191].

CP VIOLATION IN SUPERSYMMETRIC SEESAW MODELS

JUNJI HISANO

ICRR, University of Tokyo
5-1-5 Kashiwa-no-Ha
Kashiwa, 277-8582, Japan

In supersymmetric (SUSY) extensions of the seesaw mechanism the neutrino Yukawa interaction induces the flavor and CP violating sfermion mass terms via the radiative correction. In this article we review the CP violating phenomena in the SUSY seesaw models.

1. Introduction

The seesaw mechanism[1], which explains tiny neutrino masses, is one of the most promising models beyond the standard model (SM), after discovery of the neutrino oscillation. It is also compatible with the matter unification in the GUTs. In the SO(10) GUT quarks and leptons may be embedded into **16** dimensional multiplets for each generation, in cooperation with the right-handed neutrinos. In addition to it, the baryon number in the universe may be explained by the leptogenesis[2]. Hence, it is important to search for signals in the models.

Supersymmetry (SUSY) is expected in the seesaw mechanism so that the hierarchical structure is stabilized. If the SUSY breaking terms in the minimal SUSY SM (MSSM) comes from dynamics whose energy scale is above that of the seesaw mechanism or the extension, the neutrino Yukawa interaction generates the flavor and CP violating SUSY breaking terms[34]. Thus, the flavor and CP violating phenomena are sensitive to the SUSY seesaw models.

This paper is concentrated into the CP violating aspects in the SUSY seesaw model and the extension to the SUSY GUT. The lepton flavor violating (LFV) processes and the related topics in the models are reviewed by Masiero in this volume.

128

2. Minimal SUSY Seesaw Model

In the minimal SUSY seesaw model, in which only three right-handed neutrinos are introduced, the SUSY breaking terms in the lepton sector are sensitive to the neutrino Yukawa coupling. Thus, the CP violation in the minimal SUSY seesaw model may lead to the T-odd asymmetry in the lepton-flavor violating lepton decay, such as $\mu \to 3e$, and the leptonic EDMs. In this section, we review the flavor structure in the leptonic SUSY breaking terms, which is predicted in the model, and the CP violating observables.

In the minimal SUSY seesaw model the relevant leptonic part of the superpotential is

$$W_{\text{seesaw}} = f_{ij}^{\nu} L_i \overline{N}_j \overline{H}_f + f_{ij}^{l} \overline{E}_i L_j H_f + \frac{1}{2} M_{ij} \overline{N}_i \overline{N}_j \tag{1}$$

where the indexes i, j run over three generations and M_{ij} is the heavy singlet neutrino mass matrix. In addition to the three charged lepton masses, this superpotential has eighteen physical parameters, including six real mixing angles and six CP-violating phases, because the Yukawa coupling and the Majorana mass matrices are given after removing unphysical phases as $f_{ij}^{l} = f_{l_i} \delta_{ij}$, $f_{ij}^{\nu} = X_{ik}^{\star} f_{\nu_k} e^{-i\varphi_{\nu_k}} W_{kj}^{\star} e^{-i\overline{\varphi}_{\nu_k}}$, and $M_{ij} = \delta_{ij} M_{N_k}$. Here, $\sum_i \varphi_{\nu_i} = 0$ and $\sum_i \overline{\varphi}_{\nu_i} = 0$, and each W and X are unitary matrices with one phase.

At low energies the effective theory after integrating out the right-handed neutrinos is given by the effective superpotential,

$$W_{\text{eff}} = f_{l_i} \overline{E}_i L_i H_f + \frac{1}{2v^2 \sin^2 \beta} (m_\nu)_{ij} (L_i \overline{H}_f)(L_j \overline{H}_f), \tag{2}$$

where we work in a basis in which the charged lepton Yukawa couplings are diagonal. The second term in (2) leads to the light neutrino masses and mixings. The explicit form of the small neutrino mass matrix (m_ν) is given by $(m_\nu)_{ij} = \sum_k f_{ik}^{\nu} f_{jk}^{\nu} v^2 \sin^2 \beta / M_{N_k}$. The light neutrino mass matrix (m_ν) is symmetric, with nine parameters, including three real mixing angles and three CP-violating phases. It can be diagonalized by a unitary matrix Z as $Z^T m_\nu Z = m_\nu^D$. By redefinition of fields one can rewrite $Z \equiv UP$, where $P \equiv \text{diag}(e^{i\phi_1}, e^{i\phi_2}, 1)$ and U is the MNS matrix, with the three real mixing angles and the remaining CP-violating phase.

The abilities to probe the minimal seesaw model by the low-energy neutrino experiments, such as the neutrino oscillation and the double β decay, are limited. Nine parameters associated with the heavy-neutrino sector cannot be measured in a direct way. However, we may probe the model by phenomena induced by the SUSY breaking terms in the MSSM.

If the SUSY-breaking terms are generated above the right-handed neutrino mass scale, the renormalization effects may induce sizable LFV slepton mass terms, which lead to the LFV charged lepton decays. If the SUSY-breaking parameters at the GUT scale or the Planck scale are universal, off-diagonal components in the left-handed slepton mass matrix $(m_{\tilde{l}_L}^2)$ and the trilinear slepton coupling (A_l) take the approximate forms,

$$(m_{\tilde{l}_L}^2)_{ij} \simeq -\sum_k \frac{f_{ik}^{\nu\star} f_{jk}^{\nu}}{16\pi^2} \left[m_0^2 (3\log\frac{M_G^2}{M_{N_k}^2} + 1) + A_0^2 (\log\frac{M_G^2}{M_{N_k}^2} + 1) \right] ,$$

$$(A_l)_{ij} \simeq -\sum_k f_{l_i} A_0 \frac{f_{ik}^{\nu\star} f_{jk}^{\nu}}{16\pi^2} \left[\log\frac{M_G^2}{M_{N_k}^2} + 1 \right] , \tag{3}$$

where $i \neq j$, and the off-diagonal components of the right-handed slepton mass matrix are suppressed. Here, we include the one-loop finite parts[5] in the radiative correction, and ignore terms of higher order in f_l, assuming that $\tan\beta$ is not extremely large.

The non-vanishing off-diagonal components in $(m_{\tilde{l}_L}^2)$ and (A_l) predict the charged lepton-flavor violating decays, whose measurements supply the information about the seesaw model, which is independent of the low-energy neutrino experiments. Ignoring the one-loop finite corrections, the terms in Eq. (3) are proportional to $H_{ij} = \sum_k f_{ik}^{\nu\star} f_{jk}^{\nu} \log(M_G^2/M_{N_k}^2)$. Here, the Hermitian matrix H, whose diagonal terms are real and positive, is defined in terms of f^{ν} and the heavy neutrino masses M_{N_k}. This matrix has nine parameters including three phases, which are clearly independent of the parameters in (m_ν). Thus two matrices (m_ν) and H together provide the required eighteen parameters, including six CP-violating phases, by which we can parameterize the minimal SUSY seesaw model[6]. The off-diagonal terms, $H_{ij}(i \neq j)$, are related to the LFV $l_i - l_j$ transition, and they are related to the LFV charged lepton decays.

Now we discuss the CP violation in the minimal SUSY seesaw model. The CP violating observables of the charged leptons are the T-odd asymmetry in $l \to 3l'$ and the leptonic EDMs.

First, we discuss the T-odd asymmetry in $\mu^+ \to e^+ e^- e^+$. In order for CP violation to appear in any process, interference between different terms in the amplitude for the process must occur. Therefore, all possible observables in $l \to l'\gamma$ decays, such as differences between the $\mu^+ \to e^+\gamma$ and $\mu^- \to e^-\gamma$ rates, vanish in the leading order of perturbation theory. Moreover, the process $\mu^- \to e^-\gamma$ is not measurable with high accuracy because of the large backgrounds. However, when muons are polarized, a

T-odd asymmetry for final-state particles in $\mu^+ \to e^+e^+e^-$ can be defined. Since CPT is conserved, the T-odd asymmetry measures the amount of CP violation in the model.

The muon polarization vector $\vec{P}$ can be defined in the coordinate system in which the z axis is taken to be the direction of the electron momentum, the x axis the direction of the most energetic positron momentum, and the $(z \times x)$ plane defines the decay plane perpendicular to the y axis. The T-odd asymmetry is then defined[7] by

$$A_T = \frac{N(P_y > 0) - N(P_y < 0)}{N(P_y > 0) + N(P_y < 0)},$$
(4)

where $N(P_i > (<)0)$ denotes the number of events with a positive (negative) P_i component for the muon polarization. A_T is limited to be below 24%[8].

In the branching ratio for $\mu \to 3e$, the contribution from photonic penguin diagram tends to dominate due to the phase-space integral while the Z penguin and box diagrams also gives the contribution. Then, assuming that the photonic penguin diagram dominates in $Br(\mu^+ \to e^+e^+e^-)$, the T-odd asymmetry A_T, induced by non-vanishing off-diagonal terms in (m_L^2), is approximately given[9] by

$$A_T = \frac{\mathrm{Im}\left[(\Delta_{12}^l)_L(\Delta_{23}^l)_L(\Delta_{31}^l)_L\right]}{|(\Delta_{12}^l)_L|^2}$$
$$\times \frac{0.039 + 0.196 \tan\beta + 0.017/\tan\beta}{\left|(1 + 2.4\tan\beta) - \frac{(\Delta_{23}^l)_L(\Delta_{31}^l)_L}{(\Delta_{21}^l)_L}(0.64 + 1.12\tan\beta)\right|^2},$$
(5)

where $(\Delta_{ij}^l)_L \equiv ((m_L^2)_{ij}/m_S^2)$. Here, in writing Eq. (5), we have taken $(A_l)_{ij} = 0$ $(i \neq j)$, for simplicity. We see explicitly how A_T in Eq. (5) depends on the Jarlskog invariant $J_L = \mathrm{Im}[(m_{\tilde{l}_L}^2)_{12}(m_{\tilde{l}_L}^2)_{23}(m_{\tilde{l}_L}^2)_{31}]$. It is found that A_T could in principle reach $\sim 10\%$. However, if $\mathrm{Im}[(\Delta_{12}^l)_L(\Delta_{23}^l)_L(\Delta_{31}^l)_L] \ll |(\Delta_{12}^l)_L|^2$, as one might expect, or if $\tan\beta \gg 1$, A_T is suppressed.

In Fig. 1 we show (a) the branching ratios for the decays $\mu \to e\gamma$ and $\mu \to 3e$ and (b) the T-odd asymmetry A_T in $\mu^+ \to e^+e^+e^-$ decay, as functions of the common soft mass m_0. For the neutrino parameters and others in the minimal SUSY seesaw model, see Ref. [9]. When the branching ratio for $\mu \to e\gamma$ is suppressed due to the accidental cancellation, the T-odd asymmetry is enhanced. This is because the penguin contribution is suppressed and and becomes comparable to the other contributions.

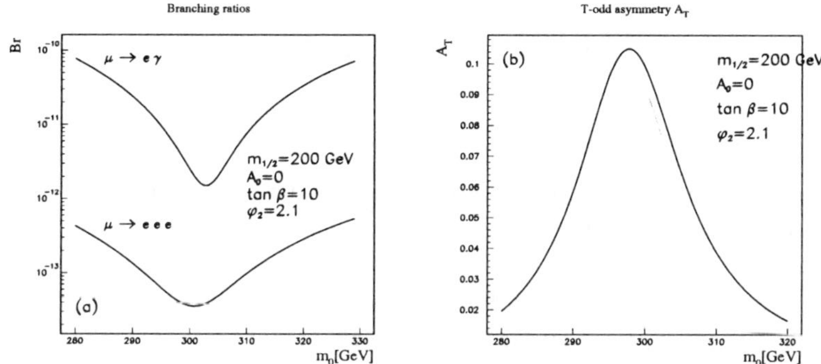

Figure 1. (a) Branching ratios for the decays $\mu \to e\gamma$ and $\mu \to 3e$ and (b) the T-odd asymmetry A_T in $\mu^+ \to e^+e^+e^-$ decay, as functions of the common soft mass m_0.

The leptonic EDMs depend on the flavor-diagonal Jarlskog invariants. If the source of the flavor violation comes from only the Yukawa interactions in the minimal SUSY seesaw model, the flavor-diagonal Jarlskog invariants, which consist of the Yukawa coupling constants, are given as

$$J_{\text{edm}}^{(i)} = \text{Im} \left[f^{l\star} \left[f^\nu f^{\nu\dagger} \left[f^{lT} f^{l\star},\ f^\nu f^{\nu\dagger} \right] f^\nu f^{\nu\dagger} \right]_{ii} \right]. \tag{6}$$

These are much suppressed by the Yukawa coupling constants. This situation is similar to the hadronic EDMs in the SM, where the CP violation comes from the CKM matrix. However, in the minimal SUSY seesaw model, when the right-handed neutrinos are not degenerate in mass, we have other flavor-diagonal Jarlskog invariants,

$$J_{\text{edm}}^{(i)\prime} = \text{Im} \left[f^{l\star} \left[f^\nu f^{\nu\dagger}, f^\nu \log \frac{M_G^2}{M_N^2} f^{\nu\dagger} \right]_{ii} \right]. \tag{7}$$

Obviously, they vanish when the right-handed neutrino masses are degenerate. Thus, it is expected that the leptonic EDMs are enhanced significantly when the right-handed neutrinos masses are not degenerate[10].

When the SUSY breaking terms are generated as in Eq. (3), the nontrivial Jarlskog invariants, $J_{\text{edm}}^{(i)\prime}$, are given as[5]

$$\text{Im} \left[\left[(A_l)(m_{\tilde{l}_L}^2) \right]_{ii} \right] = \frac{A_0 m_0^2}{(4\pi)^4} J_{\text{edm}}^{(i)\prime}, \tag{8}$$

and the leptonic EDMs are proportional to them. It is shown that the electron and muon EDMs may reach to 10^{-29} and $10^{-27} ecm$, respectively, for $m_S \sim 200\text{GeV}$[5]. The CP violating phases contributing to the leptonic

EDMs are independent of those in other CP violating observables, such as the T-odd asymmetry in $\mu \to 3e$ or the neutrino oscillation.

3. SUSY SU(5) GUT with Right-Handed Neutrinos

When the SUSY seesaw model is extended to the SUSY GUTs, the neutrino Yukawa interaction also induces the flavor violation in the hadronic sector via the radiative corrections to the squark SUSY breaking terms. In this section, we discuss the hadronic CP violation in the SUSY GUTs with the right-handed neutrinos. Here, we take the SUSY SU(5) GUT for simplicity. In this model, doublet leptons and right-handed down-type quarks are embedded in common **5**-dimensional multiplets, while doublet quarks, right-handed up-type quarks, and right-handed charged leptons are in the **10**-dimensional ones. Thus, the neutrino Yukawa coupling induces the right-handed down-type squark mixing, and this leads to rich flavor and CP violating phenomena in hadrons. They are also correlated due to the GUT relation, especially processes of the second and third generation transition[11].

First, we review the flavor structure in the squark and slepton mass matrices in the SUSY SU(5) GUT with the right-handed neutrinos. The Yukawa interactions for quarks and leptons and the Majorana mass terms for the right-handed neutrinos in this model are given by the following superpotential,

$$W = \frac{1}{4} f_{ij}^u \Psi_i \Psi_j H + \sqrt{2} f_{ij}^d \Psi_i \Phi_j \overline{H} + f_{ij}^\nu \Phi_i \overline{N}_j H + M_{ij} \overline{N}_i \overline{N}_j, \qquad (9)$$

where Ψ and Φ are for **10**- and $\overline{\mathbf{5}}$-dimensional multiplets, respectively, and $\overline{N}$ is for the right-handed neutrinos. H $(\overline{H})$ is **5**- ($\overline{\mathbf{5}}$-) dimensional Higgs multiplets. After removing the unphysical degrees of freedom, the Yukawa coupling constants in Eq. (9) are given as $f_{ij}^u = V_{ki} f_{u_k} e^{i\varphi_{u_k}} V_{kj}$, $f_{ij}^d = f_{d_i} \delta_{ij}$, and $f_{ij}^\nu = e^{i\varphi_{d_i}} U_{ij}^\star f_{\nu_j}$. Here, φ_u and φ_d are CP-violating phases inherent in the SUSY SU(5) GUT. They satisfy $\sum_i \varphi_{f_i} = 0$ ($f = u$ and d). The unitary matrix V is the CKM matrix in the extension of the SM to the SUSY SU(5) GUT, and each unitary matrices U and V have only a phase. When the Majorana mass matrix for the right-handed neutrinos is diagonal, U is the MNS matrix observed in the neutrino oscillation. In this section we assume the diagonal Majorana mass matrix in order to avoid the complexity due to the structure.

The colored Higgs multiplets H_c and $\overline{H}_c$ are introduced in H and $\overline{H}$ as SU(5) partners of the Higgs doublets in the MSSM, respectively. They have new flavor-violating interactions in Eq. (9). If the SUSY-breaking terms in the MSSM are generated by dynamics above the colored Higgs masses, such as in the gravity mediation, the sfermion mass terms may get sizable corrections by the colored Higgs interactions. In the minimal supergravity scenario the SUSY breaking terms are supposed to be given at the reduced Planck mass scale (M_G). In this case, the flavor-violating SUSY breaking mass terms at low energy are induced by the radiative correction, and they are qualitatively given in a flavor basis as

$$(m^2_{\tilde{u}_L})_{ij} \simeq -V_{i3}V^\star_{j3}\frac{f^2_b}{(4\pi)^2}\ (3m^2_0 + A^2_0)\ (2\log\frac{M^2_G}{M^2_{H_c}} + \log\frac{M^2_{H_c}}{m^2_S}),$$

$$(m^2_{\tilde{u}_R})_{ij} \simeq -e^{-i\varphi_{u_{ij}}}V^\star_{i3}V_{j3}\frac{2f^2_b}{(4\pi)^2}\ (3m^2_0 + A^2_0)\ \log\frac{M^2_G}{M^2_{H_c}},$$

$$(m^2_{\tilde{d}_L})_{ij} \simeq -V^\star_{3i}V_{3j}\frac{f^2_t}{(4\pi)^2}\ (3m^2_0 + A^2_0)\ (3\log\frac{M^2_G}{M^2_{H_c}} + \log\frac{M^2_{H_c}}{m^2_S}),$$

$$(m^2_{\tilde{d}_R})_{ij} \simeq -e^{i\varphi_{d_{ij}}}U^\star_{ik}U_{jk}\frac{f^2_{\nu_k}}{(4\pi)^2}\ (3m^2_0 + A^2_0)\ \log\frac{M^2_G}{M^2_{H_c}},$$

$$(m^2_{\tilde{l}_L})_{ij} \simeq -U_{ik}U^\star_{jk}\frac{f^2_{\nu_k}}{(4\pi)^2}\ (3m^2_0 + A^2_0)\ \log\frac{M^2_G}{M^2_{N_k}},$$

$$(m^2_{\tilde{e}_R})_{ij} \simeq -e^{i\varphi_{d_{ij}}}V_{3i}V^\star_{3j}\frac{3f^2_t}{(4\pi)^2}\ (3m^2_0 + A^2_0)\ \log\frac{M^2_G}{M^2_{H_c}}, \tag{10}$$

with $i \neq j$, where $\varphi_{u_{ij}} \equiv \varphi_{u_i} - \varphi_{u_j}$ and $\varphi_{d_{ij}} \equiv \varphi_{d_i} - \varphi_{d_i}$ and M_{H_c} is the colored Higgs mass. f_t is the top quark Yukawa coupling constant while f_b is for the bottom quark. As mentioned above, the off-diagonal components in the right-handed squarks and slepton mass matrices are induced by the colored Higgs interactions, and they depend on the CP-violating phases in the SUSY SU(5) GUT with the right-handed neutrinos[12].

When both the left-handed and right-handed squarks have the off-diagonal components in the mass matrices, the EDMs and CEDMs for the light quarks are enhanced significantly by the heavier quark mass[13][14]. The EDMs and CEDMs contribute to the hadronic EDMs, and they are constrained by the observations. In the SUSY SU(5) GUT with the right-handed neutrinos, the neutrino Yukawa coupling induces the flavor-violating mass terms for the right-handed down-type squarks. The flavor-violating mass terms for the left-handed down-type squarks are expected to be dominated by the radiative correction, which is controlled by the CKM matrix, induced by the top quark Yukawa coupling as in Eq. (10). Then, we can investigate or constrain the structure in the neutrino sector by the

134

hadronic EDMs, which is generated by the CEDMs and EDMs of the down and strange quarks.[a]

The CEDMs for the light quarks, including the strange quark, contribute to the hadronic EDMs since the CP-violating nucleon coupling is induced by them. Here, we consider only the CEDMs for the light quarks. While the EDMs for the up and down quarks contribute to the neutron EDM, it is found that the EDM contributions are comparable to or smaller than the CEDM contributions.

The CEDMs of the down-type light quarks derived by the flavor violation in the both the left-handed and right-handed quark mass matrices are given by the following dominant contribution, which is enhanced by the heavier quark masses,

$$d_{d_i}^C = \frac{\alpha_s}{4\pi} \frac{m_{\tilde{g}}}{\overline{m}_{\tilde{d}}^2} f\left(\frac{m_{\tilde{g}}^2}{\overline{m}_{\tilde{d}}^2}\right) \sum_j \mathrm{Im}\left[(\Delta_{ij}^d)_L (\Delta_j^d)_{LR} (\Delta_{ji}^d)_R\right], \qquad (11)$$

where $m_{\tilde{g}}$ and $\overline{m}_{\tilde{d}}$ are the gluino and averaged squark masses. The mass insertion parameters are defined as $(\Delta_{ij}^d)_{L/R} \equiv (m_{\tilde{d}_{L/R}}^2)_{ij}/\overline{m}_{\tilde{f}}^2$ and $(\Delta_i^d)_{LR} \equiv m_{d_i}(A_i^{(d)} - \mu\tan\beta)/\overline{m}_{\tilde{d}}^2$. The function $f(x)$ is given in Ref. [16] and $f(1) = -11/180$.

In order to translate the CEDMs of the light quarks to the EDMs of neutron and ^{199}Hg atom, we use the evaluation of the EDMs of neutron and ^{199}Hg atom in Ref. [17],

$$d_n = -1.6 \times e(d_u^C + 0.81 \times d_d^C + 0.16 \times d_s^C),$$
$$d_{\mathrm{Hg}} = -8.7 \times 10^{-3} \times e(d_u^C - d_d^C + 0.005 d_s^C), \qquad (12)$$

where d_n is generated by the charged meson loops and d_{Hg} comes from the nuclear force by the pion exchange in the chiral perturbation theory. The experimental upperbounds on the EDMs of neutron[18] and ^{199}Hg atom[19] are $|d_n| < 6.3 \times 10^{-26} e\,cm$ and $|d_{\mathrm{Hg}}| < 1.9 \times 10^{-28} e\,cm$, respectively (90%C.L.). Thus, the upperbounds on the quark CEDMs are $e|d_u^C| < 3.9(2.2) \times 10^{-26}\ e\ cm$, $e|d_d^C| < 4.8(2.2) \times 10^{-26}\ e\ cm$, and $e|d_s^C| < 2.4(44) \times 10^{-25}\ e\ cm$, from the EDM of neutron (^{199}Hg atom). Here, we assume that the accidental cancellation among the CEDMs does not suppress the EDMs. The constraint on d_s^C from ^{199}Hg atom is one-

---

[a]The up quark EDM and CEDM are also predicted in this model. Since they are suppressed by the bottom quark Yukawa coupling, they may be observable in near future when $\tan\beta$ is large[15].

order weaker than that from neutron, since the contribution to the EDM of ^{199}Hg atom is suppressed by π^0-η^0 mixing[17].

In Fig. 2 the CEDMs for the down and strange quarks in the SUSY SU(5) GUT with the right-handed neutrinos are shown[16]. The minimal supergravity scenario is assumed and $M_{H_c} = 2 \times 10^{16}$GeV. In Fig. 2(a) the strange quark CEDM is presented as a function of the right-handed tau neutrino mass. We take $m_{\nu_\tau} = 0.05$eV and $U_{\mu 3} = 1/\sqrt{2}$. For the SUSY breaking parameters, $m_0 = 500$GeV, $A_0 = 0$, $m_{\tilde{g}} = 500$GeV and $\tan\beta = 10$, which lead to $\overline{m}_{\tilde{q}} \simeq 640$GeV. The contributions from the electron and muon neutrino Yukawa interactions to the flavor violation in the right-handed down-type squark mass matrix are ignored. The contributions are bounded by the constraints from the K^0-$\overline{K}^0$ mixing and $Br(\mu \to e\gamma)$ when $|U_{e2}| \sim 1/\sqrt{2}$. From this figure, the right-handed tau neutrino mass should be smaller than $\sim 3 \times 10^{14}$GeV. In Fig. 2(b), the down quark CEDM is presented as a function of the right-handed tau neutrino mass. This comes from non-vanishing U_{e3} in our assumption that the right-handed neutrino mass matrix is diagonal. The current bound is not significant even when $U_{e3} = 0.2$.

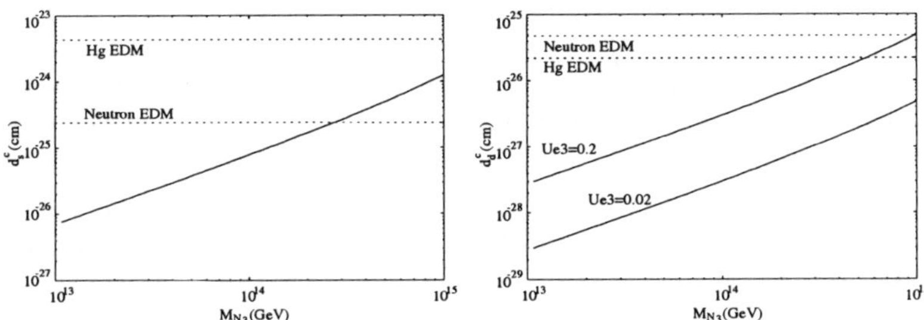

Figure 2. CEDMs for the strange quark in (a) and for the down quark in (b) as functions of the right-handed tau neutrino mass, M_{N_3}. Here, $M_{H_c} = 2 \times 10^{16}$GeV, $m_{\nu_\tau} = 0.05$eV, $U_{\mu 3} = 1/\sqrt{2}$, and $U_{e3} = 0.2$ and 0.02. For the MSSM parameters, we take $m_0 = 500$GeV, $A_0 = 0$, $m_{\tilde{g}} = 500$GeV and $\tan\beta = 10$.

The new technique for the measurement of the deuteron EDM has a great impact on the quark CEDMs if it is realized[20]. If they establish the sensitivity of $d_D \sim 10^{-27} e\,cm$, we may probe the new physics to the level of $ed_s^C \sim 10^{-26}\ e\ cm$ and $ed_d^C \sim ed_u^C \sim 10^{-28}\ e\ cm$, which are much stronger than the bounds from the neutron and ^{199}Hg atom EDMs[17]. This may imply that we may probe the structure in the neutrino sector even if $M_{N_3} \sim 10^{13}$GeV or $U_{e3} \sim 0.02$.

Finally, we discuss the correlation between the hadronic EDM and the CP asymmetry in $B \to \phi K_s$ ($S_{\phi K_s}$). The $b \to s\bar{s}s$ transition is sensitive to the new physics since it is induced at one-loop level[21]. When the right-handed down-type squarks have the flavor mixing, the b–s penguin diagram may give a sizable contribution in $B \to \phi K_s$. Thus, if the deviation from the SM prediction in $B \to \phi K_s$ is observed, it might be a signature of the effect of the neutrino Yukawa coupling in the SUSY GUTs with the right-handed neutrinos. On the other hand, the strange quark CEDM is also generated in the model, and they have a strong correlation[22].

The b–s penguin contribution induced by the right-handed squark mixing is dominated by the gluino diagram, and it is represented by the effective operator $H = -C_8^R(g_s/(8\pi^2))m_b\overline{s_R}(G\sigma)b_L$. When both the left-handed and right-handed down-type squarks have flavor violation, we get a strong correlation between C_8^R and d_s^C as

$$d_s^C = -\frac{m_b}{4\pi^2}\frac{11}{21}\mathrm{Im}\left[(\delta_{LL}^{(d)})_{23}C_8^R\right] \tag{13}$$

up to the QCD correction. Here, we take $m_{\tilde{g}} = \overline{m}_{\tilde{d}}$. The coefficient $11/21$ in Eq. (13) changes from 1 to 1/3, depending on the SUSY mass spectrum.

In Fig. 3, the correlation between d_s^C and $S_{\phi K_s}$ is shown, assuming Eq. (13). The detail of this figure is shown in Ref. [17]. Here, we take $(\Delta_{23}^d)_L = -0.04$, $\arg[C_8^R] = \pi/2$ and $|C_8^R|$ corresponding to $10^{-5} < |(\Delta_{32}^d)_R| < 0.5$. κ is a parameter for the matrix element of chromomagnetic moment in $B \to \phi K_s$, and we show the results for $\kappa = -1$ and -2. From this figure, the deviation of $S_{\phi K_s}$ from the SM prediction due to the gluon penguin contribution should be suppressed when the constraints on d_s^C from the ^{199}Hg atomic and the neutron EDMs are applied. We find that the neutron EDM gives a stronger bound on $S_{\phi K_s}$. Moreover, $S_{\phi K_s}$ may be constrained further by the future deuteron EDM measurements. Therefore, the hadronic EDMs give a very important implication to $S_{\phi K_s}$. Of cause, it should be careful to compare the EDM constraints with other low-energy observables, which are theoretically controlled better, since the hadronic EDMs may suffer from more hadronic uncertainties. However, the orders of the magnitude in the EDM constraints are still expected to have the significance.

4. Summary

In this article we review the CP violating phenomena in the SUSY seesaw models. In the SUSY extensions of the seesaw mechanism the neutrino

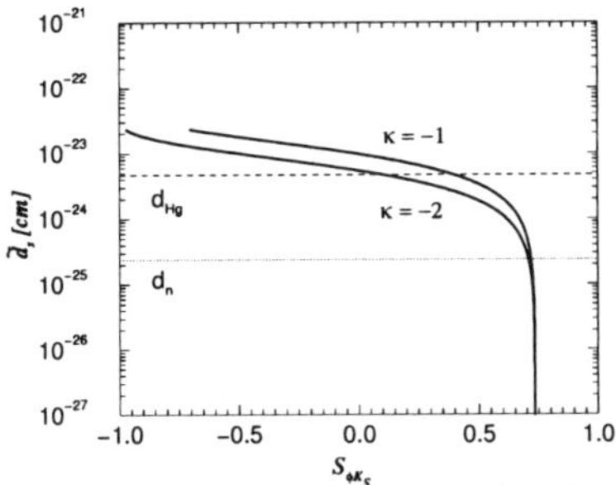

Figure 3. The correlation between d_s^C and $S_{\phi K_s}$. κ comes from the matrix element of chromomagnetic moment in $B \to \phi K_s$. The dashed (dotted) line is the upperbound on d_s^C from the EDM of ^{199}Hg atom (neutron).

Yukawa interaction induces the flavor and CP violating sfermion mass terms via the radiative correction. Thus, we may probe the models by studying the flavor and CP violating phenomena.

The recent results for the b-s penguin processes, including $B \to \phi K_s$, in the BABAR and BELLE experiments are converging. The combined results of the various b-s penguin processes are 2.7σ and 2.4σ deviated from the SM prediction in the BABAR[23] and BELLE[24] experiments, respectively. Though it is premature to judge whether they are signals of new physics, it is important to discuss the sensitivity to new physics. The deviation may be explained by introduction of the right-handed bottom and strange squarks mixing, such as in the SUSY SU(5) GUT with the right-handed neutrinos, however, the hadronic EDM constraints give bounds on the deviation. Even if the discrepancy comes from the hadronic uncertainties, the further improvements of the bound on the hadronic EDMs and the b-s penguin processes are very important.

References

1. M. Gell-Mann, P. Ramond and R. Slansky, Proceedings of the Supergravity Stony Brook Workshop, New York, 1979, eds. P. Van Nieuwenhuizen and D. Freedman (North-Holland, Amsterdam); T. Yanagida, Proceedings of the Workshop on Unified Theories and Baryon Number in the Universe, Tsukuba, Japan 1979 (edited by A. Sawada and A. Sugamoto, KEK Report No. 79-18, Tsukuba).
2. M. Fukugita and T. Yanagida, Phys. Rev. D **42** (1990) 1285.
3. L. J. Hall, V. A. Kostelecky and S. Raby, Nucl. Phys. B **267** (1986) 415.
4. F. Borzumati and A. Masiero, Phys. Rev. Lett. **57** (1986) 961; J. Hisano,

T. Moroi, K. Tobe, M. Yamaguchi and T. Yanagida, Phys. Lett. B **357** (1995) 579; J. Hisano, T. Moroi, K. Tobe and M. Yamaguchi, Phys. Rev. D **53** (1996) 2442; J. Hisano and D. Nomura, Phys. Rev. D **59** (1999) 116005.
5. Y. Farzan and M. E. Peskin, hep-ph/0405214.
6. J. R. Ellis, J. Hisano, M. Raidal and Y. Shimizu, Phys. Rev. D **66** (2002) 115013.
7. Y. Okada, K. Okumura and Y. Shimizu, Phys. Rev. D **58** (1998) 051901; Phys. Rev. D **61** (2000) 094001.
8. A. de Gouvea, S. Lola and K. Tobe, Phys. Rev. D **63** (2001) 035004.
9. J. R. Ellis, J. Hisano, S. Lola and M. Raidal, Nucl. Phys. B **621** (2002) 208.
10. J. R. Ellis, J. Hisano, M. Raidal and Y. Shimizu, Phys. Lett. B **528** (2002) 86.
11. For example, see J. Hisano and Y. Shimizu, Phys. Lett. B **565** (2003) 183; M. Ciuchini, A. Masiero, L. Silvestrini, S. K. Vempati and O. Vives, Phys. Rev. Lett. **92** (2004) 071801.
12. T. Moroi, Phys. Lett. B **493** (2000) 366.
13. S. Dimopoulos and L. J. Hall, Phys. Lett. B **344** (1995) 185.
14. R. Barbieri, L. J. Hall and A. Strumia, Nucl. Phys. B **445** (1995) 219; R. Barbieri, L. J. Hall and A. Strumia, Nucl. Phys. B **449** (1995) 437.
15. A. Romanino and A. Strumia, Nucl. Phys. B **490** (1997) 3.
16. J. Hisano, M. Kakizaki, M. Nagai and Y. Shimizu, hep-ph/0407169.
17. J. Hisano and Y. Shimizu, hep-ph/0406091.
18. P. G. Harris *et al.*, Phys. Rev. Lett. **82** (1999) 904.
19. M. V. Romalis, W. C. Griffith and E. N. Fortson, Phys. Rev. Lett. **86** (2001) 2505.
20. Y. K. Semertzidis *et al.* [EDM Collaboration], AIP Conf. Proc. **698** (2004) 200.
21. Y. Grossman and M. P. Worah, Phys. Lett. B **395** (1997) 241; R. Barbieri and A. Strumia, Nucl. Phys. B **508** (1997) 3.
22. J. Hisano and Y. Shimizu, Phys. Lett. B **581** (2004) 224.
23. Talked by M. Giorgi in ICHEP'04, August 16-22, 2004 Beijing, China, (*http://ichep04.ihep.ac.cn/*).
24. Talked by Y. Sakai in ICHEP'04, August 16-22, 2004 Beijing, China, (*http://ichep04.ihep.ac.cn/*).

SEESAW MECHANISM AND THE BARYON ASYMMETRY

M. RAIDAL

NICPB, Ravala 10, 10143 Tallinn, Estonia

I review the present understanding of the connection between non-zero neutrino masses and the baryon asymmetry of the Universe. The state-of-art results are presented for the standard thermal leptogenesis.

1. Smallness of Neutrino Masses and the Seesaw Mechanism

Non-zero neutrino masses and mixing angles provide at the moment the only convincing evidence [1] of physics beyond the standard model. A paradigm to understand the smallness of neutrino masses, involving some new heavy states which break lepton number, is called the seesaw mechanism [2]. If the heavy particles couple to the Standard Model lepton and Higgs doublets, their decoupling generates at low scale the dimension five operator

$$\frac{1}{\Lambda} LLHH. \tag{1}$$

After the electroweak symmetry breaking (1) generates small neutrino masses suppressed by the heavy scale Λ.

According to the original proposal [2], which still is by far the most popular and the most studied version of the seesaw mechanism, the heavy states are three superheavy singlet Majorana neutrinos N_i. These can be identified with the right-handed chiral fields of some grand unification gauge group such as $SO(10)$ and play a fundamental role in the anomaly cancellation. As the singlets do not have the Standard Model gauge couplings, they do not spoil the nice features of the Standard Model or its supersymmetric extension, such as the gauge coupling unification. The relevant terms for the light neutrino masses in the Lagrangian are the neutrino Yukawa couplings Y_ν and Majorana masses m_N,

$$L = \bar{L}_{Li} Y_\nu^{ij} N_{Rj} H + \frac{1}{2} \bar{N}_{Ri}^c m_N^{ij} N_{Rj} + \text{h.c.} \tag{2}$$

Integrating out the heavy neutrinos one obtains the seesaw relation for light neutrino masses

$$m_\nu = -Y_\nu^T m_N^{-1} Y_\nu v^2, \tag{3}$$

where $v = 174$ GeV. According to that, the relation between light neutrino masses and mixing in m_ν, and the structure of the heavy neutrino Yukawa couplings and mass matrix is rather complicated, and obtaining the observed neutrino mixing pattern requires complicated flavour model building [3]. As the deviation of the light neutrino mixing from bimaximal seems to be parametrized by the quark mixing matrix, this quark-lepton complementarity [4] may indicate some unification effect for the fermion Yukawa couplings. There are 9 physical degrees of freedom in the left-hand side of (3) while the right-hand side contains 18 of them. The missing 9 physical degrees of freedom at low energies can be parametrized by an orthogonal parameter matrix R [5] or by an Hermitian parameter matrix H [6]. The latter one allows to relate the missing degrees of freedom to different observables in the supersymmetric models with universal boundary condition for the soft supersymmetry breaking terms [7]. Those include the renormalization induced lepton flavour violating processes [8] and electric dipole moments [9]. The connection between low- and high-energy observables in the singlet seesaw models has been reviewed by S. Davidson in this conference [10].

The second proposal for generating light neutrino masses is to couple the Standard Model doublets to the $SU(2)_L$ triplet Higgs boson with non-zero hypercharge [11]. The relevant interaction terms are given by

$$L = \frac{1}{\sqrt{2}}(Y_T^{ij} \bar{L}_i^c i\tau_2 \mathbf{T} L_j + \lambda H^T \mathbf{T}^* i\tau_2 H + \text{h.c.}) + M_T \text{Tr}[\mathbf{T}\mathbf{T}^\dagger], \tag{4}$$

where $\mathbf{T} = \tau \cdot T$, the triplet T is in the $SU(2)_L \times U(1)_Y$ representation $T \sim (3, 1)$, and the τ_i are the three Pauli matrices. Notice that the Yukawa couplings Y_T and the Higgs self coupling λ together break lepton number explicitly. In this case the neutrino masses are suppressed by the heavy triplet mass M_T via the triplet seesaw mechanism [12]

$$m_\nu^{ij} = Y_T^{ij} \lambda \frac{v^2}{M_T}. \tag{5}$$

The triplet neutrino mass mechanism is very appealing one from the low energy neutrino phenomenology point of view because it requires introduction of the minimal number of new degrees of freedom and because the neutrino masses are directly proportional (up to the renormalization effects) to the triplet Higgs Yukawa couplings. In the triplet seesaw mechanism the low energy neutrino mass measurements determine directly, up to the overall scale, the structure of Majorana type Yukawa couplings in the fundamental Lagrangian. Indeed, comparison of (5) with (3) shows that the flavour structure of (5) is trivial. Therefore the explanation to the almost bimaximal light neutrino mixing is free of fine tunings and unnatural cancellations between numerical parameters. This simplicity also implies that the flavour violating processes in supersymmetric models are related to each other [13]. The scale of M_T can vary from almost unification scale to as low as 1 TeV [14]. In the latter case Higgs triplet can be discovered in the future collider experiments [15].

The third neutrino mass mechanism with triplet fermions [16] is considered to be somewhat exotic and has not gained much attention in the neutrino model building industry. We do not discuss this possibility in this talk any further.

2.   The Seesaw Mechanism and Leptogenesis

Another observable related to the physics of neutrino masses is the baryon asymmetry of the Universe. The observed ratio of baryon density to entrophy density [17],

$$\frac{n_B}{s} = (8.7 \pm 0.4) \times 10^{-11}, \tag{6}$$

requires the existence of physics beyond the Standard Model. To generate (6), three famous Sakharov's conditions must be satisfied [18]. Firstly, baryon number must be violated. Secondly, C and CP must be violated. Thirdly, the process must take place in out-of-equilibrium situation. In principle, those conditions can be satisfied also in the Standard Model since at non-perturbative level both B and L are separately violated [19]. However, baryogenesis in the electroweak phase transition has been extensively studied and found not to able to generate (6) because it requires very light Higgs boson mass $M_H < 40$ GeV. Therefore, currently the most widely accepted concept for generating (6) is leptogenesis.

According to the paradigm of leptogenesis [20], non-zero lepton asymmetry is generated first in out-of-equilibrium decays of some heavy states in the early universe. Thus the interactions of those heavy particles must violate lepton number and CP, and out-of-equilibrium condition is provided by the expansion of the Universe, $\Gamma < H$, where H is the Hubble parameter. Thereafter the lepton asymmetry is reprocessed to the $B - L$ asymmetry

142

by the sphaleron processes [21], generating (6). Baryogenesis via leptogenesis is the only idea which is also supported by the experimental data, namely by the non-vanishing neutrino masses and mixing. If we require the see-saw mechanism to induce simultaneously both the neutrino masses and the baryon asymmetry of the Universe, one can constrain leptogenesis from the experimental neutrino data. From that point of view different realizations of the seesaw mechanism discussed in the previous section have quite different features. In this talk I consider only thermal leptogenesis, *i.e.*, the case when all the particle species are created by thermal plasma during and after reheating of the Universe. All the following discussion applies only to that case.

In the case of the singlet neutrino seesaw mechanism [2] leptogenesis [20] is a direct, almost un-avoidable, consequence of seesaw rather that a separate mechanism. Because present neutrino data requires the existence of at least two heavy singlet neutrinos, and because in the case of Majorana particles physical CP phases exist already in the case of two generation, even the most minimal singlet seesaw model implies viable leptogenesis [22]. If the heavy and light neutrinos are hierarchical in mass so that leptogenesis comes from the decays of the lightest singlet N_1 only, there exist an upper bound on the CP asymmetry from its decay [23]

$$\epsilon_{N_1} \leq \frac{3}{16\pi} \frac{m_{N_1} m_3}{v^2},\tag{7}$$

where m_3 is the heaviest light neutrino mass. In more general case there is an upper bound on the the light neutrino mass scale from leptogenesis [24] which does not allow highly degenerate neutrino masses. This, together with (6) and with the state-of-art estimates of the thermal washout effects [25], leads to the lower bound on the leptogenesis scale, which turns out to be $m_{N_1} > 2 \times 10^9$ GeV. However, if the singlet neutrinos are partially degenerate in mass, the CP asymmetry is resonantly enhanced [26,27] and leptogenesis scale as low as $O(1)$ TeV could be viable. In principle the bound (7) can also be violated if the heavy neutrinos are not degenerate [28]. First neutrino model of that sort has been proposed in Ref. [29], and such models are quite different from the generic ones. Altogether, leptogenesis is a very natural consequence of the singlet seesaw mechanism.

On the other hand, the triplet seesaw mechanism, which in its minimal form contains just one triplet Higgs, does not provide leptogenesis in the Standard Model because of the lack of interfering amplitudes. The minimal triplet leptogenesis model must contain two triplets [30] which doubles the neutrino degrees of freedom, and the nicest phenomenological argument of simplicity in favour of this scenario is lost. However, in the minimal supersymmetric version of the triplet seesaw model leptogenesis is possible [31]. In the supersymmetric triplet seesaw model the anomaly cancellation requires introduction of two triplets with opposite $U(1)$ quantum numbers, $T \sim (3, 1)$ and $\bar{T} \sim (3, -1)$. According to the superpotential

$$W = \frac{1}{\sqrt{2}} (Y_T^{ij} L_i T L_j + \lambda_1 H_1 T H_1 + \lambda_2 H_2 \bar{T} H_2) + M_T T \bar{T}, \tag{8}$$

T and $\bar{T}$ have equal masses but only one of them couples to the lepton doublet, thus giving (5) with $\lambda = \lambda_2$, $v = v_2$. When supersymmetry breaking terms are included, T and $\bar{T}$ degeneracy is split by the soft terms, and resonant leptogenesis occurs [31] via the mechanism called "soft leptogenesis" [32]. While the scale of triplet seesaw is rather arbitrary, the scale of triplet leptogenesis is very much constrained due to the strong dependence on the size of the soft supersymmetry breaking B terms. In the following discussion we concentrate only on the singlet leptogenesis as the most interesting one.

3. Standard Thermal Leptogenesis

By the standard leptogenesis we mean the evolution of lepton asymmetry in the thermal plasma during and after reheating of the Universe. The heavy neutrinos are assumed to be hierarchical in mass so that leptogenesis occurs in the decays of the lightest heavy neutrinos N_1 only. Despite of the fact that the neutrino sector of the minimal seesaw model contains 18 free parameters, this scenario depends on only very few combinations of them. We parametrize the generated $B - L$ asymmetry via

$$Y_{B-L} = -\epsilon_{N_1} \, \eta \, Y_{N_1}^{\text{eq}}(T \gg m_{N_1}), \tag{9}$$

where ϵ_{N_1} is the CP-asymmetry in N_1 decays at zero temperature given by [27]

$$\epsilon_{N_1}(T = 0) = \frac{1}{8\pi} \sum_{j \neq 1} \frac{\text{Im}\left[(Y^\dagger Y)_{j1}^2\right]}{[Y^\dagger Y]_{11}} f\left(\frac{m_{N_j}^2}{m_{N_1}^2}\right), \tag{10}$$

with

$$f(x) = \sqrt{x}\left[\frac{x - 2}{x - 1} - (1 + x)\ln\left(\frac{1 + x}{x}\right)\right] \xrightarrow{x \gg 1} -\frac{3}{2\sqrt{x}}, \tag{11}$$

and $Y_{N_1}^{\text{eq}}(T \gg m_{N_1}) = 135\zeta(3)/(4\pi^4 g_*)$ is the neutrino equilibrium density at high temperature. Here g_* counts the effective number of spin-degrees of freedom in thermal equilibrium ($g_* = 106.75$ in the SM with no right-handed neutrinos). If the light (and also heavy) neutrino masses are taken to be hierarchical, the upper bound (7) on the CP asymmetry holds. In the following we assume that this is the case. Eq.(9) can be regarded as the definition of the leptogenesis efficiency parameter η, which contains all the finite temperature effects and dependences on the initial conditions for heavy neutrino abundances. For hierarchical neutrino masses, and for particular initial condition, the efficiency η depends on only two parameters. Those are m_{N_1} and the effective neutrino mass

$$\tilde{m}_1 = \frac{(Y_\nu Y_\nu^\dagger)_{11}}{m_{N_1}} v^2, \tag{12}$$

which is the measure on N_1 interaction strength with thermal plasma.

The obtained $B - L$ asymmetry is converted to the baryon asymmetry via sphaleron processes

$$\frac{n_B}{s} = \frac{24 + 4n_H}{66 + 13n_H} \frac{n_{B-L}}{s}, \tag{13}$$

where n_H is the number of Higgs doublets. For the SM we find numerically

$$\frac{n_B}{s} = -1.38 \times 10^{-3} \epsilon_{N_1} \eta. \tag{14}$$

Although the standard thermal leptogenesis scenario has been studied extensively [33], several important aspects of this scenario have been worked out in detail quite recently. Most important of them is adding finite temperature corrections to the decay and scattering amplitudes and to the CP asymmetry [25]. It has been found that to a good approximation the dominant effects are included if one (i) uses thermal masses of particles in the processes involved instead of zero temperature masses; (ii) renormalizes all the couplings at the first Matsubara mode,

$$E_r = 2\pi T, \tag{15}$$

using the zero-temperature renormalization group equations; (iii) uses finite temperature Feynman rules for calculation of the CP asymmetries. The important consequence of this is that at high temperatures the Higgs boson mass exceeds the sum of the singlet neutrino and lepton mass, and instead of N decay the two-body decay process is $H \to NL$. The second important refinement of the calculation is inclusion of the gauge scatterings in [25] (which in particular limit agree with the similar attempt in [34]). Thus the

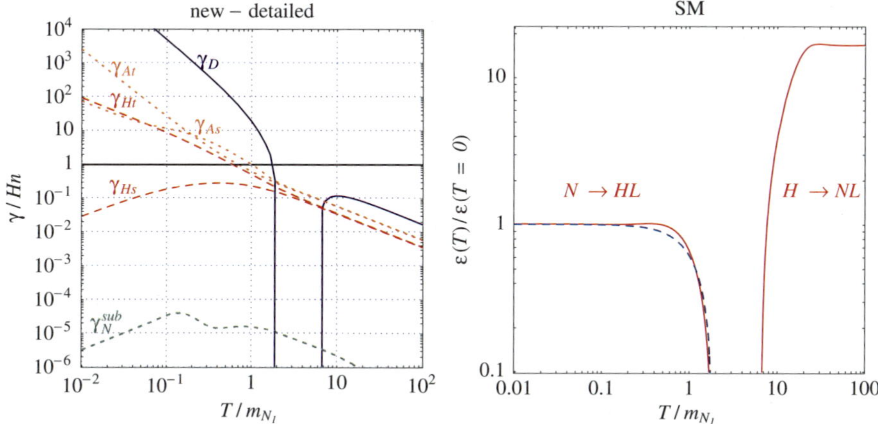

Figure 1. Evolution of the scattering densities (first plot) and the CP asymmetries (second plot) with temperature.

processes contributing to the thermal leptogenesis are

$$\Delta L = 1: \qquad D = [N_1 \leftrightarrow LH], \quad S_s = H_s + A_s, \qquad S_t = H_t + A_t,$$

$$\Delta L = 2: \qquad N_s = [LH \leftrightarrow \bar{L}\bar{H}], \qquad N_t = [LL \leftrightarrow \bar{H}\bar{H}], \tag{16}$$

where

$$H_s = [LN_1 \leftrightarrow Q_3 U_3], \; 2H_t = [N_1 \bar{U}_3 \leftrightarrow Q_3 \bar{L}] + [N_1 \bar{Q}_3 \leftrightarrow U_3 \bar{L}],$$
$$A_s = [LN_1 \leftrightarrow \bar{H}A], \quad 2A_t = [N_1 H \leftrightarrow A\bar{L}] + [N_1 A \leftrightarrow \bar{H}\bar{L}]. \tag{17}$$

We separated $\Delta L = 1$ scatterings $S_{s,t}$ into top ($H_{s,t}$) and gauge contributions ($A_{s,t}$). The evolution of scattering densities γ for the particular processes with temperature has been shown in Fig. 1 for $m_{N_1} = 10^{10}$ GeV and $\tilde{m}_1 = 0.06$ eV. It follows that there is a temperature range in which H and L thermal masses are such that all the two-body decays are kinematically forbidden. Notice that in order to avoid double counting of the two-body decays, after subtracting the resonances from the scattering $LH \leftrightarrow LH$ the corresponding γ_N^{sub} is completely negligible compared to the other processes. In the same figure we also plot the dependence of the CP asymmetries in $N \to HL$ and in $H \to NL$ decays. The solid line shows the result of full finite temperature calculation in [25] while the dashed line is obtained by approximating N_1 to be at rest in thermal plasma. Those results agree with each other with good accuracy. The results including thermal corrections plotted in Fig. 1 differ both qualitatively and quantitatively from the zero temperature calculations and change the predictions

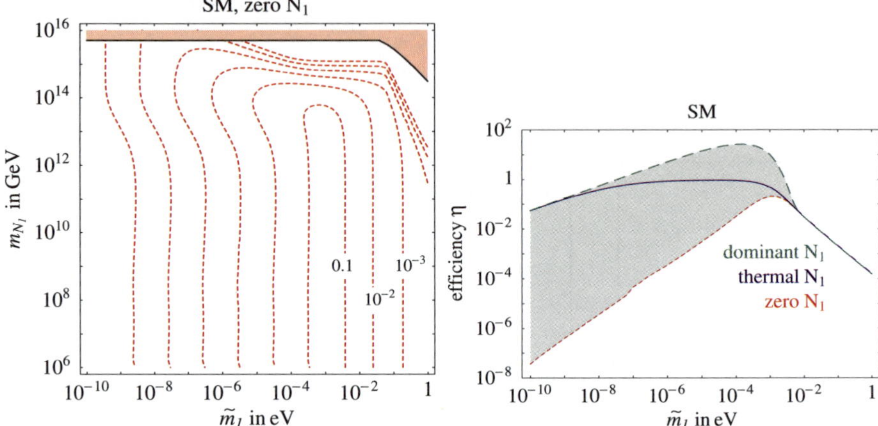

Figure 2. Isocurves of leptogenesis efficiency η. The first plot is for vanishing initial N_1 abundance, the second plot is for fixed $m_{N_1} = 10^{10}$ GeV.

for the generated $B - L$ asymmetry.

Solving the full set of Boltzmann equations for the evolution of the $B - L$ asymmetry with temperature, and parameterizing the results via Eq.(9), our results for the leptogenesis efficiency η are shown in Fig. 2. In the first plot we present the isocurves of $\eta = 10^i$, $i = -1, -2, -3, \dots$ in the $(\tilde{m}_1, m_{N_1})$ plane, assuming vanishing initial abundance for the singlet neutrinos $Y_{N_1}(T = \infty) = 0$. The maximum efficiency is found around $\tilde{m}_1 \sim 10^{-3}$ eV. For smaller values of $\tilde{m}_1$ the Yukawa interactions of N_1 are too weak for N_1 to reach thermal abundance, which leads to the suppression of the $B - L$ asymmetry. For larger values of $\tilde{m}_1$ the Yukawa interaction becomes too strong so that the N_1 decays occur not sufficiently in out-of-equilibrium, which again leads to the suppression of the $B - L$ asymmetry. Our result can be summarized with a simple analytical fit [25]

$$\frac{1}{\eta} \approx \frac{3.3 \times 10^{-3}\text{eV}}{\tilde{m}_1} + \left(\frac{\tilde{m}_1}{0.55 \times 10^{-3}\text{eV}}\right)^{1.16}, \qquad (18)$$

valid for $m_{N_1} \ll 10^{14}$ GeV. This enables the reader to study leptogenesis in neutrino mass models without setting up and solving the complicated Boltzmann equations.

The dependence of the efficiency on the initial conditions for N_1 is demonstrated on the second plot in Fig. 2 where we assume $m_{N_1} = 10^{10}$ GeV and study three different cases, zero initial abundance for N_1, thermal initial abundance for N_1 and when N_1 dominates the Universe. For

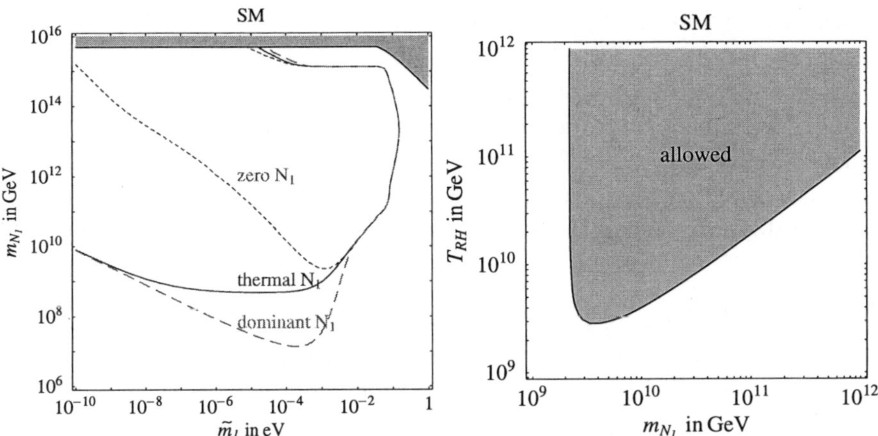

Figure 3.   Regions in $(\tilde{m}_1, m_{N_1})$ and (m_{N_1}, T_{RH}) planes allowed by successful leptogenesis.

small values of $\tilde{m}_1$ the results depend on the initial conditions. However, if $\tilde{m}_1 = \sqrt{\Delta m^2_{sol}}$ or $\tilde{m}_1 = \sqrt{\Delta m^2_{atm}}$, the prediction of thermal leptogenesis is practically independent of any pre-existing initial condition. This makes the thermal leptogenesis predictions so robust.

In order to study in which parameter space the thermal leptogenesis is capable to yield the observed baryon asymmetry of the Universe, one has to combine the results for the efficiency η with the maximal value of the CP asymmetry (corrections to Eq.(7) from non-zero m_1 have to be taken into account). This has been done in Fig. 3. In the first plot in this figure we present the contours for successful leptogenesis in the $(\tilde{m}_1, m_{N_1})$ plane for the three initial conditions for N_1 as before. There exist lower bounds on the N_1 mass from the requirement of successful leptogenesis which depend on the initial conditions. For the vanishing initial N_1 abundance, which is probably the most appropriate assumption for singlets, the bound is [25]

$$m_{N_1} > 2 \times 10^9 \ \text{GeV}. \tag{19}$$

To derive the results presented so far we have assumed that the reheating temperature of the Universe exceeds the singlet neutrino mass, $T_{RH} \gg m_{N_1}$. This may not be the case in all the scenarios, especially in supersymmetric ones because supersymmetry sets an upper bound on the reheating temperature of the Universe from the overproduction of gravitinos [35]. Therefore one has to study how our predictions depend on the reheating of the Universe. We have set up and solved Boltzmann equations which include T_{RH} as a free parameter in Ref. [25]. The result is presented on the second plot in Fig. 3 where we show the region in the (m_{N_1}, T_{RH}) plane allowed by thermal leptogenesis. We have taken $\tilde{m}_1 = 10^{-3}$ eV and assumed vanishing initial abundance for N_1. The results show that T_{RH} can be as low as m_{N_1} without considerable loss of the efficiency. However, if $T_{RH} < m_{N_1}$ successful leptogenesis requires large values of m_{N_1}, considerably larger than the minimally allowed one.

4. Conclusions

Baryogenesis via leptogenesis, the idea of generating the observed baryon asymmetry of the Universe in out-of-equilibrium decays of heavy particles which violate lepton number, is supported by the light neutrino data. Remarkably, the standard thermal leptogenesis works in particularly robust way in the neutrino parameter space suggested by the seesaw mechanism. Currently it is not possible to make exact predictions between the low energy neutrino measurements and the generated baryon asymmetry because of too many free parameter in the seesaw model. Nevertheless the experimental and theoretical success in understanding the seesaw meshanism during its first 25 years has been impressive. Hopefully in next 25 years, if next generation experiments will find new lepton flavour violating observables, leptogenesis can be experimantally established as the mechanism of baryogenesis.

Acknowledgments

I would like to thank all my collaborators with whom I have studied the issues presented in this talk over several past years. This work has been supported by the ESF Grants 5135 and 5935, by the EC MC contract MERG-CT-2003-503626, and by the Ministry of Education and Research of the Republic of Estonia.

References

1. M. H. Ahn *et al.* [K2K Collab.], Phys. Rev. Lett. **90**, 041801 (2003) [arXiv:hep-ex/0212007]; K. Eguchi *et al.* [KamLAND Collab.], Phys. Rev. Lett. **90**, 021802 (2003) [arXiv:hep-ex/0212021]; S. N. Ahmed *et al.* [SNO Collab.], arXiv:nucl-ex/0309004; M. Ishitsuka [Super-Kamiokande Collaboration], arXiv:hep-ex/0406076.
2. P. Minkowski, Phys. Lett. B **67** (1977) 421; M. Gell-Mann, P. Ramond and R. Slansky, Proceedings of the Supergravity Stony Brook Workshop, New

York, 1979, eds. P. Van Nieuwenhuizen and D. Freedman (North-Holland, Amsterdam); T. Yanagida, Proceedings of the Workshop on Unified Theories and Baryon Number in the Universe, Tsukuba, Japan 1979 (eds. A. Sawada and A. Sugamoto, KEK Report No. 79-18, Tsukuba); S.L. Glashow, in Quarks and Leptons, Cargèse, eds. M. Lévy et al., (Plenum, 1980, New-York), p. 707; T. Yanagida, in Proceedings of the Workshop on the Unified Theory and the Baryon Number in the Universe, edited by O. Sawada and A. Sugamoto (KEK Report No. 79-18, Tsukuba, 1979), p. 95; R.N. Mohapatra and G. Senjanović, Phys. Rev. Lett. **44**, (1980) 912.

3. For reviews see, G. Altarelli and F. Feruglio, Phys. Rept. **320**, 295 (1999); G. Altarelli and F. Feruglio, arXiv:hep-ph/0306265; M. C. Gonzalez-Garcia and Y. Nir, Rev. Mod. Phys. **75**, 345 (2003) [arXiv:hep-ph/0202058]; A. Y. Smirnov, arXiv:hep-ph/0402264; F. Feruglio, arXiv:hep-ph/0410131.

4. M. Raidal, Phys. Rev. Lett. **93** (2004) 161801 [arXiv:hep-ph/0404046]; H. Minakata and A. Y. Smirnov, arXiv:hep-ph/0405088; P. H. Frampton and R. N. Mohapatra, arXiv:hep-ph/0407139.

5. J. A. Casas and A. Ibarra, Nucl. Phys. B **618** (2001) 171 [arXiv:hep-ph/0103065].

6. S. Davidson and A. Ibarra, JHEP **0109** (2001) 013 [arXiv:hep-ph/0104076]; J. R. Ellis, J. Hisano, M. Raidal and Y. Shimizu, Phys. Rev. D **66** (2002) 115013 [arXiv:hep-ph/0206110].

7. F. Borzumati and A. Masiero, Phys. Rev. Lett. **57** (1986) 961; L. J. Hall, V. A. Kostelecky and S. Raby, Nucl. Phys. B **267** (1986) 415.

8. J. Hisano, T. Moroi, K. Tobe, M. Yamaguchi and T. Yanagida, Phys. Lett. B **357** (1995) 579 [arXiv:hep-ph/9501407]; J. Hisano, T. Moroi, K. Tobe and M. Yamaguchi, Phys. Rev. D **53** (1996) 2442 [arXiv:hep-ph/9510309].

9. J. R. Ellis, J. Hisano, S. Lola and M. Raidal, Nucl. Phys. B **621** (2002) 208 [arXiv:hep-ph/0109125]; J. R. Ellis, J. Hisano, M. Raidal and Y. Shimizu, Phys. Lett. B **528** (2002) 86 [arXiv:hep-ph/0111324]; Y. Farzan and M. E. Peskin, arXiv:hep-ph/0405214.

10. S. Davidson, arXiv:hep-ph/0409339, these proceedings.

11. J. Schechter and J. W. F. Valle, Phys. Rev. **D22** (1980) 2227; T. P. Cheng and L. F. Li, Phys. Rev. **D22** (1980) 2860.

12. M. Magg and C. Wetterich, Phys. Lett. **B94** (1980) 61; C. Wetterich, Nucl. Phys. **B187** (1981) 343; G. Lazarides, Q. Shafi and C. Wetterich, Nucl Phys. **B181** (1981) 287; R.N. Mohapatra and G. Senjanović, Phys. Rev. **D23** (1981) 165.

13. A. Rossi, Phys. Rev. D **66** (2002) 075003 [arXiv:hep-ph/0207006].

14. E. Ma, M. Raidal and U. Sarkar, Phys. Rev. Lett. **85** (2000) 3769 [arXiv:hep-ph/0006046].

15. K. Huitu, J. Maalampi, A. Pietila and M. Raidal, Nucl. Phys. B **487** (1997) 27 [arXiv:hep-ph/9606311]; F. Cuypers and M. Raidal, Nucl. Phys. B **501** (1997) 3 [arXiv:hep-ph/9704224].

16. E. Ma, Phys. Rev. Lett. **81** (1998) 1171 [arXiv:hep-ph/9805219].

17. H. V. Peiris *et al.*, Astrophys. J. Suppl. **148** (2003) 213 [arXiv:astro-ph/0302225].

18. A. D. Sakharov, Pisma Zh. Eksp. Teor. Fiz. **5** (1967) 32 [JETP Lett. **5** (1967 SOPUA,34,392-393.1991 UFNAA,161,61-64.1991) 24].

19. G. 't Hooft, Phys. Rev. Lett. **37** (1976) 8.

20. M. Fukugita and T. Yanagida, Phys. Lett. B **174** (1986) 45.

21. V. A. Kuzmin, V. A. Rubakov and M. E. Shaposhnikov, Phys. Lett. B **155** (1985) 36.

22. P. H. Frampton, S. L. Glashow and T. Yanagida, Phys. Lett. **B548** (2002) 119 [arXiv:hep-ph/0208157]; M. Raidal and A. Strumia, Phys. Lett. **B553** (2003) 72 [arXiv:hep-ph/0210021].

23. S. Davidson and A. Ibarra, Phys. Lett. **B535** (2002) 25 [arXiv:hep-ph/0202239].

24. W. Buchmuller, P. Di Bari and M. Plumacher, Phys. Lett. B **547** (2002) 128 [arXiv:hep-ph/0209301].

25. G. F. Giudice, A. Notari, M. Raidal, A. Riotto and A. Strumia, Nucl. Phys. **B685** (2004) 89 [arXiv:hep-ph/0310123].

26. M. Flanz, E.A. Paschos and U. Sarkar, Phys. Lett. **B345** (1995) 248; M. Flanz, E.A. Paschos, U. Sarkar and J. Weiss, Phys. Lett. **B389** (1996) 693; A. Pilaftsis, Nucl. Phys. **B504** (1997) 61; A. Pilaftsis, Phys. Rev. **D56** (1997) 5431, [arXiv:hep-ph/9707235]; J. R. Ellis, M. Raidal and T. Yanagida, Phys. Lett. B **546** (2002) 228 [arXiv:hep-ph/0206300].

27. L. Covi, E. Roulet and F. Vissani, Phys. Lett. B **384** (1996) 169 [arXiv:hep-ph/9605319].

28. S. Davidson and R. Kitano, JHEP **0403** (2004) 020 [arXiv:hep-ph/0312007]; T. Hambye, Y. Lin, A. Notari, M. Papucci and A. Strumia, arXiv:hep-ph/0312203.

29. M. Raidal, A. Strumia and K. Turzynski, arXiv:hep-ph/0408015.

30. E. Ma and U. Sarkar, Phys. Rev. Lett. **80** (1998) 5716 [arXiv:hep-ph/9802445]; T. Hambye, E. Ma, U. Sarkar, Nucl. Phys. **B602** (2001) 23 [hep-ph/0011192].

31. G. D'Ambrosio, T. Hambye, A. Hektor, M. Raidal and A. Rossi, arXiv:hep-ph/0407312.

32. Y. Grossman, T. Kashti, Y. Nir and E. Roulet, arXiv:hep-ph/0307081; G. D'Ambrosio, G. F. Giudice and M. Raidal, Phys. Lett. **B575** (2003) 75 [arXiv:hep-ph/0308031].

33. M. A. Luty, Phys. Rev. D **45** (1992) 455; M. Plumacher, Z. Phys. C **74** (1997) 549 [arXiv:hep-ph/9604229]; M. Plumacher, Nucl. Phys. B **530** (1998) 207 [arXiv:hep-ph/9704231]; R. Barbieri, P. Creminelli, A. Strumia and N. Tetradis, Nucl. Phys. B **575** (2000) 61 [arXiv:hep-ph/9911315]; W. Buchmuller, P. Di Bari and M. Plumacher, Nucl. Phys. B **643** (2002) 367 [arXiv:hep-ph/0205349].

34. A. Pilaftsis and T. E. J. Underwood, Nucl. Phys. B **692** (2004) 303 [arXiv:hep-ph/0309342].

35. M. Y. Khlopov and A. D. Linde, Phys. Lett. B **138** (1984) 265; J. R. Ellis, D. V. Nanopoulos and S. Sarkar, Nucl. Phys. **B259** (1985) 175; J. R. Ellis, D. V. Nanopoulos, K. A. Olive and S. J. Rey, Astropart. Phys. **4** (1996) 371; M. Kawasaki and T. Moroi, Prog. Theor. Phys. **93** (1995) 879.

VARIOUS REALIZATIONS OF LEPTOGENESIS AND NEUTRINO MASS CONSTRAINTS

T. HAMBYE

Rudolf Peierls Center for Theoretical Physics
University of Oxford
1, Keble Road, Oxford OX1 3NP, UK
E-mail: hambye@thphys.ox.ac.uk

Seven types of leptogenesis models which can lead to a successful explanation of baryogenesis are presented. Emphasis is put on the conditions which need to be fulfilled by the neutrino masses as well as by the heavy state masses. The model dependence of these conditions is discussed.

1. Introduction

Following the recent convincing evidence for neutrino oscillations, leptogenesis[1] has become a well motivated possible explanation of the origin of the baryon asymmetry of the universe. In these proceedings, after a short introduction on the three basic ingredients of leptogenesis, we will extend the discussion of Ref.[2] along two main directions. First, in the framework of the usual seesaw model with three right-handed neutrinos ("type-I" seesaw model[3]), we will study the neutrino mass constraints, in particular the neutrino mass upper bound, which exist in order that leptogenesis can successfully explain the baryon asymmetry of the universe. We will show how this neutrino mass upper bound can be relaxed by no longer assuming that the right-handed neutrinos have a hierarchical mass spectrum. Secondly, other models of leptogenesis which might also be attractive for various reasons will be introduced: "type-II" seesaw model with a scalar Higgs triplet, "type-III" seesaw model with fermion triplets, and combinations of them such as the "type-I" plus "type-II" model which is motivated by left-right or SO(10) models. The corresponding neutrino mass bounds will be discussed for all these models. Finally, we will also briefly consider the case of radiative neutrino mass models.

152

2. The Three Basic Ingredients of Leptogenesis

There are essentially three ingredients to take into account for leptogenesis to work. In this section we will discuss only the case of the "type-I" seesaw model with three heavy right-handed neutrinos N_i which is based on the following Lagrangian

$$L = L_{\mathrm{SM}} + \bar{N}_i i \partial\!\!\!/ N_i + \left(\lambda^{ij} \, H^\dagger N_i L_j + \frac{M_{N_i}}{2} N_i N_i + \mathrm{h.c.} \right), \qquad (1)$$

with $L_j = (\nu_{jL}, \, l_{jL})^T$, $H = (H^0, H^-)^T$. In Eq. (1) we made the choice to work in the basis where the N_i mass matrix is real and diagonal. This model has 18 parameters; 9 combinations of them enter in the neutrino mass matrix $M_\nu^I = -\lambda^T M_N^{-1} \lambda v^2$, and nine of them decouple from it. A very useful parametrization (in term of a orthogonal complex R matrix) which allows to span easily all the parameter space of this model in agreement with the data, and which is very useful to determine the neutrino mass bounds below, can be found in Ref.[4,5]. For the following we order the neutrino masses as: $m_{\nu_3} > m_{\nu_2} > m_{\nu_1} \geq 0$ and $M_{N_3} \geq M_{N_2} \geq M_{N_1} \geq 0$.

2.1. *The CP Asymmetry*

At a temperature well above their masses one can expect the right-handed neutrinos to be in thermal equilibrium in the universe thermal bath, due to Yukawa interactions or other possible interactions (such as gauge interactions in the right-handed sector). However, once the temperature drops below their masses the right-handed neutrinos disappear from the thermal bath by decaying to leptons and Higgs bosons. The crucial quantity for leptogenesis is the CP-asymmetry, that is to say the averaged ΔL which is produced each time one N_i decays. At lowest order, that is to say at one-loop order, for N_1 (and similarly for $N_{2,3}$), it is given by

$$\varepsilon_{N_1} \equiv \frac{\Gamma(N_1 \to lH^*) - \Gamma(N_1 \to \bar{l}H)}{\Gamma(N_1 \to lH^*) + \Gamma(N_1 \to \bar{l}H)} = - \sum_{j=2,3} \frac{3}{2} \frac{M_{N_1}}{M_{N_j}} \frac{\Gamma_{N_j}}{M_{N_j}} I_j \frac{2S_j + V_j}{3}$$

$$(2)$$

where

$$I_j = \frac{\mathrm{Im}\left[(\lambda\lambda^\dagger)^2_{1j} \right]}{|\lambda\lambda^\dagger|_{11} |\lambda\lambda^\dagger|_{jj}}, \qquad \frac{\Gamma_{N_j}}{M_{N_j}} = \frac{|\lambda\lambda^\dagger|_{jj}}{8\pi} \equiv \frac{\tilde{m}_j M_{N_j}}{8\pi v^2}, \qquad (3)$$

and where

$$S_j = \frac{M^2_{N_j} \Delta M^2_{1j}}{(\Delta M^2_{1j})^2 + M^2_{N_1} \Gamma^2_{N_j}}, \qquad V_j = 2\frac{M^2_{N_j}}{M^2_{N_1}} \left[\left(1 + \frac{M^2_{N_j}}{M^2_{N_1}}\right) \log\left(1 + \frac{M^2_{N_1}}{M^2_{N_j}}\right) - 1 \right],$$

$$(4)$$

with $\Delta M_{ij}^2 = M_{N_j}^2 - M_{N_i}^2$. The factors S_j (V_j) comes from the one-loop self-energy (vertex) contribution to the decay widths, Fig. 1. The I_j factors are the CP-violating coupling combinations entering in the asymmetry.

2.2. *The Efficiency Factor*

Once the averaged ΔL produced per decay has been calculated, the second ingredient to consider is the efficiency factor η. This factor allows to calculate the lepton asymmetry produced from the CP-asymmetry,

$$\frac{n_L}{s} = \varepsilon_{N_i} Y_{N_i}|_{T>>M_{N_i}} \eta, \tag{5}$$

where $Y_{N_i} = n_{N_i}/s$ is the number density of N_i over the entropy density, with $Y_{N_i}|_{T>>M_{N_i}} = 135\zeta(3)/(4\pi^4 g_*)$ where $g_* = 112$ is the number of degrees of freedom in thermal equilibrium in the "type-I" model before the N_i decayed. If all right-handed neutrinos decay out-of-equilibrium, the lepton asymmetry produced is just given by the CP asymmetry times the number of N_i over the entropy density before the N_i decayed, i.e. $\eta = 1$. However, the efficiency factor can be much smaller than one, if they are not fully out-of-equilibrium while decaying, and/or if there are at this epoch L-violating processes partly in thermal equilibrium. The processes which can put the N_i in thermal equilibrium and/or violate L are the inverse decay process and $\Delta L = 1, 2$ scatterings. To avoid a large damping effect, it is necessary that these processes are not too fast with respect to the Hubble constant. For the inverse decay process (which is the most dangerous process, see e.g. the discussion of Ref.[6]), this gives the condition: $\Gamma_{N_i}/H(T \simeq M_{N_i}) \leq 1$ with $H(T) = \sqrt{4\pi^3 g_*/45}\, T^2/M_{\text{Planck}}$. In practice to calculate η we need to put all these processes in the Boltzmann equations[7,8] which allow a precise calculation of the produced lepton asymmetry as a function of the temperature T. The corresponding efficiency factor including finite temperature effects can be found in Ref.[8] in the limit where the right-handed neutrinos have a hierarchical spectrum $M_{N_1} << M_{N_{2,3}}$. In this limit only the

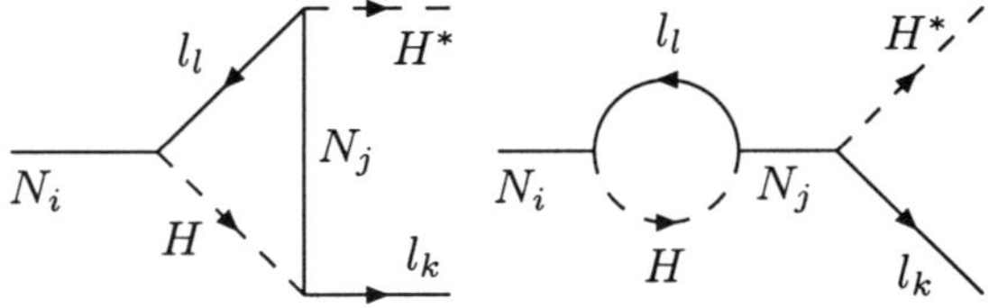

Figure 1. One-loop diagrams contributing to the asymmetry from the N_i decay.

asymmetry produced by the decay of the lightest right-handed neutrino N_1 survives and is important, which simplifies greatly the calculations.

2.3. *The B to L Asymmetry Conversion*

Once the L asymmetry n_L/s has been produced and determined, the last step is to calculate the corresponding baryon asymmetry produced due to the partial conversion of the L asymmetry to a B asymmetry by the Standard Model non-perturbative sphaleron processes. The conversion factor is $n_B/s = -(28/79)n_L/s$.[9] It comes from taking into account the fact that these processes conserve $B - L$, violate $B + L$ and are fast (i.e. in thermal equilibrium) above the electroweak scale. From this conversion factor we finally get the total baryon asymmetry produced $n_B/s = -1.38 \cdot 10^{-3} \varepsilon_{N_1} \eta$ which has to be compared with the experimental BBN and WMAP value for this ratio: $n_B/s = (8.7 \pm 0.4) \cdot 10^{-11}$.

3. The Neutrino Mass Constraints

In addition to the three ingredients above there is a fourth crucial ingredient which makes the all interest of the leptogenesis mechanism: the neutrino mass constraints. These constraints come from the fact these are the same interactions which produce both the neutrino masses and leptogenesis. There are two types of neutrino mass constraints on leptogenesis.

3.1. *The Neutrino Mass Constraint on the Efficiency*

The first constraint is the neutrino mass constraint on the size of the washout, that is to say on η. It comes from the fact that, in full generality, the ratio Γ_{N_i}/H, which has to be smaller than unity to have no washout suppression, is always larger than the ratio of the lightest neutrino mass m_{ν_1} over the $m_* = 16\pi^2 v^2 \sqrt{g_* \pi/45}/M_{\mathrm{Planck}} \sim 10^{-3}$ eV scale:[10]

$$\frac{\Gamma_{N_i}}{H} \geq \frac{m_{\nu_1}}{10^{-3}\,\mathrm{eV}}. \tag{6}$$

This inequality comes simply from Eq. (3) and the $v^2|\lambda\lambda^\dagger|^2_{jj}/M_{N_j} \geq m_{\nu_1}$ inequality. It means that, if $m_{\nu_1} > 10^{-3}$ eV, there will be some washout and the larger is m_{ν_1} above this scale the larger is the washout, i.e. the smaller is η. For example for $m_{\nu_1} = \sqrt{\delta m^2_{\mathrm{atm}}} \simeq 0.05$ eV the value of η is ~ 0.01 for $M_{N_1} \leq 10^{14}$ GeV. For $m_{\nu_1} = 0.5$ eV this value becomes few 10^{-4} for $M_{N_1} \leq 10^{12}$ GeV and decreases for larger values of M_{N_1}. The fact that m_*, which is a function of the electroweak and Planck scales, is of order the neutrino masses is a quite remarkable fact since it means that for leptogenesis the N_i are naturally only slightly in thermal equilibrium or out-of-equilibrium. Note that Eq. (6) is often rewritten in term of the so-called effective masses of Eq. (3) as $\tilde{m}_i \geq 10^{-3}$ eV.

3.2. *The Neutrino Mass Constraint on the Size of ε_{N_i}*

The second neutrino mass constraint is on the size of the CP-asymmetries ε_{N_i}. It generally applies but, contrary to the first constraint above, it do not always applies. This depends on the type of mass spectrum the right-handed neutrino have: very hierarchical, "normally" hierarchical or quasi-degenerate.

3.2.1. *Very hierarchical right-handed neutrinos: $M_{N_1} << M_{N_{2,3}}$*

If right-handed neutrino masses differ by several orders of magnitude, the size of the ε_{N_1} asymmetry is quite constrained by the size of the neutrino masses. There exists an upper bound[11,12,13] which in its exact form was given by Ref.[13]:

$$|\varepsilon_{N_1}| \leq \frac{3}{16\pi} \frac{M_{N_1}}{v^2}(m_{\nu_3} - m_{\nu_1}) = \frac{3}{16\pi} \frac{M_{N_1}}{v^2} \frac{\Delta m_{\mathrm{atm}}^2}{m_{\nu_3} + m_{\nu_1}}. \qquad (7)$$

Since $\Delta m_{\mathrm{atm}}^2 \simeq 2 \cdot 10^{-3}~eV^2$ is fixed experimentally, this bound decreases as the neutrino masses increase. Therefore as the neutrino masses increase there are two suppression effects arising: the washout effect increases and the upper bound on the asymmetry decreases. Successful leptogenesis leads therefore to the upper bound:[14,8,5]

$$m_{\nu_3} < 0.12 - 0.15~\mathrm{eV}. \qquad (8)$$

For more details see Ref.[2] in these proceedings. Note that in order to derive this bound Eq. (7) is not sufficient. As the final produced lepton asymmetry depends not only on m_{ν_3} and M_{N_1} but also, through the efficiency factor, on Γ_{N_1} (or $\tilde{m}_1$), it is necessary[14,5] to have an upper bound for fixed values of these 3 parameters and not only as a function of m_{ν_3} and M_{N_1} as in Eq. (7). This bound can be found in Ref.[5]. Note also that, due to the fact that the upper bound on the CP-asymmetry is proportional to M_{N_1}, successful leptogenesis with hierarchical right-handed neutrinos implies also a lower bound on this mass:[13,11,12,8]

$$M_{N_1} > 5 \times 10^8~\mathrm{GeV}. \qquad (9)$$

This result holds for the case where the N_1 are in thermal equilibrium before decaying. Starting instead (due to inflation dynamics) from a universe with no (with only) right-handed neutrinos at a temperature above their mass, this bound becomes[8]: $M_{N_1} > 2 \times 10^9$ GeV, $(2 \times 10^7$ GeV$)$.

3.2.2. *"Normally" hierarchical right-handed neutrinos*

If right-handed neutrinos have a hierarchy similar to the ones of the charged leptons or quarks, that is to say if $M_{N_1} \simeq (10 - 100)M_{N_2}$ with $M_{N_3} > M_{N_2}$, the L-asymmetry production is still dominated by the decays of the lightest right-handed neutrino N_1. In this case the upper bound on the CP-asymmetry is the same as for very-hierarchical neutrinos, except that there are extra corrections[5] in $M_{N_1}^2/M_{N_{2,3}}^2$ to be added in Eq. (7):

$$\delta\varepsilon_1 \simeq \frac{3}{16\pi} \frac{M_{N_1}}{v^2} \tilde{m}_{2,3} \frac{M_{N_1}^2}{M_{N_{2,3}}^2} . \tag{10}$$

These corrections generically are small so that all the bounds obtained in the previous section are still valid, but not always. There exist configurations of the Yukawa couplings which lead to large corrections. The point is that, contrary to the leading term in Eq. (7), the corrections do not decrease when the neutrino masses decrease and they do not necessarily vanish for degenerate light-neutrino masses. As a result[5], for special configurations giving large $\tilde{m}_{2,3}$ but small neutrino masses, one can have successful leptogenesis with M_{N_1} well below the lower bound of Eq. (9) and with neutrino masses well above the bound of Eq. (8). An explicit example of such Yukawa coupling configuration leading to successful leptogenesis with $M_{N_1} \simeq 10^6$ GeV has been recently considered in Ref.[15].

3.2.3. *Quasi-degenerate right-handed neutrinos*

If at least two right-handed neutrinos have masses very close to each other, $M_{N_1} \sim M_{N_2}$, the situation is changing drastically with respect to the two previous cases. This is due to the fact that the one loop self-energy diagram of Fig.1 displays in this case a resonance behaviour[16,5,17] coming from the propagator of the virtual right-handed neutrino in this diagram. This effect can be seen from the S_j factors in Eq. (2). Since in the seesaw model the decay widths of the N_i are generically much smaller than their masses, this resonance effect can lead to a several order of magnitude enhancement of the asymmetry. At the resonance, that is to say for $M_{N_2} - M_{N_1} = \Gamma_{N_2}/2$, the S_2 factor, which is unity in the $M_{N_1} << M_{N_{2,3}}$ limit, is as large as

$\frac{1}{2} M_{N_2}/\Gamma_{N_2}$. In this case, up to a 1/2 factor, the S_2 factor compensates the Γ_{N_2}/M_{N_2} prefactor in Eq. (2) which gives $\varepsilon_{N_1} = \frac{1}{2} I_2$. Moreover in this case one can show that the asymmetry is not bounded anymore by an expression depending on the neutrino masses. In particular the upper bound is not proportional to $\Delta m_{\mathrm{atm}}^2/(m_{\nu_3} + m_{\nu_1})$ as in the hierarchical case,[a] Eq. (7). The Yukawa coupling factor I_2 turns out to be bounded just by unity so that the upper bound on ε_{N_1} is just $\frac{1}{2}$ independently of light and heavy neutrino masses[5]. Together with ε_{N_2}, which is equal to ε_{N_1} in this case and has also to be taken into account, CP-violation is just bounded by unity. As a result, since neither the maximal asymmetry nor the washout effect (coming from inverse decay[b]) depend on M_{N_1}, successful leptogenesis can be obtained at any scale except that the L to B conversion from sphalerons still needs to be effective. This requires M_{N_1} to be above the electroweak scale (i.e. typically above ~ 1 TeV). Moreover since the upper bound on the CP asymmetry is independent of neutrino masses, there is no more suppression of the asymmetry for large neutrino masses. The only remaining suppression effect arising for large neutrino masses is the one of section 3.1 above coming from the washout. Therefore the upper bound on the neutrino masses in this case gets considerably relaxed. This is shown in Fig.2.a where is plotted, as a function of m_{ν_3}, the level of degeneracy which is needed to have successful leptogenesis. Values far above the eV are possible which means that *in full generality there is no more relevant upper bound on neutrino masses coming from leptogenesis.* The value $m_{\nu_3} \simeq 1$ eV can lead to successful leptogenesis with a level of degeneracy of order $(M_{N_2} - M_{N_1})/M_{N_2} \simeq 4 \cdot 10^{-2}$ which is quite moderate.

One could argue that such a degeneracy of right-handed neutrinos is unnatural and that we would anyway generally expect a hierarchical spectrum for the right-handed neutrino by analogy with charged leptons or quarks. However, since the bound on neutrino masses from leptogenesis is relevant only for a quasi-degenerate spectrum of light neutrinos, when calculating this bound one can wonder what would be the most natural right-handed neutrino mass spectrum to explain such correlations between the light-neutrino masses. Presumably, three quasi-degenerate right-handed neutri-

---

[a]One can check[13,5] that this neutrino mass factor comes from the fact that for hierarchical right-handed neutrinos $S_2 = S_3$. This equality has no reason to be true anymore with quasi-degenerate right-handed neutrinos.

[b]Note that there is a M_{N_i} mass dependence in the washout coming from $\Delta L = 2$ scatterings. This effect however can be large only for large values of the M_{N_i} close to the GUT scale.

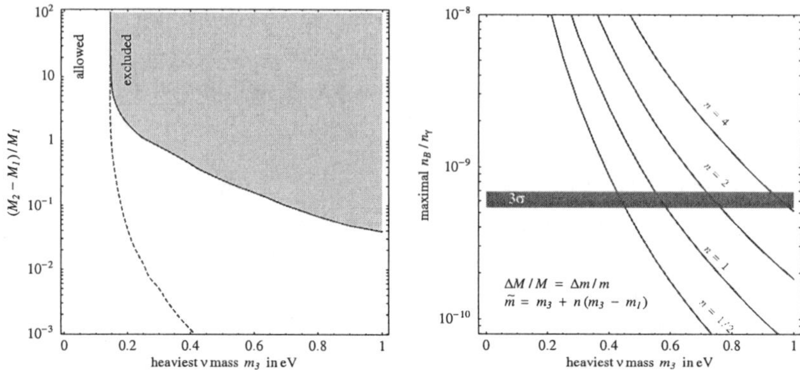

Figure 2. Fig.2.a shows[5] how much the maximal value of neutrino mass compatible with thermal leptogenesis increases when right-handed neutrinos are allowed to be quasi-degenerate. The dashed line is what we obtain if we take into account only the resonance enhancement effect, not the neutrino mass effect, see text and Ref.[5]. Fig.1.b holds assuming Eqs. (11)-(12) as a generic example that everything is as degenerate as neutrinos, considering the natural possibility that the $\tilde{m}_i$ should be quasi-degenerate with the neutrino masses, taking $\tilde{m}_{1,2,3} < m_{\nu_1} + n(m_{\nu_3} - m_{\nu_1})$.

nos are the most natural spectrum in this case (with Yukawa couplings correlated among each other). Such a quasi degenerate spectrum, unlike a hierarchical spectrum, wouldn't require in addition precise correlations between the Yukawa couplings and the right-handed neutrino masses (in order that the ratio between the M_{N_i} is compensated in the seesaw mass formula by the same ratio between the Yukawa couplings). Therefore the bound given in Fig. 2.a is the fully relevant one.

To complete this section note that there exists nevertheless one specific pattern, which is probably the most realistic one, which leads to more stringent bounds.[5] In fact, if neutrinos were quasi-degenerate, the degeneracy would presumably not be accidental but due to some reason: a broken SO(3) flavour symmetry is probably the simplest possibility. One expects that in such framework all quantities, and not only neutrino masses, are close to the ideal limit where three degenerate right-handed neutrinos give equal masses to three orthogonal combinations of left-handed neutrinos. Therefore one expects something like $\tilde{m}_i - \tilde{m}_j \approx m_{\nu_i} - m_{\nu_j}$ and

$$\frac{M_{N_2} - M_{N_1}}{M_{N_1}} \sim \frac{m_{\nu_2} - m_{\nu_1}}{m_1} \approx \frac{\tilde{m}_2 - \tilde{m}_1}{\tilde{m}_{\nu_1}} \approx \frac{\Delta m^2_{\text{sun}}}{2m^2_{\nu_1}} \approx 0.5 \ 10^{-4} \left(\frac{\text{eV}}{m_{\nu_2}}\right)^2 \quad (11)$$

$$\frac{M_{N_3} - M_{N_2}}{M_{N_2}} \sim \frac{m_{\nu_3} - m_{\nu_2}}{m_{\nu_2}} \approx \frac{\tilde{m}_3 - \tilde{m}_2}{\tilde{m}_2} \approx \frac{\Delta m^2_{\text{atm}}}{2m^2_{\nu_3}} \approx 10^{-3} \left(\frac{\text{eV}}{m_{\nu_2}}\right)^2. \quad (12)$$

Right-handed neutrinos can be more degenerate than in the above estimates if only the neutrino Yukawa couplings deviate from the symmetric limit, and can be less degenerate only if there are accidental cancellations between non-universal Yukawa couplings and non-degenerate $M_{N_{1,2,3}}$ in the see-saw prediction for neutrino masses.

Assuming the relations of Eqs. (11)-(12), the bound on the neutrino masses can be read off from Fig. 2.b where we give the maximal baryon asymmetry we obtain as a function of m_{ν_3} for values of $\tilde{m}_{1,2,3} < m_{\nu_1} + n(m_{\nu_3} - m_{\nu_1})$ with $n = \{1/2, 1, 2, 4\}$. Taking $\tilde{m}_{1,2,3} < m_{\nu_3}$ $(n = 1)$, as the generic example for the case that the $\tilde{m}$ would be precisely of order the neutrino masses, gives the constraint

$$m_{\nu_3} < 0.6 \text{ eV}, \tag{13}$$

which is stronger than in the fully general case of Fig. 2.a. The suppression effect comes essentially from the fact that for values of $\tilde{m}_{1,2,3}$ close to the neutrino masses the Yukawa coupling factors I_j are suppressed[5,14]. The result is quite sensitive to how these quantities are close to each other. For example taking n=4 (which could be considered as a quite moderately fine-tuned case) leads already to an upper bound as large as 1 eV.

In summary without a predictive flavour model which would show how the correlations between the seesaw parameters at the origin of the degenerate spectrum occur, in order to have a safe bound we must consider the fully general case of Fig. 2.a (where n was left as a free parameter in order to maximize the asymmetry). Even in a very constrained situation the neutrino masses can be as large as 0.6 eV, Eq. (13), or 1 eV.[c]

4. Leptogenesis in the Framework of Other Seesaw Models

The type-I seesaw mechanism is probably the most direct extension of the Standard Model we can consider in order to explain the neutrino masses and is in this sense the most attractive. However it is not the only attractive seesaw model. Beside the type I seesaw, one can think about two other basic seesaw mechanisms. The first one is the type-II seesaw model[18] where neutrino masses are due to the exchange of an heavy scalar Higgs triplet. The second one is from the exchange of three heavy self-conjugated $SU(2)_L$ triplets of fermions[19], a model which according to us should be called type-III seesaw model as it induces the neutrino mass from a third type of

---

[c]Stronger constraints will arise if supersymmetry exists and if right-handed neutrinos lighter than 10^{10} GeV will be needed to avoid gravitino overproduction.

heavy particles. In addition to these three seesaw basic mechanisms one can also think about combinations of them. In the following we consider these various alternatives and see whether they can lead to successful leptogenesis and what are the corresponding mass bounds.

4.1. *The Type-II Seesaw Model*

The type-II seesaw model with just one heavy scalar triplet Δ_L which couples to 2 leptons doublets and to two Higgs doublets is a particularly minimal model. It is based on the Lagrangian

$$L \ni -M_\Delta^2 Tr\Delta_L^\dagger \Delta_L - (Y_\Delta)_{ij} L_i^T C i\tau_2 \Delta_L L_j + \mu H^T i\tau_2 \Delta_L H + h.c. , \quad (14)$$

with

$$\Delta_L = \begin{pmatrix} \frac{1}{\sqrt{2}}\delta^+ & \delta^{++} \\ \delta^0 & -\frac{1}{\sqrt{2}}\delta^+ \end{pmatrix} . \quad (15)$$

It leads to the neutrino mass matrix: $M_\nu^{II} = 2Y_\Delta v_\Delta \simeq 2Y_\Delta \mu^* v^2 / M_\Delta^2$. This model in full generality has only 11 parameters, the triplet mass, its μ coupling and 6 real parameters plus 3 phases in the Y_Δ Yukawa coupling matrix. This model has the attractive property that the knowledge of the full low energy neutrino mass matrix M_ν would allow to determine the full flavour high energy structure in Y_Δ, both matrices being just proportional to each other. However for leptogenesis it turns out that this model is too minimal. As explained in Ref.[20,21,5,22], since the triplet is not a self-conjugated particle, there is no vertex diagram and leptogenesis could come only from a self-energy diagram involving two leptons in the final state and two Higgs doublets in the self-energy, third diagram of Fig.3. This diagram with just one triplet is real and therefore doesn't bring any CP-violation. At two loops the asymmetries are too suppressed. Therefore the standard model extended by just one scalar triplet leads in a "minimal" way to neutrino masses but do not lead to successful leptogenesis.

However, based on the type-II model with just one triplet coupling to leptons, there is one framework which can work. It is the supersymmetric version of this model, i.e. the MSSM model extended by a pair of scalar triplets with opposite hypercharges. In a way similar to the soft leptogenesis mechanism with right-handed neutrinos[23,2], the supersymmetry breaking terms involving the triplets can remove the mass degeneracy between the two triplets and lead to resonant leptogenesis from the self-energy diagram involving the two triplets[22]. This requires triplet mass between $\sim 10^3$ GeV and $\sim 10^9$ GeV.[d]

4.2. *The Heavy Triplet of Fermion Model*

The third seesaw basic way to induce neutrino masses is by adding to the Standard Model 3 self-conjugated $SU(2)_L$ triplets of fermions.[19] The Lagrangian keeps the same structure as the one of the type-I seesaw model but with different $SU(2)_L$ contractions:

$$L = L_{\rm SM} + \bar{N}_i^a i\!\!\not{D} N_i^a + (\lambda^{ij}\tau^a_{\alpha\beta} N_i^a L_j^\alpha H^{\dagger\beta} + \frac{M_{N_i}}{2} N_i^a N_i^a + {\rm h.c.}) , \quad (16)$$

with $a = 1,2,3$; $\alpha,\beta = 1,2$. As a result it leads to the same seesaw formula than with singlets: $M_\nu = -\lambda^T M_N^{-1}\lambda v^2$. For leptogenesis, the one-loop diagrams are exactly the same as with right-handed neutrinos, Fig. 1, which lead to the same asymmetries up to $SU(2)_L$ factors of order unity,[5]

$$\varepsilon_{N_1} = \sum_{j=2,3} \frac{3}{2}\frac{M_{N_1}}{M_{N_j}}\frac{\Gamma_{N_j}}{M_{N_j}} I_j \frac{V_j - 2S_j}{3} , \quad (17)$$

and is therefore 3 times smaller in the hierarchical limit where $V_j = S_j = 1$, see Eq. (2). The final amount of baryon asymmetry is given by the CP-asymmetry times the efficiency factor η times a numerical coefficient which is 3 times bigger than in the singlet case because now N_1 has three components: $\frac{n_B}{s} = -4.1 \cdot 10^{-3}\varepsilon_{N_1}\eta$. The decay width of each of the 3 components of N_1 is given by the same expression as in the singlet case, Eq. (3). Scatterings are same as in the type-I model except for $SU(2)_L$ factors[5]. The only important difference is that the triplets have $SU(2)_L \times U(1)$ gauge scatterings the singlets do not have. This reduces the efficiency factor[5]. As a result, in the hierarchical limit $M_{N_1} << M_{N_{2,3}}$, the bounds for successful leptogenesis are slightly more stringent than for singlets:

$$M_{N_1} > 1.5 \cdot 10^{10} \text{ GeV}, \quad m_{\nu_3} < 0.12\text{eV} . \quad (18)$$

In the quasi-degenerate case the discussion is similar to the one of the type-I model, see section 3.2.3 above. A value of M_{N_1} as low as ~ 1 TeV is possible, very close to the resonance.

---

[d]Note that triplets, unlike right-handed neutrinos have gauge scatterings, which tend to put them in closer thermal equilibrium, reducing the efficiency. The exact efficiency for a scalar triplet is yet to be calculated but the one for a fermion triplet is known[5]. All lower bounds on the scalar triplet masses we give here are obtained using the fermion triplet efficiency, assuming that both efficiencies are same, as gauge scatterings are expected to be similar up to factors of order one. This should be checked explicitly.

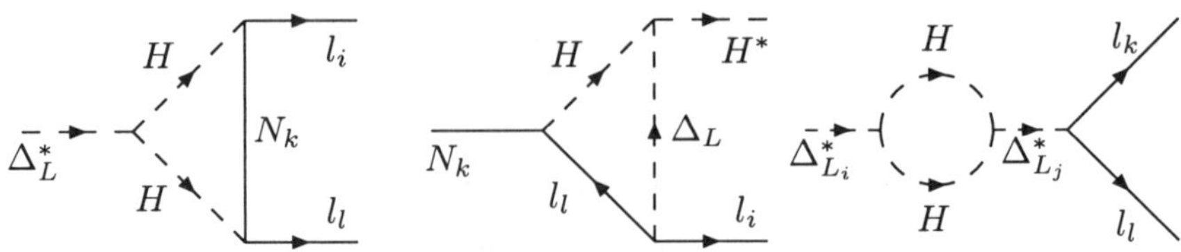

Figure 3.   One-loop diagrams contributing to the asymmetry from the Δ_L and N decays.

4.3. *The Type-I plus Type-II Model*

The case where we add to the standard model three right-handed neutrinos and one scalar triplet is quite interesting because it is the situation of the ordinary left-right models and of the renormalizable $SO(10)$ models such as defined in Ref.[24]. In this case the relevant Lagrangian is just the sum of the Lagrangians of Eqs. (1) and (14). To discuss this possibility it is necessary to consider two cases, depending on which particle is the lightest one, the scalar triplet Δ_L or the lightest right-handed neutrino N_1.

4.3.1. *The $M_\Delta < M_{N_1}$ case*

If the triplet is lighter than N_1, the production of the asymmetry will be naturally dominated by the decay of the triplet to two leptons. The CP-asymmetry comes in this case from the difference between the decay width of Δ_L^* to two leptons and of Δ_L to two anti-leptons. The leptogenesis one loop diagram is a vertex diagram involving both the decaying triplet and a virtual right-handed neutrino, first graph of Fig. 3. This diagram was first displayed in Ref.[25]. Calculating explicitly its contribution we get[26]:

$$\varepsilon_\Delta = 2 \cdot \frac{\Gamma(\Delta_L^* \to l + l) - \Gamma(\Delta_L \to \bar{l} + \bar{l})}{\Gamma_{\Delta_L^*} + \Gamma_{\Delta_L}} \tag{19}$$

$$= -\frac{1}{8\pi} \frac{M_\Delta^2}{\left(\sum_{ij} |(Y_\Delta)_{ij}|^2 M_\Delta^2 + |\mu|^2\right)} \frac{1}{v^2} Im[(M_\nu^{I*})_{il}(Y_\Delta)_{il}\mu^*], \tag{20}$$

where M_ν^I is the type-I contribution to the neutrino mass matrix (given in section 2). For each of the 3 triplet components the total decay width is:

$$\Gamma_\Delta = \frac{1}{8\pi} M_\Delta \left(\sum_{ij} |(Y_\Delta)_{ij}|^2 + \frac{|\mu|^2}{M_\Delta^2} \right). \tag{21}$$

In this case leptogenesis works in a way similar to the type-I model apart from 3 important differences. The first is that, like for triplet of fermions above, there are gauge scatterings which tend to put the Δ_L closer to thermal equilibrium. The effect is similar to the one encountered for triplet of fermions above. The second difference is that here there is no one-loop self-energy diagram, so there is no possible resonance effect. The third one is that the interplay between the neutrino masses and the size of the washout and the size of the asymmetry is completely different from the type-I case. The decay width is proportional to the triplet couplings but the asymmetry is essentially proportional to the right-handed neutrino couplings (taking into account the fact that there are two triplet couplings in both numerator and denominator of the asymmetry). As a result one may consider the possibility that the type-II contribution to neutrino masses is small enough to avoid large washout and that the type-I contribution to neutrino masses is large. By increasing in this way the type-I neutrino mass contribution, the washout remains unchanged but the asymmetry increases proportionally to the neutrino masses[26]. Therefore, there is no upper bound on neutrino masses in this model to have successful leptogenesis[26]. Note that there is nevertheless a lower bound on the triplet mass because the asymmetry is proportional to this mass. For a hierarchical spectrum of light neutrinos (i.e. $m_{\nu_3} \sim \sqrt{\Delta m_{\text{atm}}^2}$) the bound is essentially the same[d] as for the triplet of fermions, Eq. (18). As there is no possible resonance effect this bound is an absolute bound. For larger values of m_{ν_3} this bound decreases because in this case the asymmetry increases. However, assuming m_{ν_3} below 1 eV, this bound cannot decrease by more than one order of magnitude.

4.3.2. *The $M_\Delta > M_{N_1}$ case*

If the triplet is sizeably heavier than at least one right-handed neutrino, then it is the decay of the right-handed neutrino to a lepton and a Higgs boson which dominates the production of the L asymmetry. In this case leptogenesis can be produced from the pure type-I model just as in sections 2 and 3. However here, in addition to this pure type-I contribution, there is a new contribution to leptogenesis coming from a diagram[25,27,26,28] with a real right-handed neutrino and a virtual triplet, second diagram of Fig. 3. Disregarding the pure type-I contribution (assuming that it has a small contribution because the type-I Yukawa couplings and/or their phases are small) this new diagram can perfectly lead to successful leptogenesis alone. For the asymmetry defined in Eq. (2) one obtains from this diagram[26,28]

$$\varepsilon_{N_1}^\Delta = \frac{3}{16\pi} \frac{M_{N_1}}{v^2} \frac{\sum_{il} Im[\lambda_{1i}\lambda_{1l}(M_\nu^{II*})_{il}]}{\sum_i |\lambda_{1i}|^2}, \tag{22}$$

where M_ν^{II} is the neutrino mass matrix induced by the type-II contribution (given in section 4.1 above). The discussion is similar to the one of the $M_\Delta < M_{N_1}$ case above inverting the role of type-I and type-II two contributions. It is now the decay of the right-handed neutrino to lepton and Higgs, induced by the type-I couplings, Eq. (3), which essentially determines the washout. To increase the asymmetry without increasing the washout, one can then consider the possibility of keeping the type-I contribution small, increasing the type-II contribution which increases the asymmetry but not the washout. As a result, here too, there is no more upper bound on neutrino masses for leptogenesis[26]. For hierarchical light neutrinos the lower bound on the lightest right-handed neutrino mass is to a good approximation the same as in the pure type-I model[26], Eq. (9). As the asymmetry is linear in both M_{N_1} and the neutrino masses, for larger values of m_{ν_3} the asymmetry linearly increases, so the bound on M_{N_1} linearly decreases. But here too assuming m_{ν_3} below 1 eV, this bound cannot decrease by more than one order of magnitude. The precise bound is given in Ref.[28] as a function of the efficiency factor η.

Note that, if instead of considering a dominant type-II contribution, we consider the case where both type-I and type-II contributions are important this lower bound on M_{N_1} doesn't get relaxed. In this case both contributions appear to be just proportional to their respective contributions to the neutrino masses so that, baring a possible cancellation of CP-violating phases, leptogenesis is expected to be dominated by the contribution which dominates the neutrino masses[26]. From Eq. (2) the pure type-I contribution to leptogenesis for hierarchical right-handed neutrinos turns out to be given by Eq. (22) replacing M_ν^{II} by M_ν^{I}, a fact which can be nicely understood by using effective dimension five neutrino mass operators.[28]

For a quasi-degenerate spectrum of right-handed neutrino the pure type-I contribution to leptogenesis is expected dominant and the triplet contribution which doesn't display any resonant behaviour is in this case negligible.

4.4. *The Multiple Type-II Models*

If there is more than one heavy scalar triplet, leptogenesis can be easily induced by the decay of the triplets to two leptons with a one-loop self-energy diagram involving two different triplets[20,21], third diagram of Fig.3. The asymmetry in this case is given by[20]

$$\varepsilon_{\Delta_i} = -\frac{1}{\pi} M_{\Delta_i} \frac{Im[(\mu_i^* \mu_j (Y_{\Delta_i})_{kl} (Y_{\Delta_j}^*)_{kl}]}{|(Y_{\Delta_i})_{kl}|^2 M_{\Delta_i}^2 + |\mu_i|^2} \frac{M_{\Delta_j}^2 \Delta M_{1j}^2}{(\Delta M_{1j}^2)^2 + M_{\Delta_1}^2 \Gamma_{\Delta_j}^2}, \qquad (23)$$

where there is now a triplet scalar index on the couplings of Eqs. (14) and (21). For hierarchical triplets, under the assumption of footnote d above, successful leptogenesis leads to a triplet mass bound similar to the one of Eq. (18) for hierarchical fermion triplets, in accordance with the estimate of Ref.[21]. This bound is higher than for right-handed neutrinos due to gauge scatterings. Similarly for quasi-degenerate triplets, one can go down to 1-10 TeV (very close to the resonance). And, here too, even with hierarchical triplets, there is no more upper bound on neutrino masses for successful leptogenesis because one can always increase the neutrino mass contribution of the virtual triplet in the self-energy diagram leading to a larger asymmetry without increasing the washout.

4.5. *The Multiple Type-I Seesaw Case*

To add extra fermion singlets S_i on top of the 3 right-handed neutrinos N_i is also a possibility one might consider,[29,30,31] especially if these extra $SU(2)_L \times U(1)$ singlets are also singlets of SO(10) in case one can build a renormalizable SO(10) model[29,30] without a 126 scalar multiplet to give mass to the N_i's. In this case, as there is no 126, there is no triplet. This may lead to successful leptogenesis just from the same diagrams as in the type-I model, Fig. 1, except that there are now more than 3 singlets to be put in these diagrams. There are diagrams where both real and virtual singlets are N_i's or are S_i's. Clearly, for example if the lightest heavy particle is a S, one can increase the Yukawa couplings of the N leaving the S Yukawa couplings unchanged. In this way the neutrino masses increase but not the washout which is due the S couplings, so that, here too, there is no relevant upper bound on neutrino masses for successful leptogenesis.

Note that extra diagrams with a real N and a virtual S in Fig.1 (or viceversa) are also possible if they couple to the same scalar multiplets. In the SO(10) models of Ref.[29,30] the S and N do not couple to the same multiplets because the S which is a singlet of SO(10) couples to the 16 of matter ψ_{16} and a scalar $H_{\overline{16}}$ but the N which is in ψ_{16} couples to an other ψ_{16} and a scalar H_{10}. However through mixing of the $H_{\overline{16}}$ and H_{10} from a coupling involving the vev of another $H_{\overline{16}}$ such a diagram may exist, a possibility still to consider. It is straightforward to check that such a diagram can lead to successful leptogenesis and here too without leading to a relevant upper bound on the neutrino masses.

Note also that there is an other possibility with extra singlets[31], simply by assuming more than three generations of fermions and by assuming a huge hierarchy of Yukawa couplings (to have $m_{\nu_4} > 45$ GeV). This may also lead to successful leptogenesis (e.g. at a low scale).

5. The Radiative Models

An other class of models one might consider to explain the neutrino masses and mixings is the one where neutrino masses are induced by loop diagrams. This is quite interesting as in most of these models the scale where they are generated is not far beyond the reach of present particle accelerators. However to generate leptogenesis at the $\sim$ TeV scale in this framework turns out to be a quite hard task for various reasons. A first reason is that in these models, such as the ones based on violation of R-parity[32] or based on the presence of a charged scalar $SU(2)_L$ singlet (Zee model[33]), the heavy states whose decays could generate the lepton asymmetry are not gauge singlets. As a result there is a large washout suppression coming from gauge scatterings involving these heavy states which are very fast at low scale. A second reason is that at the TeV scale the Hubble constant is much smaller than for example at 10^{10} GeV since it is proportional to the square of the temperature. Since the decay width is only linear in the mass, the condition $\Gamma < H$ requires therefore couplings much smaller at the TeV scale than at 10^{10} GeV. Since the asymmetry is proportional to these tiny couplings, this leads in general to too small lepton asymmetry, see Ref.[12].

One solution to these problems is to consider three body decays instead of two-body decays.[12] An other solution is to consider the seesaw extended MSSM, that is the say the MSSM extended by three right-handed (s)neutrinos. In this framework it has been shown[34] that a large enough asymmetry can be obtained from the L-violating soft supersymmetry breaking terms involving the right-handed sneutrinos. With sneutrino masses of order a few TeV, successful leptogenesis together with light neutrino masses (induced radiatively from the same soft terms) can be induced.

6. Summary

In summary there are quite a few models which from the same interactions can lead to successful generation of neutrino masses and leptogenesis: type-I model with right-handed neutrinos, type-II model (although only if involving the soft terms), type-III model with triplet of fermions, type-I plus type-II model, multiple type-I or type-II models, or even a radiative model (in the seesaw extended MSSM from soft terms).

The upper bound on neutrino masses is quite sensitive to the model considered as well as on the assumptions made on the heavy mass spectrum in each model. In the type-I model (and similarly in the type-III model) a stringent bound, Eq. (8), can be found only assuming a hierarchical spectrum of right-handed neutrinos. However this assumption doesn't appear to be the most natural assumption one can make when considering this bound, for which the light neutrinos have a quasi-degenerate spectrum. A probably more natural quasi-degenerate spectrum of right-handed neutrinos leads instead in full generality to an upper bound on neutrino masses far beyond the eV scale. Even in a highly constrained model, such as the one based on a SO(3) symmetry in section 3.2.3, a value around 1 eV appears to be possible for successful leptogenesis. Only with extra assumptions (such as low reheating temperature) one might get a more stringent bound. In the other models (multiple type-I and/or type-II) the upper bound is far above the eV scale even for a hierarchical spectrum of heavy states.

Similarly the lower bound on the scale of leptogenesis is sensitive to the model and/or the heavy mass spectrum considered. However for this scale there is at least one firm conclusion one can draw: in all the seesaw models we considered here, if the masses of the heavy states differ by orders of magnitude, this scale has to be orders of magnitude above the TeV scale, that is to say above $10^7 - 10^{10}$ GeV depending on the model.

Acknowledgments

It is a pleasure to thank L. Boubekeur, G. D'Ambrosio, A. Hektor, Y. Lin, E. Ma, A. Notari, M. Papucci, M. Raidal, A. Rossi, U. Sarkar, G. Senjanovic and A. Strumia with whom part of the work presented here was done. This work was supported by the Marie Curie HPMF-CT-01765 and EU HPRN-CT-2000-00148 (Across the Energy Frontier) contracts.

References

1. M. Fukugita and T. Yanagida, *Phys. Lett.* **B174** (1986) 45.
2. M. Raidal, these proceedings.
3. P. Minkowski, *Phys. Lett.* **B67**, 421 (1977). M. Gell-Mann, P. Ramond and R. Slansky, in *Supergravity*, eds. P. van Nieuwenhuizen et al., (North-Holland, 1979), p. 315; S.L. Glashow, in Quarks and Leptons, Cargèse, eds. M. Lévy et al., (Plenum, 1980), p. 707; T. Yanagida, in *Proceedings of the Workshop on the Unified Theory and the Baryon Number in the Universe*, eds. O. Sawada et al., (KEK Report 79-18, Tsukuba, 1979), p. 95; R.N. Mohapatra and G. Senjanović, Phys. Rev. Lett. 44, (1980) 912. See also these proceedings.
4. J.A. Casas, A. Ibarra, *Nucl. Phys.* **B618**, 171 (2001).

5. T. Hambye, Y. Lin, A. Notari, M. Papucci and A. Strumia, *Nucl. Phys.* **B695**, 169 (2004).

6. P. Di Bari, hep-ph/0406115.

7. See e.g. M. Plumacher, Ph.D. thesis, hep-ph/9807557.

8. G.F. Giudice, A. Notari, M. Raidal, A. Riotto, A. Strumia, *Nucl. Phys.* **B685**, 89 (2004).

9. J. A Harvey and M.S. Turner, *Phys. Rev.* **D42**, 3344 (1990).

10. W. Buchmüller and M. Plumacher, *Phys. Rept.* **320**, 329 (1999).

11. R. Barbieri, P. Creminelli, A. Strumia, N. Tetradis, *Nucl. Phys.* **B575**, 61 (2000); W. Buchmüller and T. Yanagida, *Phys. Lett.* **B445**, 399 (1999).

12. T. Hambye, *Nucl. Phys.* **B633**, 171 (2000).

13. K. Hamaguchi, H. Murayama, T. Yanagida, *Phys. Rev.* **D65**, 043512 (2002); S. Davidson, A. Ibarra, *Phys. Lett.* **B535**, 25 (2002).

14. W. Buchmuller, P. Di Bari, M. Plümacher, *Nucl. Phys.* **B665**, 445 (2003).

15. M. Raidal, A. Strumia and K. Turzyński, hep-ph/0408015.

16. M. Flanz, E.A. Paschos, U. Sarkar, Phys. Lett. B345 (1995) 248; M. Flanz, E.A. Paschos, U. Sarkar, J. Weiss, Phys. Lett. B389 (1996) 693; L. Covi, E. Roulet, F. Vissani, Phys. Lett. B384 (1996) 169; A. Pilaftsis, Phys. Rev. D56 (1997) 5431; A. Pilaftsis, T.E.J. Underwood, hep-ph/0309342.

17. T. Hambye, J. March-Russell and S. West, *JHEP* **0497**, 070 (2004).

18. M. Magg and C. Wetterich, Phys. Lett. **B94** (1980) 61; C. Wetterich, Nucl. Phys. **B187** (1981) 343; G. Lazarides, Q. Shafi and C. Wetterich, Nucl Phys. **B181** (1981) 287; R.N. Mohapatra and G. Senjanović, Phys. Rev. **D23** (1981) 165. R.N. Mohapatra; G. Senjanovic; C. Wetterich; these proceedings.

19. R. Foot, H. Lew, X.-G. He and G.C. Joshi, Z. Phys. C44 (1989) 441; E. Ma, Phys. Rev. Lett. 81 (1998) 1171.

20. E. Ma and U. Sarkar, Phys. Rev. Lett. 80 (1998) 5716.

21. T. Hambye, E. Ma and U. Sarkar, Nucl. Phys. B602 (2001) 23.

22. G. D'Ambrosio, T. Hambye, A. Hektor, M. Raidal and A. Rossi, hep-ph/0407312.

23. G. D'Ambrosio, G.F. Giudice and M. Raidal, *Phys. Lett.* **B575**, 75 (2003).

24. B. Bajc, G. Senjanovic and F. Vissani, *Phys. Rev. Lett.* **90**, 051802 (2003); R.N. Mohapatra and S.P. Ng, *Phys. Lett.* **B570**, 215 (2003).

25. P. O'Donnell and U. Sarkar, Phys. Rev. **D49** (1994) 2118.

26. T. Hambye and G. Senjanovic, *Phys. Lett.* **B582**, 73 (2004).

27. G. Lazarides and Q. Shafi, Phys. Rev. **D58** (1998) 071702.

28. S. Antusch and S.F. King, *Phys. Lett.* **B597**, 199 (2004).

29. C.H. Albright and S.M. Barr, *Phys. Rev.* **D70**, 033013 (2004).

30. U. Sarkar, hep-ph/0409019.

31. A. Abada, H. Aissaoui and M. Losada, hep-ph/0409343.

32. M.A. Diaz, M. Hirsch, W. Porod, J.C. Romao and J.W.F. Valle, *Phys. Rev.* **D68**, 013009 (2003) and refs. therein.

33. A. Zee, *Phys. Lett.* **B93**, 389 (1980); *Phys. Lett.* **B161**, 141 (1985).

34. L. Boubekeur, T. Hambye and G. Senjanovic, *Phys. Rev. Lett.* **93**, 111601 (2004). See also Ref.[17].

NEUTRINOS IN EXTRA DIMENSIONS

EMILIAN DUDAS[1,2]

[1] *LPT, Univ. de Paris-Sud, Bât. 210, F-91405 Orsay Cedex, France*
[2] *CPhT, École Polytechnique, F-91128 Palaiseau, France*

We review alternatives and generalisations of the standard seesaw mechanism of generating neutrino masses in theories with large extra dimensions and low scale strings. We first review the models in which the sterile neutrinos are bulk fields propagating in gravitational type extra dimensions for flat and warped space. We then present a multiple seesaw mechanism which provide small neutrino masses in low scale strings, even without very large extra dimensions.

1. Millimeter and TeV^{-1} large extra dimensions

The seesaw mechanism of producing neutrino masses [1] is one of the cornerstones of physics beyond the standard model. The experimental evidence for neutrino masses and mixings, in conjunction with the seesaw mechanism, can be considered as the first indirect hint for a new scale in physics, close to the grande unification scale. As such, neutrino physics is a unique window into the high energy physics, inaccessible to accelerator experiments.

The presence of branes in String Theory opens new perspectives for particle physics phenomenology. Indeed, in Type I strings the string scale is not necessarily tied to the Planck scale. In view of the new D-brane picture, let us take a closer look at the simplest example of compactified Type I string, with only D9 branes present. The string scale can be in the TeV range if the string coupling is extremely small, $\lambda_I \sim 10^{-32}$. One can see that in this case the compact volume is very small $V M_I^6 \sim 10^{-32}$. Let us split the compact volume into two parts, $V = V^{(1)} V^{(2)}$, where $V^{(1)}$, of dimension $6 - n$, is of order one in string units and $V^{(2)}$, of dimension n, is very small. The Kaluza-Klein states of the brane fields along $V^{(2)}$ are much heavier than the string scale and therefore are difficult to excite. The physics is then better captured in this case performing T-dualities along

$V^{(2)}$, which read

$$\lambda_I' = \frac{\lambda_I}{V^{(2)} M_I^n} \quad , \quad V_\perp = \frac{1}{V^{(2)} M_I^{2n}} \cdot \tag{1}$$

In the T-dual picture, neglecting numerical factors, we find

$$M_P^2 \sim \frac{1}{\alpha_{GUT} \lambda_I'} V_\perp M_I^{2+n} \quad , \quad \frac{1}{\alpha_{GUT}} \sim \frac{V_{||} M_I^{6-n}}{\lambda_I'} \, , \tag{2}$$

where for transparency of notation we redefined $V^{(1)} \equiv V_{||}$. After the n T-dualities, the D9 brane becomes a D(9-n) brane, since the T-dual winding modes of the bulk (orthogonal) compact space are very heavy and therefore the brane fields cannot propagate in the bulk. As seen from (2), for a very large bulk volume the string scale can be very low $M_I << M_P$. The geometric picture here is that we have a D-brane with some compact radii parallel to it, of order M_I^{-1}, and some very large, orthogonal compact radii. In particular, if the full compact space is orthogonal to the brane ($n = 6$), from (2) the T-dual string coupling is fixed by the unified coupling $\lambda_I' \sim \alpha_{GUT}$, and therefore we find [5]

$$M_P^2 \sim \frac{1}{\alpha_{GUT}^2} V_\perp M_I^{2+n} \, , \tag{3}$$

a relation similar to that proposed in the field-theoretical scenario of [3].

Let us now imagine a "brane-world" picture[a], in which the Standard Model gauge group and charged fields are confined to the D-brane under consideration. We can then ask a very important question: what are the present experimental limits on parallel $R_{||}$ and perpendicular $R_\perp$ type radii ? The Standard Model fields have light KK states in the parallel directions $R_{||}$. Their possible effects in accelerators were studied in detail [7] and the present limits are $R_{||}^{-1} \geq 4 - 5$ TeV. On the other hand, Standard Model excitations related to $R_\perp$ are very heavy and are thus irrelevant at low energy. The main constraints on $R_\perp$ come from the presence of very light Kaluza-Klein gravitational excitations, which can therefore generate effects in colliders [8] and deviations from the Newton law of gravitational attraction. The actual experimental limits on such deviations are limited to the mm range and experiments in the near future are planned to improve them [9]. There are also collider effects coming from string physics at energies close to M_I [10]. For $M_I \sim$ TeV in (3), the case of only one extra dimension is clearly excluded, since it asks for $R_\perp^{-1} \sim 10^8$ Km. However, for two

---

[a]For earlier proposals of such a "brane-world" picture, see [6].

extra dimensions, we find $R_\perp^{-1} \sim 1$mm, only marginally excluded by the present experimental data. On the other hand, if all compact dimensions are perpendicular and large, one finds $R_\perp^{-1} \sim$ fm, distance scale completely inaccessible for Newton law measurements. Such a physical picture with $M_I \sim$ TeV provides in principle a new solution to the gauge hierarchy problem, i.e. of why the Higgs mass M_h is much lower than the 4d Planck mass M_P, provided the physical cutoff M_I has similar values $M_I \simeq M_h$.

In Type I strings [11], the brane we considered can be a D9 or a D5 brane, up to T-dualities. Our brane world can live on any of the branes; let us choose for concrctcness that our Standard Model gauge group be on a D9 brane. Notice that, while D9 branes fill (before T-dualities) the full 10d space, D5 branes fill only six dimensions. The D5 degrees of freedom can of course propagate in what we called previously bulk space, and can change slightly our previous picture. The relation between the corresponding D9 and D5 gauge couplings is

$$g_9^2/g_5^2 = V_\perp \quad , \tag{4}$$

where $V_\perp$ denotes here (before T-dualities) the compact volume perpendicular to the D5 brane. If $V_\perp >> 1$ in string units, then D5 branes live in (at least part of) the bulk and, by (4) their gauge coupling is very suppressed compared to our (D9) gauge coupling. In particular, if $V_\perp$ in (4) is as in (3), the D5 gauge couplings are of gravitational strength. The fields in mixed 95 representations are charged under both gauge groups. Then, due to their very small gauge couplings, the D5 gauge groups manifest themselves as global symmetries on our D-brane, and could be used for protecting baryon and lepton number nonconservation processes. Indeed, global symmetries are presumably violated by nonrenormalizable operators suppressed by the fundamental scale M_I and, since M_I can be very low, we need suppression of many higher-dimensional operators.

There are clearly many challenging questions that such a scenario must answer in order to be seriously considered as an alternative to the conventional "desert picture" of supersymmetric unification at energies of the order of 10^{16} GeV. The gauge hierarchy problem still has a counterpart here, understanding the possible mm size of the compact dimensions (perpendicular to our brane) in a theory with a fundamental length (energy) in the 10^{-16} mm (TeV) range. There are several ideas concerning this issue in the literature, which however need further studies in realistic models in order to prove their viability. A serious theoretical question concerns gauge coupling unification, that in this case, if it exists, must be completely different from the conventional MSSM (Minimal Supersymmetric Standard Model) one. Moreover, there is more and more convincing evidence for neutrino masses and mixings, and the conventional picture provides an elegant explanation of their pattern via the seesaw mechanism with a mass scale of the order of the $10^{12} - 10^{15}$ GeV, surprisingly close to the usual GUT scale. Cosmology, astrophysics, accelerator physics and flavor physics put additional strong constraints on the low-scale string scenario.

2. Bulk physics: Neutrino masses with large extra dimensions

There is more and more convincing evidence for the existence of neutrino masses and mixings. Any extension of the Standard Model should therefore address this question, at least at a qualitative level. The most elegant mechanism for explaining the smallness of neutrino masses postulates the existence of right-handed neutrinos with very large Majorana masses 10^{11} GeV $\leq M \leq 10^{15}$ GeV. Via the seesaw mechanism very small neutrino masses, of the order of $m_\nu \sim v^2/M$, are generated, where $v \simeq 246$ GeV is the vev of the Higgs field. This suggests the presence of a large (intermediate or GUT) scale in the theory, related to new physics. On the other hand, low-scale string models do not have such a large scale and therefore superficially have problems to acommodate neutrino masses.

The scenario we present here is based on the observation that right-handed neutrinos can be put in the bulk of a very large (mm size) compact space [14,15], perpendicular to the brane where we live. Consider for simplicity the case of one family of neutrinos. The model consists of our brane with the left-handed neutrino ν_L and Higgs field confined to it and one bulk Dirac neutrino, $\Psi = (\psi_1, \bar{\psi}_2)^T$ in Weyl notation, invading a space with (again for simplicity) one compact perpendicular direction y. The compact direction is taken here to be an orbifold S^1/Z_2, since as is well known circle compactifications are not phenomenologically realistic. The Z_2 orbifold acts on the spinors as $Z_2\Psi(y) = \pm\gamma_5\Psi(-y)$, so that one of the two-component Weyl spinors, ψ_1, is even under the Z_2 action $y \to -y$, while the other spinor ψ_2 is odd. If the left-handed neutrino ν_L is restricted to a brane located at the orbifold fixed point $y = 0$, ψ_2 vanishes at this point and so ν_L couples only to ψ_1. This then results in a Lagrangian of the form

$$L = -\frac{1}{2} \int d^4x \, dy \, M_s \left\{ \bar{\psi} i \bar{\gamma}^M \partial_M \psi - \partial_M \bar{\psi} i \bar{\gamma}^M \psi \right\}$$
$$- \int d^4x \left\{ \bar{\nu}_L i \bar{\sigma}^\mu D_\mu \nu_L + (\hat{m} \nu_L \psi_1 |_{y=0} + \text{h.c.}) \right\} . \qquad (5)$$

Here M_s is the mass scale of the higher-dimensional fundamental theory (a reduced Type I string scale) and the spacetime index M runs over all five dimensions: $x^M \equiv (x^\mu, y)$. The first line describes the kinetic-energy term for the 5d Ψ field, while the second line describes the kinetic energy of the 4d two-component neutrino field ν_L, as well as the coupling between ν_L and ψ_1. Note that in 5d, a bare Dirac mass term for Ψ would not have been invariant under the action of the Z_2 orbifold, since $\bar{\Psi}\Psi \sim \psi_1 \psi_2 +$ h.c.

Now compactify the Lagrangian (5) down to 4d, expanding the 5d Ψ field in Kaluza-Klein modes. The orbifold relations $\psi_{1,2}(-y) = \pm \psi_{1,2}(y)$ imply that the Kaluza-Klein decomposition takes the form

$$\psi_1(x,y) = \frac{1}{\sqrt{2\pi R}} \sum_{n=0}^{\infty} \psi_1^{(n)}(x) \cos(ny/R) ,$$
$$\psi_2(x,y) = \frac{1}{\sqrt{2\pi R}} \sum_{n=1}^{\infty} \psi_2^{(n)}(x) \sin(ny/R) . \qquad (6)$$

However, a more general possibility emerges naturally from the Scherk-Schwarz compactification. Recall that our original 5d Dirac spinor field Ψ is decomposed in the Weyl basis as $\Psi = (\psi_1, \bar{\psi}_2)^T$, where ψ_1 and ψ_2 have the mode expansions given in (6). Let us consider performing a local rotation in (ψ_1, ψ_2) space of the form

$$\begin{pmatrix} \hat{\psi}_1 \\ \hat{\psi}_2 \end{pmatrix} \equiv U \begin{pmatrix} \psi_1 \\ \psi_2 \end{pmatrix} \qquad \text{where} \quad U \equiv \begin{pmatrix} \cos(\omega y/R) & -\sin(\omega y/R) \\ \sin(\omega y/R) & \cos(\omega y/R) \end{pmatrix} , \qquad (7)$$

with ω an (for the moment) arbitrary real number. The effect of the matrix U in (7) is to twist the fermions after a $2\pi R$ rotation on y. Such twisted boundary conditions, as we have seen, are allowed in field and in string theory if the higher-dimensional theory has a suitable $U(1)$ symmetry. The 4d Lagrangian of the component fields coming from the 5d Lagrangian is found from (5) by replacing everywhere $\psi_i \to \hat{\psi}_i$.

For convenience, let us define $M_0 = \omega/R$ and the linear combinations $N^{(n)} \equiv (\psi_1^{(n)} + \psi_2^{(n)})/\sqrt{2}$ and $M^{(n)} \equiv (\psi_1^{(n)} - \psi_2^{(n)})/\sqrt{2}$ for all $n > 0$. Then integrating over the compactified dimension yields We also define in the

174

following the effective Dirac neutrino mass couplings

$$m \equiv \frac{\hat{m}}{\sqrt{2}\sqrt{\pi M_s R}} \ . \tag{8}$$

In the Lagrangian (5), the Standard-Model neutrino ν_L mixes with the entire tower of Kaluza-Klein states of the higher-dimensional Ψ field. Indeed, if for simplicity we restrict our attention to the case of only one extra dimension, define

$$N^T \equiv (\nu_L, \psi_1^{(0)}, N^{(1)}, M^{(1)}, N^{(2)}, M^{(2)}, ...) \ , \tag{9}$$

and integrate over the compactified dimension, we see that the mass terms in the Lagrangian (5) take the form $(1/2)(N^T M N + \text{h.c.})$, where the mass matrix is

$$M \ = \ \begin{pmatrix} 0 & m & m & m & m & m & \cdots \\ m & M_0 & 0 & 0 & 0 & 0 & \cdots \\ m & 0 & M_0 + 1/R & 0 & 0 & 0 & \cdots \\ m & 0 & 0 & M_0 - 1/R & 0 & 0 & \cdots \\ m & 0 & 0 & 0 & M_0 + 2/R & 0 & \cdots \\ m & 0 & 0 & 0 & 0 & M_0 - 2/R & \cdots \\ \vdots & \vdots & \vdots & \vdots & \vdots & \vdots & \ddots \end{pmatrix} \ . \tag{10}$$

Let us start for simplicity by disregarding the possible bare Majorana mass term, setting $M_0 = 0$. In this case, the characteristic polynomial which determines the eigenvalues λ of the mass matrix (10) can be worked out exactly and takes the form

$$\left[\prod_{k=1}^{\infty} \left(\frac{k^2}{R^2} - \lambda^2 \right) \right] \left[\lambda^2 - m^2 + 2\lambda^2 m^2 R^2 \sum_{k=1}^{\infty} \frac{1}{k^2 - \lambda^2 R^2} \right] \ = \ 0 \ , \tag{11}$$

clearly invariant under $\lambda \to -\lambda$. From this we immediately see that all eigenvalues fall into *degenerate*, pairs of opposite sign. In order to solve this eigenvalue equation, it is convenient to note that $\lambda = k/R$ is never a solution (unless of course $m = 0$), as the cancellation that would occur in the first factor in (11) is offset by the divergence of the second factor. We are therefore free to disregard the first factor entirely, and focus on solutions for which the second factor vanishes. The summation in second factor can be performed exactly, resulting in the transcendental equation

$$\lambda R \ = \ \pi (mR)^2 \cot(\pi \lambda R) \ . \tag{12}$$

All the eigenvalues can be determined from this equation, as functions of the product mR. This equation can be analyzed graphically [14], and in the limit $mR \to 0$ (corresponding to $m \to 0$), the eigenvalues are k/R, $k \in Z$, with a double eigenvalue at $k = 0$. Conversely, in the limit $mR \to \infty$, the eigenvalues with $k > 0$ shift smoothly toward $(k + 1/2)/R$, while those with $k < 0$ shift smoothly toward $(k - 1/2)/R$. Finally, the double zero eigenvalue splits toward the values $\pm 1/(2R)$. The overlap between the light mass eigenstates and the neutrino gauge eigenstate is generically less than half in this scenario. The important prediction of this scenario is that the gauge neutrino and the (lightest) sterile neutrino are degenerate in mass, a possibility that can be experimentally tested.

Let us now return to the more general case $M_0 \neq 0$. To this end, it is useful to define

$$k_0 \equiv [M_0 R], \qquad \epsilon \equiv M_0 - \frac{k_0}{R}, \tag{13}$$

where $[x]$ denotes here the integer nearest to x. Thus, ϵ is the smallest diagonal entry in the mass matrix (10), corresponding to the excited Kaluza-Klein state $M^{(k_0)}$. In other words, $\epsilon \equiv M_0$ (modulo R^{-1}) satisfies $-1/(2R) < \epsilon \leq 1/(2R)$. The remaining diagonal entries in the mass matrix can then be expressed as $\epsilon \pm k'/R$, where $k' \in Z^+$. Reordering the rows and columns of our mass matrix, we can therefore cast it into the form

$$M = \begin{pmatrix} 0 & m & m & m & m & m & \cdots \\ m & \epsilon & 0 & 0 & 0 & 0 & \cdots \\ m & 0 & \epsilon + 1/R & 0 & 0 & 0 & \cdots \\ m & 0 & 0 & \epsilon - 1/R & 0 & 0 & \cdots \\ m & 0 & 0 & 0 & \epsilon + 2/R & 0 & \cdots \\ m & 0 & 0 & 0 & 0 & \epsilon - 2/R & \cdots \\ \vdots & \vdots & \vdots & \vdots & \vdots & \vdots & \ddots \end{pmatrix} . \tag{14}$$

While this is of course nothing but the original mass matrix (10), the important consequence of this rearrangement is that the *heavy* mass scale M_0 has been replaced by the *light* mass scale ϵ. Unlike M_0, we see that $|\epsilon| \sim O(R^{-1})$. Thus, the heavy Majorana mass scale M_0 completely *decouples* from the physics. Indeed, the value of M_0 enters the results only through its determinations of k_0 and the precise value of ϵ. Therefore, interestingly enough, the presence of the infinite tower of regularly-spaced Kaluza-Klein states ensures that only the value of M_0 modulo R^{-1} plays a role.

The easiest way to solve (14) for the eigenvalues $\lambda_\pm$ is to integrate out the Kaluza-Klein modes. It turns out that there are two relevant cases to consider, depending on the value of ϵ. If $|\epsilon| \gg m$ (which can arise when

$mR \ll 1$), *all* of the Kaluza-Klein modes are extremely massive relative to m, and we can integrate them out to obtain an effective $\nu_L \nu_L$ mass term of size

$$|\epsilon| \gg m : \quad m_\nu = m^2/\epsilon + m^2 \sum_{k'=1}^{\infty} \left(\frac{1}{\epsilon + k'/R} + \frac{1}{\epsilon - k'/R} \right) = \pi m^2 R \cot(\pi R \epsilon) .$$

(15)

We shall discuss the special case $\epsilon = 1/(2R)$ later on. Alternatively, if $|\epsilon| \not\gg m$, the lightest Kaluza-Klein mode $M^{(k_0)}$ should not be integrated out, and the end result is an effective $\nu_L \nu_L$ mass term of size μ, where

$$|\epsilon| \not\gg m : \quad \mu \equiv -m^2 \sum_{k'=1}^{\infty} \left(\frac{1}{\epsilon + k'/R} + \frac{1}{\epsilon - k'/R} \right) = \frac{m^2}{\epsilon} - \pi m^2 R \cot(\pi R \epsilon) .$$

(16)

Note that $\mu \to 0$ smoothly as $\epsilon \to 0$, with μ otherwise of size $O(m^2 R)$. Diagonalizing the final 2×2 mass matrix mixing ν_L and $M^{(k_0)}$ in the presence of this mass term then yields

$$|\epsilon| \not\gg m : \qquad \lambda_\pm = \frac{1}{2} \left[(\mu + \epsilon) \pm \sqrt{(\mu - \epsilon)^2 + 4m^2} \right] .$$

(17)

We therefore conclude that, although we may have started with a bare Majorana mass $M_0 >> R^{-1}$, in all cases the final neutrino mass remains of order $m^2 R$. Even though we might have expected a neutrino mass of order m^2/M_0 from the mixing between ν_L and the original zero-mode $\psi_1^{(0)}$, the contribution m^2/M_0 from the zero-mode is completely canceled by the summation over the Kaluza-Klein tower, while the seesaw between ν_L and $M^{(k_0)}$ becomes dominant.

In string theory, however, there are additional topological constraints that permit only *discrete* values of ω. Taking $\omega = 1/2$ then implies $\psi_{1,2}(2\pi R) = -\psi_{1,2}(0)$, which shows that lepton number is broken globally (although not locally) as the spinor is taken around the compactified space. It is straightforward to show that when $\epsilon = 1/2R$, the characteristic eigenvalue equation $\det(M - \lambda I) = 0$ for the mass matrix (10),(14) becomes

$$\lambda R \left[\prod_{k=1}^{\infty} (\lambda^2 R^2 - (k - \tfrac{1}{2})^2) \right] \left[1 - 2m^2 R^2 \sum_{k=1}^{\infty} \frac{1}{\lambda^2 R^2 - (k - 1/2)^2} \right] = 0 .$$

(18)

This has an exact trivial solution $\lambda = 0$, corresponding to an exactly massless neutrino. Indeed, the characteristic polynomial for the mass matrix in this case has the form

$$\lambda R = -\pi (mR)^2 \tan(\pi \lambda R) .$$

(19)

It is then clear than the zero eigenvalue is always present, irrespective of the value of the radius. In fact, by changing the value of M_0, we see that it is possible to smoothly *interpolate* between the scenario with $M_0 = 0$ and the scenario we are discussing here [14]. This also provides another explanation of why only the value $\epsilon \sim M_0$ (modulo R^{-1}) is relevant physically. The regular, repeating aspect of the infinite towers of Kaluza-Klein states is now manifested graphically in the periodic nature of the cotangent function.

The scenario(s) presented have also other interesting consequences. The neutrino eigenstate can now oscillate into an infinite tower of right-handed KK neutrinos with a probability that can be reliably estimated and experimentally tested. Moreover, even if in the last scenario presented the physical neutrino is massless, its probability of oscillation into the tower of KK states is nonvanishing. In particular, a neutrino mass difference $\Delta m \sim 10^{-2} eV$, that fits the experimental data, could well be explained by an oscillation of the massless neutrino into the first KK state, for a radius $R^{-1} \sim 10^{-2} eV$, precisely in the sub-mm region we are interested in ! Finally, let us mention that recent SNO data strongly disfavors oscillations of an active neutrino into a sterile one. The scenario mentioned above involves oscillations into several sterile neutrinos and, as such, is not directly excluded by the recent data. More effort is needed, however, in order to check the viability of this scenario in light of the new solar neutrino data.

There is another way to break lepton number, similar but slightly different than the breaking by compactification, by using standard Majorana masses M_0. Here there are two possibilities, depending if the Majorana masses are brane localized or of bulk type. In the first case, by considering again only one generation and by denoting by m_D the Dirac neutrino mass, we find, analogously to the previous examples, an eigenvalue equation

$$\lambda, \tan(\pi \lambda R) = \pi R \ (m_D^2 - \lambda M_0) \ . \tag{20}$$

Here, in the relevant limit $M_0 >> m_D$, we find the standard seesaw formula $\lambda_0 \sim m_D^2/M_0$ for the lowest mass eigenvalue. Notice that, even if both M_0 and m_D are volume suppressed due to the large wave function spreading in the extra dimension, their ratio m_D^2/M_0 is volume independent. This possibility realizes therefore the standard seesaw mechanism.

If the Majorana masses are of bulk type, in the case of "vector" type bulk masses [16] we find the eigenvalue equation

$$\lambda \tan[\pi(M_0 - \lambda)R] = -\pi R m_D^2 \ . \tag{21}$$

Here, as explained before, M_0 only matters modulo $1/R$, and therefore the lighest state has always a mass of the order of $1/R$.

For more than one relevant dimension, the active/sterile mixing is not dominated by the lightest KK modes and the mixing with the heaviest modes does not lead to oscillations. In this case, after integrating out heavy KK modes, we generically get a dimension-six operator of the type

$$\epsilon_{ij} G_F (H^+ \bar{L}_i)(i\, \gamma^\mu \partial_\mu)(H \bar{L}_j) \ , \tag{22}$$

generating flavour-changing transitions which allow for more experimental signatures in such models. This and also more phenomenological aspects of these models can be found, e.g. in [17].

In the context of warped compactifications, there are new ways of getting small neutrino masses, by using various ways of localizing bulk fields. For example, in the Randall-Sundrum scenario, if the Standard Model is localized on the TeV brane and the sterile neutrino is a bulk field, by adding a bulk mass term of the type $m\epsilon(y)\bar{\Psi}\Psi$ the right-handed neutrino acquires a wave-function profile $\nu_R(y) \sim \exp(k - 2m)|y|$, where k is the AdS radius. If $m > k/2$, the right-handed neutrino is localised on the Planck brane and the wave-function overlap between the left and right-handed neutrino produce very small Dirac neutrino masses [18].

In GUT orbifold models, where the fundamental scale is close to the standard MSSM unification scale $2 \times 10^{16} GeV$ and the compact radii are of the order of $10^{14} - 10^{15} GeV$, the simplest way to get realistic neutrino masses is by adding bulk sterile neutrinos with large Majorana masses. Then, as described previously, the relevant seesaw scale is the compactification scale and the resulting neutrino masses $m_\nu \sim m_D^2 / R^{-1}$ are of the right order of magnitude [19].

3. Multiple seesaw in low scale strings

Let us consider one active neutrino (let us say the electron one) in a theory whose fundamental mass scale M is low, of the order of $10 - 100 \ TeV$. A successful generation of neutrino masses ask in this case for two conditions. The first is to forbid the operator responsible for the seesaw mass $(1/M) \ HH\nu_L\nu_L$. The second is to generate a small neutrino mass by some other mechanism.

A minimal proposal proposed in [20] involves a minimal set of two singlet fermions Ψ_1, Ψ_2 and two $SU(2)_L$ doublets Ψ_4, Ψ_3 of hypercharges $Y = \pm 1/2$, respectively. The symmetries of the model, whose discussion (in a particular realisation) is postponed for the next section, give mass terms in the lagrangian of the form

$$\frac{1}{2} \, V \, M \, V^T \, , \tag{23}$$

where the vector V denotes the collection of fermionic fields $V = (\nu_L, \Psi_1, \Psi_2, \Psi_3, \Psi_4)$. The mass matrix M in (23) we are interesting in is given by

$$M = \begin{pmatrix} 0 & m_1 & 0 & 0 & 0 \\ m_1 & M_1 & M_2 & m_2 & m_3 \\ 0 & M_2 & 0 & m_4 & m_5 \\ 0 & m_2 & m_4 & 0 & M_3 \\ 0 & m_3 & m_5 & M_3 & 0 \end{pmatrix} \, . \tag{24}$$

The entries m_i denote electroweak-type mass terms, given by the vacuum expectation value(s) of the Higgs field(s), whereas the capital letters M_a denote Majorana mass terms (independent of the electroweak scale), generically of the order of the fundamental scale, in our case assumed to be of the order $10-100 \, TeV$. The eigenvalues/eigenvectors of the mass matrix (24) are obtained rather easily in the relevant approximation $M_a >> m_i$. In this limit, the mass matrix has a block-diagonal form of a 3×3 matrix times a 2×2 mass matrix. There are four heavy eigenstates of mass approximately given by

$$\lambda_{2,3} \simeq \frac{1}{2} \, (M_1 \pm \sqrt{M_1^2 + 4M_2^2}) \, ,$$
$$\lambda_{4,5} \simeq \pm M_3 \, . \tag{25}$$

In this block-diagonal limit, the gauge neutrino stays massless, despite the presence of the Majorana mass term $M_1 \Psi_1 \Psi_1$ in (24). This is easily explained by considering the 3×3 upper block in (24), which has zero determinant. Indeed, by inverting the 2×2 matrix block for the singlets Ψ_1, Ψ_2

$$M_2 = \begin{pmatrix} M_1 & M_2 \\ M_2 & 0 \end{pmatrix} \, , \tag{26}$$

we get in the seesaw diagram entries in the singlet propagators of the type $S_{12}(p=0) = (1/M_2)$ and $S_{22}(p=0) = (M_1/M_2)^2$, where p is the momentum flowing into the fermion propagator. The main point here is that the $S_{11}(p=0)$ entry *is zero*. Consequently, the usual seesaw diagram obtained by integrating out the heavy Ψ_1 field is forbidden. The neutrino mass appears only by taking into account the small 3×2 block off-diagonal terms in

(24). In this case, the light eigenvalue is most easily obtained by computing the determinant of the matrix (24)

$$\det M \;=\; -2\, m_4\, m_5\, m_1^2\, M_3 \;. \tag{27}$$

By combining (25) and (27), we obtain the value of the light neutrino eigenstate, which in our simplified model is mostly given by the electron neutrino

$$m_\nu = \lambda_1 \simeq \frac{2m_1^2 m_4 m_5}{M_2^2 M_3} \;. \tag{28}$$

The final result (28) can be easily understood as a result of a multiple seesaw-type diagram obtained by integrating out the heavy fields $\Psi_1, \Psi_2, \Psi_3, \Psi_4$ and inserting the appropriate Higgs vev(s) (see the figure). The first step is to integrate out the weak fermion doublets ψ_3, ψ_4. The Majorana mass matrix (26) is then modified to

$$M_2 \;=\; \begin{pmatrix} M_1 & M_2 \\ M_2 & \frac{2m_4 m_5}{M_3} \end{pmatrix} \;, \tag{29}$$

The second step involves integrating out the singlet ψ_1, ψ_2 fermion fields in the remaining 3×3 mass matrix

$$M_2 \;=\; \begin{pmatrix} 0 & m_1 & 0 \\ m_1 & M_1 & M_2 \\ 0 & M_2 & \frac{2m_4 m_5}{M_3} \end{pmatrix} \;, \tag{30}$$

leading to the eiganvalues $\lambda_{2,3}$ in (25) and $\lambda_1 = m_\nu$ in (28).

The lagrangian leading to the mass matrix (24) breaks the $U(1)$ lepton number which we define to act on the various fields as

$$\nu_L \to e^{i\alpha}\nu_L \;, \psi_1 \to e^{-i\alpha}\psi_1 \;,\; \psi_2 \to e^{i\alpha}\psi_2 \;,$$
$$\psi_3 \to e^{-i\alpha}\psi_3 \;,\; \psi_4 \to e^{-i\alpha}\psi_4 \;. \tag{31}$$

The breaking of the lepton number (by $\Delta L = 2$) is due to the Majorana mass parameters M_1 and M_3. However, only M_3 is relevant for the neutrino mass generation. The smalness of the neutrino mass (28) is then understood by the fact ψ_3 and ψ_4 do not directly couple to the gauge neutrino. The transmission of the lepton number breaking goes therefore also through the heavy fermions ψ_1, ψ_2 and generated a cubic supression in the heavy masses (28).

We can as well include the other (muon and tau) neutrinos. With the minimal field content (24) only one neutrino linear combination acquires a mass at tree level. In order to give a tree-leel mass of the type (28) for a second neutrino, we need a second pair of singlet fields ψ_5, ψ_6 and a second pair of weak fermion doublets ψ_7, ψ_8. The second gauge neutrino couples to ψ_1 and ψ_5, but not to ψ_2 and ψ_6. The need of doubling the exotic fermion spectrum can be easily shown by following the integrating out procedure outlined above. Indeed, first of all we need to double the singlet fermion content in order to provide two (linear combinations of) gauge neutrinos to couple to two different singlets. On the other hand, by integrating out the weak doublets we generate the Majorana mass matrix for the singlet fields. With only one pair of weak doublets, the Majorana mass matrix for the singlets will have zero eigenvalues and as a result one zero mass eigenstate will survive in the physical spectrum. Adding a second pair of weak doublets will·solve this problem by generating large Majorana masses for all singlets $\psi_1, \psi_2, \psi_5, \psi_6$.

There exist already in the literature other mechanisms for getting small neutrino masses in models with low energy fundamental scale [14]. All of mechanisms studied in [14] take advantage of the presence of large extra dimensions in models with a low string scale. The mechanism we put forward in our paper does not rely crucially on large extra dimensions. Its phenomenology is different in nature from the previous ones and future experiments can easily distinguish it from other mechanisms.

The model just presented was based on a minimal field content and a specific mass matrix (24), which contained a certain number of important zeros. Their presence must be assured by some symmetry of the miscroscopic theory. An explicit example based on an abelian symmetry which provides precisely the required zero structure can be found in [20].

Acknowledgments

I would like to thank the organizers of Seesaw25 for providing a stimulating atmosphere. Work supported in part by the RTN European Program HPRN-CT-2000-00148.

References

1. M. Gell-Mann, P. Ramond and R. Slansky, in *Supergravity* (P. van Nieuwenhuisen and D.Z. Freedman eds.), North-Holland, Amsterdam, 1979; T. Yanagida, in *Proceedings of the Workshop on Unified Theories and Baryon Number in the Universe* (A. Sawada and A. Sugamoto eds.), KEK preprint 79-18, 1979; R. N. Mohapatra and G. Senjanovic, Phys. Rev. Lett. **44** (1980) 912.

2. E. Witten, *Nucl. Phys.* **B471** (1996) 135; J.D. Lykken, *Phys. Rev.* **D54** (1996) 3693.

3. N. Arkani-Hamed, S. Dimopoulos and G. Dvali, *Phys. Lett.* **B429** (1998) 263; *Phys. Rev.* **D59** (1999) 086004.

4. K.R. Dienes, E. Dudas and T. Gherghetta, *Phys. Lett.* **B436** (1998) 55; *Nucl. Phys.* **B537** (1999) 47; hep-ph/9807522.

5. I. Antoniadis, N. Arkani-Hamed, S. Dimopoulos and G. Dvali, *Phys. Lett.* **B436** (1998) 257.

6. V.A. Rubakov and M.E. Shaposhnikov, *Phys. Lett.* **B125** (1983) 139 and *Phys. Lett.* **B125** (1985) 372; K. Akama, in Lecture Notes in Physics, 176, 1982, K. Kikkawa (N. Nakanishi and H. Nariai eds.), Springer-Verlag, page 267 and *Prog. Theor. Phys.* **78** (1987) 184; M. Visser, *Phys. Lett.* **B159** (1985) 22; P. Horava and E. Witten, *Nucl. Phys.* **B460** (1996) 506; *Nucl. Phys.* **B475** (1996) 94.

7. I. Antoniadis, Phys. Lett. B **246** (1990) 377; I. Antoniadis, K. Benakli and M. Quiros, Phys. Lett. B **331** (1994).

8. G.F. Giudice, R. Rattazzi and J.D. Wells, *Nucl. Phys.* **B544** (1999) 3; E.A. Mirabelli, M. Perelstein and M.E. Peskin, *Phys. Rev. Lett.* **82** (1999) 2236; T. Han, J.D. Lykken and R.J. Zhang, *Phys. Rev.* **D59** (1999) 105006.

9. J.C. Long, H.W. Chan and J.C. Price, *Nucl. Phys.* **B539** (1999) 23.

10. E. Dudas and J. Mourad, Nucl. Phys. B **575** (2000) 3 ; S. Cullen, M. Perelstein and M. E. Peskin, Phys. Rev. D **62** (2000) 055012 ; I. Antoniadis, K. Benakli and A. Laugier, JHEP **0105**, 044 (2001) [arXiv:hep-th/0011281].

11. A. Sagnotti, in: Cargese '87, Non-Perturbative Quantum Field Theory, eds. G. Mack et al. (Pergamon Press, Oxford, 1988) p. 521; G. Pradisi and A. Sagnotti, *Phys. Lett.* **B216** (1989) 59; M. Bianchi, G. Pradisi and A. Sagnotti, *Nucl. Phys.* **B376** (1992) 365.

12. T.R. Taylor and G. Veneziano, *Phys. Lett.* **B212** (1988) 147.

13. D. Ghilencea and G.G. Ross, *Phys. Lett.* **B442** (1998) 165 and Nucl. Phys. B **569** (2000) 391; Z. Kakushadze, *Nucl. Phys.* **B548** (1999) 205; Z. Kakushadze and T.R. Taylor, Nucl. Phys. B **562** (1999) 78.

14. K.R. Dienes, E. Dudas and T. Gherghetta, *Nucl. Phys.* **B557** (1999) 25 .

15. S. Dimopoulos, talk at SUSY98 conference (July 1998); N. Arkani-Hamed, S. Dimopoulos, G. Dvali and J. March-Russell, Phys. Rev. D **65** (2002) 024032.

16. A. Lukas, P. Ramond, A. Romanino and G. G. Ross, Phys. Lett. B **495**, 136 (2000) [arXiv:hep-ph/0008049].

17. A. De Gouvea, G. F. Giudice, A. Strumia and K. Tobe, Nucl. Phys. B **623**, 395 (2002) [arXiv:hep-ph/0107156] ; G. Bhattacharyya, H. V. Klapdor-Kleingrothaus, H. Paes and A. Pilaftsis, Phys. Rev. D **67** (2003) 113001 [arXiv:hep-ph/0212169] ; A. Ioannisian and A. Pilaftsis, Phys. Rev. D **62** (2000) 066001 [arXiv:hep-ph/9907522].

18. Y. Grossman and M. Neubert, Phys. Lett. B **474** (2000) 361 [arXiv:hep-ph/9912408].

19. H. D. Kim and S. Raby, JHEP **0307** (2003) 014 [arXiv:hep-ph/0304104].

20. E. Dudas and C. A. Savoy, Acta Phys. Polon. B **33**, 2547 (2002) [arXiv:hep-ph/0205264].

COSMOLOGICAL BOUNDS ON MASSES OF NEUTRINOS AND OTHER THERMAL RELICS

STEEN HANNESTAD

*Physics Department, University of Southern Denmark, Campusvej 55,
DK-5230 Odense M, Denmark, E-mail: hannestad@fysik.sdu.dk*

With the advent of precision data, cosmology has become an extremely powerful
tool for probing particle physics. The prime example of this is the cosmological
bound on light neutrino masses. Here I review the current status of cosmological neutrino mass bounds as well as the various uncertainties involved in deriving
them. From WMAP, SDSS, and Lyman-α forest data an upper bound on the sum
of neutrino masses of 0.65 eV (95% C.L.) can be derived without any assumptions
about bias. I also present new limits on other light, thermally produced particles. For example, a hypothetical new Majorana fermion decoupling around the
electroweak phase transition must have $m \lesssim 5$ eV.

1. Introduction

Neutrinos are the among the most abundant particles in the universe. This
means that they have a profound impact on many different aspects of cosmology, from the question of leptogenesis in the very early universe, over
big bang nucleosynthesis, to late time structure formation.

At late times $(T \lesssim T_{\mathrm{EW}})$ neutrinos mainly influence cosmology because
of their energy density and, even later, their mass.

The absolute value of neutrino masses are very difficult to measure experimentally. On the other hand, mass differences between neutrino mass
eigenstates, (m_1, m_2, m_3), can be measured in neutrino oscillation experiments.

The combination of all currently available data suggests two important
mass differences in the neutrino mass hierarchy. The solar mass difference of
$\delta m_{12}^2 \simeq 8 \times 10^{-5}$ eV^2 and the atmospheric mass difference $\delta m_{23}^2 \simeq 2.6 \times 10^{-3}$
eV^2 [1,2,3,4].

In the simplest case where neutrino masses are hierarchical these results
suggest that $m_1 \sim 0$, $m_2 \sim \delta m_{\mathrm{solar}}$, and $m_3 \sim \delta m_{\mathrm{atmospheric}}$. If the hierarchy is inverted [5,6,7,8,9,10] one instead finds $m_3 \sim 0$, $m_2 \sim \delta m_{\mathrm{atmospheric}}$,

and $m_1 \sim \delta m_{\text{atmospheric}}$. However, it is also possible that neutrino masses are degenerate [11,12,13,14,15,16,17,18,19,20,21], $m_1 \sim m_2 \sim m_3 \gg \delta m_{\text{atmospheric}}$, in which case oscillation experiments are not useful for determining the absolute mass scale.

Experiments which rely on kinematical effects of the neutrino mass offer the strongest probe of this overall mass scale. Tritium decay measurements have been able to put an upper limit on the electron neutrino mass of 2.3 eV (95% conf.) [22]. However, cosmology at present yields a much stronger limit which is also based on the kinematics of neutrino mass.

Very interestingly there is also a claim of direct detection of neutrinoless double beta decay in the Heidelberg-Moscow experiment [23,24], corresponding to an effective neutrino mass in the $0.1 - 0.9$ eV range. If this result is confirmed then it shows that neutrino masses are almost degenerate and well within reach of cosmological detection in the near future.

Neutrinos are not the only possibility for stable eV-mass particles in the universe. There are numerous other candidates, such as axions and majorons, which might be present. As will be discussed later, the same cosmological mass bounds can be applied to any generic light particle which was once in thermal equilibrium.

Here I focus mainly on the issue of cosmological mass bounds. Much more detailed reviews of neutrino cosmology can for instance be found in [25,26].

2. Neutrino Decoupling

In the standard model neutrinos interact via weak interactions with e^+ and e^-. In the absence of oscillations neutrino decoupling can be followed via the Boltzmann equation for the single particle distribution function [27]

$$\frac{\partial f}{\partial t} - Hp\frac{\partial f}{\partial p} = C_{\text{coll}}, \tag{1}$$

where C_{coll} represents all elastic and inelastic interactions. In the standard model all these interactions are $2 \leftrightarrow 2$ interactions in which case the collision integral for process i can be written

$$C_{\text{coll,i}}(f_1) = \frac{1}{2E_1} \int \frac{d^3\mathbf{p_2}}{2E_2(2\pi)^3} \frac{d^3\mathbf{p_3}}{2E_3(2\pi)^3} \frac{d^3\mathbf{p_4}}{2E_4(2\pi)^3}$$
$$\times (2\pi)^4\delta^4(p_1 + p_2 - p_3 + p_4)\Lambda(f_1, f_2, f_3, f_4)S|M|^2_{12\to34,i} \tag{2}$$

where $S|M|^2_{12\to34,i}$ is the spin-summed and averaged matrix element including the symmetry factor $S = 1/2$ if there are identical particles in initial or final states. The phase-space factor is $\Lambda(f_1, f_2, f_3, f_4) = f_3 f_4 (1 - f_1)(1 - f_2) - f_1 f_2 (1 - f_3)(1 - f_4)$.

The matrix elements for all relevant processes can for instance be found in Ref. [28]. If Maxwell-Boltzmann statistics is used for all particles, and neutrinos are assumed to be in complete scattering equilibrium so that they can be represented by a single temperature, then the collision integral can be integrated to yield the average annihilation rate for a neutrino

$$\Gamma = \frac{16 G_F^2}{\pi^3} (g_L^2 + g_R^2) T^5, \tag{3}$$

where

$$g_L^2 + g_R^2 = \sin^4 \theta_W + (\frac{1}{2} + \sin^2 \theta_W)^2 \ \text{for} \ \nu_e$$

$$\sin^4 \theta_W + (-\frac{1}{2} + \sin^2 \theta_W)^2 \ \text{for} \ \nu_{\mu,\tau}. \tag{4}$$

This rate can then be compared with the Hubble expansion rate

$$H = 1.66 g_*^{1/2} \frac{T^2}{M_{\text{Pl}}} \tag{5}$$

to find the decoupling temperature from the criterion $H = \Gamma|_{T=T_D}$. From this one finds that $T_D(\nu_e) \simeq 2.4$ MeV, $T_D(\nu_{\mu,\tau}) \simeq 3.7$ MeV, when $g_* = 10.75$, as is the case in the standard model.

This means that neutrinos decouple at a temperature which is significantly higher than the electron mass. When $e^+ e^-$ annihilation occurs around $T \sim m_e/3$, the neutrino temperature is unaffected whereas the photon temperature is heated by a factor $(11/4)^{1/3}$. The relation $T_\nu/T_\gamma = (4/11)^{1/3} \simeq 0.71$ holds to a precision of roughly one percent. The main correction comes from a slight heating of neutrinos by $e^+ e^-$ annihilation, as well as finite temperature QED effects on the photon propagator [29,30,31,32,33,34,28,35,36,37,38,39,40,41].

3. Neutrinos in Structure Formation

Given that the neutrino temperature is known, the present contribution to the matter density from massive neutrinos is

$$\Omega_\nu h^2 = N_\nu \frac{m_\nu}{92.5 \ \text{eV}}. \tag{6}$$

Thus, even a sub-eV neutrino mass gives a significant neutrino contribution to the energy density and therefore has an effect on structure formation.

Perturbations in the neutrino distribution can be followed by solving the Boltzmann equation [42] through the epoch of structure formation. There are publicly available codes, such as CMBFAST [43], which can do this.

The main difference between neutrinos and cold dark matter is that neutrinos are light and only become non-relativistic around the epoch of recombination, significantly later than matter-radiation equality. While they are relativistic, neutrinos free-stream a distance roughly given by $\lambda = 2\pi/k = c\tau$, where τ is conformal time. The total free-streaming length is therefore roughly given by $\lambda_f \sim c\tau(T = m_\nu)$. At scales smaller than this, the solution to the Boltzmann equation is exponentially damped. For scales larger than the free-streaming length neutrino perturbations are unaffected. It is therefore possible to divide the solution into two distinct regimes: 1) $k > \tau(T = m)$: Neutrino perturbations are exponentially damped 2) $k < \tau(T = m)$: Neutrino perturbations follow the CDM perturbations. Calculating the free streaming wavenumber in a flat CDM cosmology leads to the simple numerical relation (applicable only for $T_{\text{eq}} \gg m \gg T_0$)

$$\lambda_{\text{FS}} \sim \frac{20 \text{ Mpc}}{\Omega_x h^2} \left(\frac{T_x}{T_\nu}\right)^4 \left[1 + \log\left(3.9\frac{\Omega_x h^2}{\Omega_m h^2}\left(\frac{T_\nu}{T_x}\right)^2\right)\right]. \tag{7}$$

In Fig. 1 transfer functions for various different neutrino masses in a flat ΛCDM universe ($\Omega_m + \Omega_\nu + \Omega_\Lambda = 1$) are plotted. The parameters used were $\Omega_b = 0.04$, $\Omega_{\text{CDM}} = 0.26 - \Omega_\nu$, $\Omega_\Lambda = 0.7$, $h = 0.7$, and $n = 1$.

When measuring fluctuations it is customary to use the power spectrum, $P(k,\tau)$, defined as

$$P(k,\tau) = |\delta_k(\tau)|^2. \tag{8}$$

The power spectrum can be decomposed into a primordial part, $P_0(k)$, and a transfer function $T(k,\tau)$,

$$P(k,\tau) = P_0(k)T(k,\tau). \tag{9}$$

The transfer function at a particular time is found by solving the Boltzmann equation for $\delta(\tau)$.

At scales much smaller than the free-streaming scale the present matter power spectrum is suppressed roughly by the factor [44]

$$\frac{\Delta P(k)}{P(k)} = \frac{\Delta T(k,\tau = \tau_0)}{T(k,\tau = \tau_0)} \simeq -8\frac{\Omega_\nu}{\Omega_m}, \tag{10}$$

as long as $\Omega_\nu \ll \Omega_m$. The numerical factor 8 is derived from a numerical solution of the Boltzmann equation, but the general structure of the

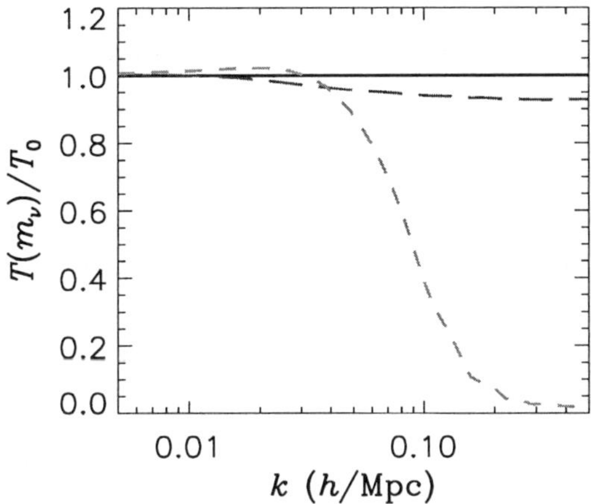

Figure 1. The transfer function $T(k, t = t_0)$ for various different neutrino masses. The solid (black) line is for $m_\nu = 0$, the long-dashed for $m_\nu = 0.3$ eV, and the dashed for $m_\nu = 1$ eV.

equation is simple to understand. At scales smaller than the free-streaming scale the neutrino perturbations are washed out completely, leaving only perturbations in the non-relativistic matter (CDM and baryons). Therefore the *relative* suppression of power is proportional to the ratio of neutrino energy density to the overall matter density. Clearly the above relation only applies when $\Omega_\nu \ll \Omega_m$, when Ω_ν becomes dominant the spectrum suppression becomes exponential as in the pure hot dark matter model. This effect is shown for different neutrino masses in Fig. 1.

The effect of massive neutrinos on structure formation only applies to the scales below the free-streaming length. For neutrinos with masses of several eV the free-streaming scale is smaller than the scales which can be probed using present CMB data and therefore the power spectrum suppression can be seen only in large scale structure data. On the other hand, neutrinos of sub-eV mass behave almost like a relativistic neutrino species for CMB considerations. The main effect of a small neutrino mass on the CMB is that it leads to an enhanced early ISW effect. The reason is that the ratio of radiation to matter at recombination becomes larger because a sub-eV neutrino is still relativistic or semi-relativistic at recombination. With the WMAP data alone it is very difficult to constrain the neutrino mass, and to achieve a constraint which is competitive with current experimental bounds it is necessary to include LSS data from 2dF or SDSS. When this is done the bound becomes very strong, somewhere in the range of 1 eV for the sum of neutrino masses, depending on assumptions about priors. This bound can be strengthened even further by including data from the Lyman-α forest and assumptions about bias. In this case the bound on the sum of neutrino masses becomes as low as 0.4-0.6 eV.

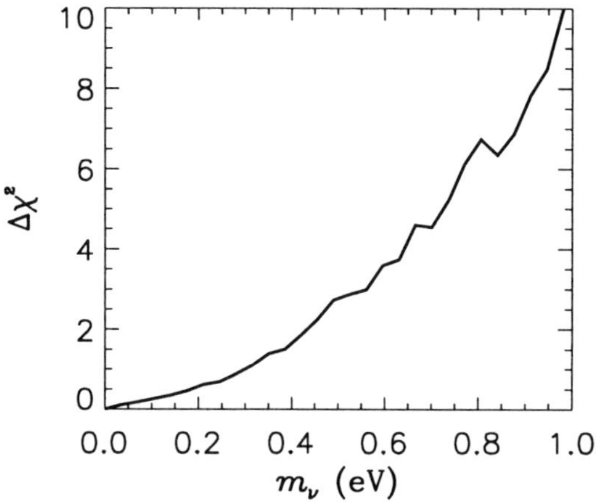

Figure 2. $\Delta\chi^2$ as a function of the the sum of neutrino masses m_ν.

In Fig. 2 $\Delta\chi^2$ is shown for an analysis which includes the WMAP CMB data [45], the SDSS galaxy survey data [46,47], the Riess *et al.* SNI-a "gold" sample [49], and the Lyman-α forest data from Croft *et al.* [50]. For the Lyman-α forest data, the error bars on the last three data points have been increased in the same fashion as was done by the WMAP collaboration [45], in order to make them compatible with the analysis of Gnedin and Hamilton [51]. In addition to the neutrino mass, which I take to be distributed in three degenerate species, I take the minimum standard model with 6 parameters: Ω_m, the matter density, Ω_b, the baryon density, H_0, the Hubble parameter, and τ, the optical depth to reionization. The normalization of both CMB, LSS, and Ly-α spectra are taken to be free and unrelated parameters. The priors used are given in Table 1.

In this particular analysis, the 95% C.L. upper bound on the sum of

Table 1. Priors on cosmological parameters used in the likelihood analysis.

Parameter	Prior	Distribution
$\Omega = \Omega_m + \Omega_X$	1	Fixed
h	0.72 ± 0.08	Gaussian [48]
$\Omega_b h^2$	0.014–0.040	Top hat
n_s	0.6–1.4	Top hat
τ	0–1	Top hat
Q	—	Free
b	—	Free

neutrino masses is 0.65 eV.

In Table 2 the present upper bound on the neutrino mass from various analyses is quoted, as well as the assumptions going into the derivation. As can be gauged from this table, a fairly robust bound on the sum of neutrino masses is at present somewhere around 0.5-1 eV, depending on the specific priors and data sets used.

Table 2. Various recent limits on the neutrino mass from cosmology and the data sets used in deriving them. 1: WMAP data, 2: Other CMB data, 3: 2dF data, 4: Constraint on σ_8 (different in 4^a, 4^b, and 4^c), 5: SDSS data, 6: Constraint on H_0, 7: Constraint from Lyman-α forest.

Ref.	Bound on $\sum m_\nu$	Data used
Spergel et al. (WMAP) [45]	0.69 eV	$1,2,3,4^a,6, 7$
Hannestad [52]	1.01 eV	1,2,3,6
Allen, Smith and Bridle [53]	$0.56^{+0.3}_{-0.26}$ eV	$1,2,3,4^b,6$
Tegmark et al. (SDSS) [47]	1.8 eV	1,5
Barger et al. [54]	0.75 eV	1,2,3,5,6
Crotty, Lesgourgues and Pastor [55]	1.0 (0.6) eV	1,2,3,5 (6)
Seljak et al. [56]	0.42 eV	$1,2,4^c,5,6,7$
The present work	0.65 eV	1,5,6,7

4. General Thermal Relics

While light neutrinos are the canonical light, thermally produced particles, there are other species which can have the same properties. At $T \sim M_{\mathrm{Pl}}$ gravitons are presumably in thermal equilibrium. However, since inflation occurs at a lower temperature, there should be no thermal graviton background present.

Other examples include majorons, axions, etc. Any such species which was once in equilibrium is characterized by only two quantities, its mass,

m_X, and the temperature at which it decoupled from thermal equilibrium, T_D. At this temperature the number of degrees of freedom was g_*. The relevant masses are $\ll$ MeV so that the particles decouple when they are relativistic, i.e. at decoupling they are characterized by a Fermi-Dirac or Bose-Einstein distribution of temperature T_D.

In the present-day universe, the new particles will be non-relativistic and contribute a matter fraction

$$\Omega_X h^2 = \frac{m_X g_X}{183 \text{ eV}} \frac{g_{*\nu}}{g_*} \times 1 \text{ (fermions)} \quad \frac{4}{3} \text{ (bosons)} \tag{11}$$

where g_X is the number of the particle's internal degrees of freedom while $g_{*\nu}$ is the effective number of thermal degrees of freedom when ordinary neutrinos freeze out with $g_{*\nu} = 10.75$ in the absence of new particles.

In the late epochs that are important for structure formation, the momentum distribution of the new particles is characterized by a thermal distribution with temperature T_X that is given by

$$\frac{T_X}{T_\nu} = \left(\frac{g_{*\nu}}{g_*} \right)^{1/3}. \tag{12}$$

In Ref. [57] CMBFAST was modified to incorporate a such thermal relics, both fermionic and bosonic.

Here I present an updated analysis for the fermionic case with $g_X = 2$ (i.e. a single Majorana fermion), using the same data sets as described above. Fig. 3 shows the 68% and 95% likelihood contours for $\Omega_X h^2$ and g_*. Overlayed are isocontours for particle masses.

As an example, a species freezing out at the electroweak transition temperature has $g_* = 106.75$, and from the figure it can be seen that the upper bound on the mass of such a particle is around 5 eV. For a corresponding scalar the mass bound will be slightly higher. For a species decoupling after the QCD phase transition where $g_* \lesssim 20$ the mass bound is roughly $m \lesssim 1$ eV.

5. Conclusion

I have reviewed the present status of cosmological mass bounds on neutrinos and other light, thermally produced particles. Already now these bounds are about an order of magnitude stronger than current laboratory limits on the neutrino mass, albeit more model dependent. In the coming years a wealth of new cosmological data will become available from such experiments as the Planck Surveyor CMB satellite. This is likely to allow

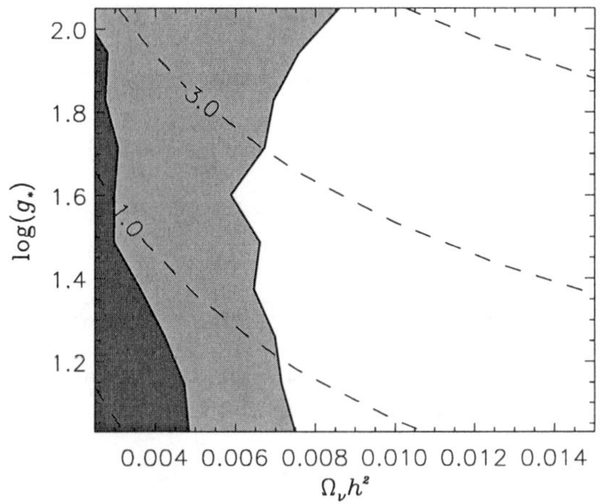

Figure 3. The full lines show 68% and 95% confidence regions in the $(\Omega_\nu h^2, g_*)$ plane for the case where the thermal relic is a fermion with $g = 2$.

for a measurement of the (sum of) neutrino masses in the 0.1 eV regime [58,59,60,61]. Together with direct detection experiments like KATRIN [62] and future neutrinoless double beta decay experiments, cosmology will provide the answer to whether neutrino masses are hierarchical.

References

1. M. Maltoni, T. Schwetz, M. A. Tortola and J. W. F. Valle, arXiv:hep-ph/0405172.
2. M. Maltoni, T. Schwetz, M. A. Tortola and J. W. F. Valle, Phys. Rev. D **68**, 113010 (2003) [arXiv:hep-ph/0309130].
3. P. Aliani, V. Antonelli, M. Picariello and E. Torrente-Lujan, arXiv:hep-ph/0309156.
4. P. C. de Holanda and A. Y. Smirnov, arXiv:hep-ph/0309299.
5. V. A. Kostelecky and S. Samuel, Phys. Lett. B **318**, 127 (1993).
6. G. M. Fuller, J. R. Primack and Y. Z. Qian, Phys. Rev. D **52**, 1288 (1995) [arXiv:astro-ph/9502081].
7. D. O. Caldwell and R. N. Mohapatra, Phys. Lett. B **354**, 371 (1995) [arXiv:hep-ph/9503316].
8. S. M. Bilenky, C. Giunti, C. W. Kim and S. T. Petcov, Phys. Rev. D **54**, 4432 (1996) [arXiv:hep-ph/9604364].
9. S. F. King and N. N. Singh, Nucl. Phys. B **596**, 81 (2001) [arXiv:hep-ph/0007243].
10. H. J. He, D. A. Dicus and J. N. Ng, arXiv:hep-ph/0203237.

11. A. Ioannisian and J. W. Valle, Phys. Lett. B **332**, 93 (1994) [arXiv:hep-ph/9402333].
12. P. Bamert and C. P. Burgess, Phys. Lett. B **329**, 289 (1994) [arXiv:hep-ph/9402229].
13. R. N. Mohapatra and S. Nussinov, Phys. Lett. B **346**, 75 (1995) [arXiv:hep-ph/9411274].
14. H. Minakata and O. Yasuda, Phys. Rev. D **56**, 1692 (1997) [arXiv:hep-ph/9609276].
15. F. Vissani, arXiv:hep-ph/9708483.
16. H. Minakata and O. Yasuda, Nucl. Phys. B **523**, 597 (1998) [arXiv:hep-ph/9712291].
17. J. R. Ellis and S. Lola, Phys. Lett. B **458**, 310 (1999) [arXiv:hep-ph/9904279].
18. J. A. Casas, J. R. Espinosa, A. Ibarra and I. Navarro, Nucl. Phys. B **556**, 3 (1999) [arXiv:hep-ph/9904395].
19. J. A. Casas, J. R. Espinosa, A. Ibarra and I. Navarro, Nucl. Phys. B **569**, 82 (2000) [arXiv:hep-ph/9905381].
20. E. Ma, J. Phys. G **25**, L97 (1999) [arXiv:hep-ph/9907400].
21. R. Adhikari, E. Ma and G. Rajasekaran, Phys. Lett. B **486**, 134 (2000) [arXiv:hep-ph/0004197].
22. C. Kraus *et al.* European Physical Journal C (2003), proceedings of the EPS 2003 - High Energy Physics (HEP) conference.
23. H. V. Klapdor-Kleingrothaus, A. Dietz, H. L. Harney and I. V. Krivosheina, Mod. Phys. Lett. A **16**, 2409 (2001) [arXiv:hep-ph/0201231].
24. H. V. Klapdor-Kleingrothaus, I. V. Krivosheina, A. Dietz and O. Chkvorets, Phys. Lett. B **586**, 198 (2004) [arXiv:hep-ph/0404088].
25. A. D. Dolgov, Phys. Rept. **370**, 333 (2002) [arXiv:hep-ph/0202122].
26. S. Hannestad, New Journal of Physics **6**, 108 (2004) [arXiv:hep-ph/0404239].
27. E. W. Kolb and M. S. Turner, "The Early Universe,", Addison-Wesley (1990).
28. S. Hannestad and J. Madsen, Phys. Rev. D **52**, 1764 (1995) [arXiv:astro-ph/9506015].
29. D. A. Dicus, E. W. Kolb, A. M. Gleeson, E. C. Sudarshan, V. L. Teplitz and M. S. Turner, Phys. Rev. D **26**, 2694 (1982).
30. N. C. Rana and B. Mitra, Phys. Rev. D **44**, 393 (1991).
31. M. A. Herrera and S. Hacyan, Astrophys. J. **336**, 539 (1989).
32. A. D. Dolgov and M. Fukugita, Phys. Rev. D **46**, 5378 (1992).
33. S. Dodelson and M. S. Turner, Phys. Rev. D **46**, 3372 (1992).
34. B. D. Fields, S. Dodelson and M. S. Turner, Phys. Rev. D **47**, 4309 (1993) [arXiv:astro-ph/9210007].
35. A. D. Dolgov, S. H. Hansen and D. V. Semikoz, Nucl. Phys. B **503**, 426 (1997) [arXiv:hep-ph/9703315].
36. A. D. Dolgov, S. H. Hansen and D. V. Semikoz, Nucl. Phys. B **543**, 269 (1999) [arXiv:hep-ph/9805467].
37. N. Y. Gnedin and O. Y. Gnedin, Astrophys. J. **509**, 11 (1998).
38. S. Esposito, G. Miele, S. Pastor, M. Peloso and O. Pisanti, Nucl. Phys. B **590**, 539 (2000) [arXiv:astro-ph/0005573].
39. G. Steigman, arXiv:astro-ph/0108148.

40. G. Mangano, G. Miele, S. Pastor and M. Peloso, arXiv:astro-ph/0111408.

41. S. Hannestad, Physical Review D **65**, 083006 (2002).

42. C. P. Ma and E. Bertschinger, Astrophys. J. **455**, 7 (1995) [arXiv:astro-ph/9506072].

43. U. Seljak and M. Zaldarriaga, Astrophys. J. **469** (1996) 437 [astro-ph/9603033]. See also the CMBFAST website at http://cosmo.nyu.edu/matiasz/CMBFAST/cmbfast.html

44. W. Hu, D. J. Eisenstein and M. Tegmark, Phys. Rev. Lett. **80** (1998) 5255 [astro-ph/9712057].

45. C. L. Bennett *et al.*, Astrophys. J. Suppl. **148** (2003) 1 [astro-ph/0302207]; D. N. Spergel *et al.*, Astrophys. J. Suppl. **148** (2003) 175 [astro-ph/0302209].

46. M. Tegmark *et al.* [SDSS Collaboration], arXiv:astro-ph/0310725.

47. M. Tegmark *et al.* [SDSS Collaboration], arXiv:astro-ph/0310723.

48. W. L. Freedman *et al.*, Astrophys. J. **553** (2001) 47 [astro-ph/0012376].

49. A. G. Riess *et al.*, Astrophys. J. **607**, 665 (2004).

50. R. A. Croft *et al.*, Astrophys. J. **581**, 20 (2002) [arXiv:astro-ph/0012324].

51. N. Y. Gnedin and A. J. S. Hamilton, arXiv:astro-ph/0111194.

52. S. Hannestad, JCAP **0305** (2003) 004 [astro-ph/0303076].

53. S. W. Allen, R. W. Schmidt and S. L. Bridle, astro-ph/0306386.

54. V. Barger, D. Marfatia and A. Tregre, arXiv:hep-ph/0312065.

55. P. Crotty, J. Lesgourgues and S. Pastor, arXiv:hep-ph/0402049.

56. U. Seljak *et al.*, astro-ph/0407372.

57. S. Hannestad and G. Raffelt, JCAP **0404**, 008 (2004).

58. S. Hannestad, Phys. Rev. D **67** (2003) 085017 [astro-ph/0211106].

59. J. Lesgourgues, S. Pastor and L. Perotto, arXiv:hep-ph/0403296.

60. M. Kaplinghat, L. Knox and Y. S. Song, astro-ph/0303344.

61. K. N. Abazajian and S. Dodelson, Phys. Rev. Lett. **91** (2003) 041301 [astro-ph/0212216].

62. See the contribution by Christian Weinheimer to the present volume.

LOW ENERGY NEUTRINO DETECTION AND DIRECT DARK MATTER SEARCH

S. SCHÖNERT

Max-Planck-Institut für Kernphysik
Saupfercheckweg 1
69117 Heidelberg, Germany
email: stefan.schoenert@mpi-hd.mpg.de

Physics goals, target materials and detection techniques for particle physics and astrophysics experiments are discussed with focus on low-energy neutrino and dark matter detection. Despite the very specific experimental requirements for the various physics goals, there exist strong technological links amongst the detection techniques as well as common challenges related to radioactive backgrounds. Emphasis in this paper is given on detectors using organic liquid scintillation, liquid noble gas and high-purity germanium diodes.

1. Introduction and focus of this paper

In the past, progress in neutrino physics has been driven mostly by experiments. Though the concept of massive neutrinos and flavor mixing has been early proposed, as well a mechanism that explains the smallness of their mass, as celebrated at this workshop, theory hardly constrained the parameter space of neutrino properties. After more than three decades of experimental study, neutrino masses and mixing are now well established with the observation of neutrino oscillations from the sun, from cosmic rays as well as from nuclear power plants. Future oscillation experiments will map with high precision the elements of the PMNS mixing matrix as well as the neutrino mass differences. Of highest priority is to measure, or further constrain, the admixture of the heaviest neutrino state to the electron described by Θ_{13}. Its observation or non-observation will decide whether the experimental study of CP properties in the lepton sector is within reach.

Yet open is the very fundamental question whether the neutrino is a Majorana particle, i.e. its own anti-particle, as required by many theories and as discussed at this conference. The observation of neutrinoless double beta decay is the most sensitive method to answer this question.

Moreover, one can derive the effective mass and probe the neutrino mass hierarchy. Recently, part of the Heidelberg-Moscow collaboration[1] claimed evidence for first observation of this decay. This claim however is discussed controversial and needs experimental scrutiny.

Beyond the study of particle properties[2], low-energy neutrinos are being used to probe the interior of the Earth, the sun and Super-Novae. At the high energy end of the experimental spectra, neutrinos will be used to study the highest energetic astrophysical objects, e.g. active galactic nuclei.

Closely related to neutrino experiments because of similar experimental techniques and challenges, is the direct search for dark matter particles. There is strong observational evidence that most of the mass density in the universe consists of dark non-baryonic matter. Unobserved weakly interacting massive particles (WIMPs) could be an explanation. Super-symmetric extensions of the standard model predict such particles.

Both neutrino and dark matter experiments encounter similar experimental problems: the event rate is extremely low thus large target masses are required; the energy deposition in the detector occurs at keV to MeV energies such that backgrounds from radioactive decays need to be strongly reduced and the residual signals actively discriminated. Despite many similarities of the experimental challenges, specific requirements apply addressing different physics objectives as itemized below:

- **Neutrinoless double beta decay (DBD):** high energy resolution to discriminate against background events; single site energy depositions versus Compton events, tracking of the decay electrons as well as tagging of the progeny isotope to discriminate neutrinoless DBD events from backgrounds.
- **Dark matter (DM) search:** low energy threshold for detection of nuclear recoils; event-by-event discrimination of nuclear recoils versus ionizing events; large target mass for the observation of annual modulations.
- **Solar neutrinos:** impurities at concentrations of μBq/m^3 for electron scattering experiments; targets mass of 10 tons or more; tag for charged current interaction.
- **Reactor neutrinos:** high performance Gd-loaded organic scintillators for tagging delayed neutrons for Θ_{13} studies with backgrounds at concentrations of mBq/m^3; unloaded scintillators for Δm_{sol} measurements;

Future experiments are confronted with a large parameter space to investigate; e.g. the effective Majorana neutrino mass can be as low as 1 meV and the WIMP interaction cross section at 10^{-10} pb or below. Therefore, most of the longer term experimental projects plan a phased approach with a step-by-step increase of target mass and reduction of the specific background event rate.

Given the context of this workshop and the limited space (and time) available, I choose not to follow the standard approach to organize the paper according to the physics goals or the experimental projects, but rather to focus on detector materials and techniques applied. This unavoidably leads to a personal bias of the projects discussed. It is intended to illustrate the underlying strategies pursued by the experimentalists and to provide some orientation for our colleagues from theory.

Though some of the target materials can be employed for different physics goals, they need a particular technical implementation to be competitive. Thus, true multi-purpose experiments which address more than one of the fundamental questions at a competitive level appear not to be at hand. However, there are 'multi-purpose target materials' which can be used for different physics goals. Since techniques to reduce internal backgrounds are material specific, they can be applied in experiments which use the same material, but address different physics goals. Table 1 lists a selection of target materials together with possible experimental implementation and the pursued physics goals. Obviously, other interesting target materials and techniques exist, however, they are not discussed here for the aforementioned reasons.

Table 1. Detector targets, detection techniques and physics goals. Org. liquid scint.: organic liquid scintillator; Scint. photons: scintillation photon detection; EL: electro-luminescence; Tag: tagging of the progeny isotope; ^{enr,nat}Ge: high purity (HP) germanium diodes enriched in ^{76}Ge; ^{nat}Ge: natural HP-germanium diodes.

Target material	Technique/Operation	Physics goals
Liquid scint. (unloaded)	Scint. photons	Solar-, SN-, geo-, reac.-ν (Δm_{sol})
(loaded)	Scint. photons	Solar-, reactor-ν (Θ_{13})
Liquefied rare gas	Scint. photons	DM, solar-ν (Θ_{sol}), DBD
	Scintillation & EL	DM
	Scint. & Charge & Tag	DBD
HP-Germanium	^{enr,nat}Ge at LN/LAr tmp.	DBD, DM
	^{nat}Ge at tens of mK tmp.	DM

2. Organic liquid scintillators

Organic liquid scintillator detectors have been built for neutrino spectroscopy at MeV and sub-MeV energies using target masses up to one kton. After years of developments, ultra purities can now be achieved if the liquid scintillators are carefully handled and purified to remove trace amounts of radioactive trace elements; concentrations of 10^{-16}g of uranium (thorium) per gram of scintillator and below have been achieved resulting in background event rates of $< 1\mu$Bq/ton. This is the bench mark value for measuring solar neutrinos via neutrino electron scattering. The main physics objectives comprise the detection of solar ^{7}Be, pep and CNO neutrinos. The pioneering experiment (c.f. Fig. 1) in this field is BOREXINO[3]. The KAMLAND[4] experiment plans to address this goals after completion of the reactor neutrino phase.

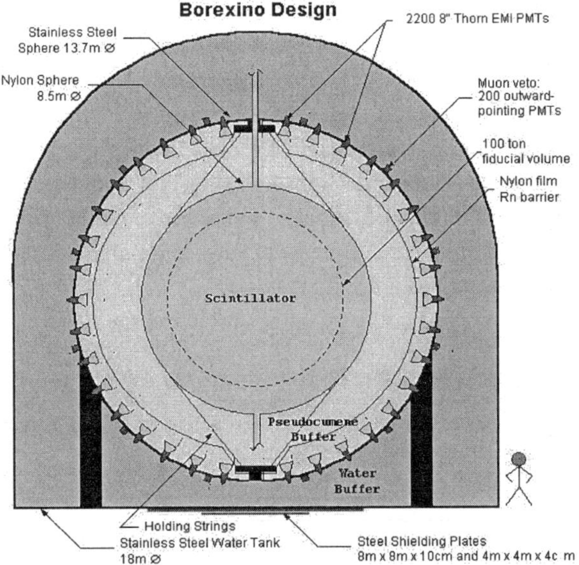

Figure 1. Schematic view of the Borexino detector at Gran Sasso. A stainless steel sphere confines the inner detector containing 300 tons of liquid scintillator and 1 kton of transparent buffer liquid.

^{14}C/^{12}C ratios at 10^{-18} as determined in the BOREXINO Counting Test Facility[5] prohibits rare event detection at energies below the 156 keV endpoint energy of the ^{14}C beta decay. Thus, high-purity organic liquid scintillators are used mainly for solar ^{7}Be, CNO and pep neutrino spectroscopy, for Super-Novae and reactor neutrino detection.

Gd-loaded scintillators are required for future reactor experiments which will probe Θ_{13} down to values of $\sin^2 2\Theta_{13} \simeq 0.03$ or even below [6,7]. The neutrons produced in $\bar{\nu}_e + p \rightarrow e^+ + n$ are captured on Gd under emission of gammas with a total energy of 8 MeV. Because of this high energy release, compared to the capture on free protons, background suppression and cut efficiency uncertainties are superior to unloaded scintillators. The challenge is to synthesize high performance Gd-loaded liquid scintillators which have high light yield, long attenuation length and are chemically stable over long periods of time[8].

3. Liquefied noble gases

Liquefied noble gases emit scintillation light from the de-excitation of the excimers. Depending on the ionization density in the liquid the singlet and triplet states are contributing differently to the scintillation light. The life time of singlet and triplet states are very different in the case of Ne and Ar indicating excellent pulse shape discrimination properties which can be applied for background suppression.

Table 2. Properties of liquefied noble gas relevant for detector design. b.p.: boiling point; ρ: density in the liquid phase; $d_{1/2}$: 1/e attenuation of 2.6 MeV photons; w.l.: wave length of scintillation peak emission; ph/MeV: scintillation photons per MeV; $\tau_{s,t}$: fluorescence decay time for singlet and triplet states.

	b.p. [K]	ρ [g/cm^3]	$d_{1/2}$ [cm]	w.l. [nm]	ph/MeV	τ_s	τ_t
He	4	0.13	202	73	22,000	N.D.	N.D.
Ne	27	1.2	22	80	15,000	13 ns	2.4 μs
Ar	87	1.4	20	128	40,000	6 ns	1.6 μs
Xe	165	3.1	9	175	42,000	4 ns	21 ns

If electro-negative impurities are removed carefully, electrons can be drifted over long distances and extracted into the gas phase where they can produce secondary electro-luminescence light. The ratio between primary light and secondary light depends on the ionization density of the interaction. This method provides an interesting discrimination technique especially for the case of xenon where pulse shape discrimination is limited. Application are under study for event-by-event discrimination for DM search in two phase liquid xenon as well as liquid argon detectors.

As evidenced in table 2 the properties of liquefied noble gas strongly vary with the atomic number. For example, to design a compact detector with good self-shielding against external radiation, a high density material as liquefied xenon is superior. On the contrary, radio-purities are superior in liquid He. In general, noble gases can be purified by distillation as well as by chromatography. Contaminations of concern are the radio-active isotopes ^{85}Kr, ^{39}Ar, ^{42}Ar and ^{222}Rn.

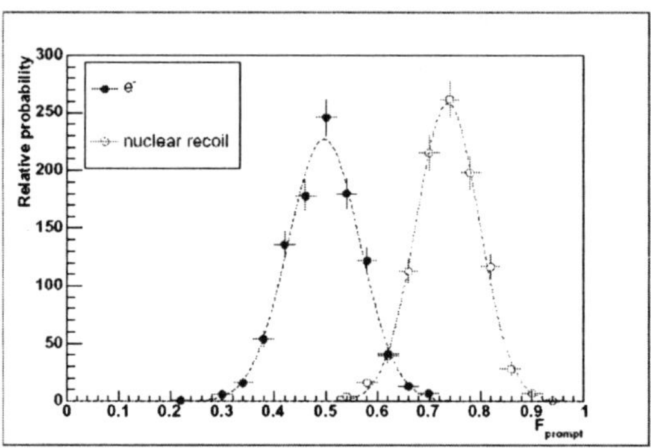

Figure 2. MC estimations of pulse shape discrimination of electrons and nuclear recoils in a large liquid neon detector.

The use of liquefied neon for the detection of dark matter, solar neutrinos as well as Super-Novae neutrinos has been proposed in the framework of the CLEAN project[9]. Given its low boiling point, neon can be readily purified from krypton, argon and radon. Recently, the sensitivity of dark matter detection has been emphasized and the potential to discriminate ionizing from nuclear recoil events by pulse shape discrimination pointed out (Fig. 2). Based on MC studies, CLEAN projects a dark matter sensitivity down to 10^{-10} pb.

A two-phase detector for dark matter detection with liquid argon has been proposed recently by the WARP collaboration[10]. The experimental concept is similar to the two-phase xenon projects. The lower cost of argon with respect to xenon permits to implement a massive active liquid argon veto system which simultaneously serves as the cryogenic bath. The scintillation photons at 128 nm are shifted to $\sim$ 440 nm and detected with bi-alkali photomultiplier submerged in liquid argon. Wavelength shifting is done by using reflector foils coated with wavelength shifting fluors. Photo electron (pe) yields of 3 pe/keV have been reported. The main challenge using argon for dark matter detection is the beta-instable isotope ^{39}Ar which can not be removed by chemical purification methods. With a specific activity of approximately 1 Bq/kg, a reduction factor of $> 10^6$ is required to achieve a sensitivity of 10^{-10} pb.

Detectors using liquid xenon are being pursued for various applications at several laboratories[11]. The present emphasis concerns the development of dark matter detectors either using single-phase (scintillation light) or dual-phase detectors (scintillation and electro luminescence), as well as the the development of new detectors for neutrinoless double-beta decay of ^{136}Xe.

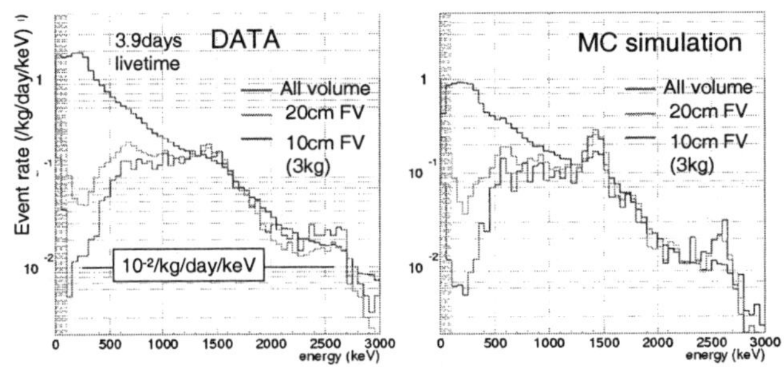

Figure 3. Data and simulations of the XMASS 100 kg prototype experiment showing the powerful concept of self shielding at low energies.

The XMASS project pursues a phased approach using single-phase detector: currently they operate a 100 kg prototype experiment with about 3 kg fiducial volume (FV) to investigate the self shielding performance, purification techniques and low-background technologies[12]. In the second phase they plan to setup a 1 ton detector (100 kg FV) with the emphasis on dark matter search, and in a third phase to operate a 20 ton detector (10 ton FV) for the spectroscopy of solar pp-neutrinos to test solar models and to improve the knowledge of the mixing angle Θ_{sol} as displayed in Fig. 4.

The XMASS experimental concept relies on self-shielding, high intrinsic purity (removal of ^{85}Kr by distillation) and on reduction of external background sources (high purity photomultiplier).

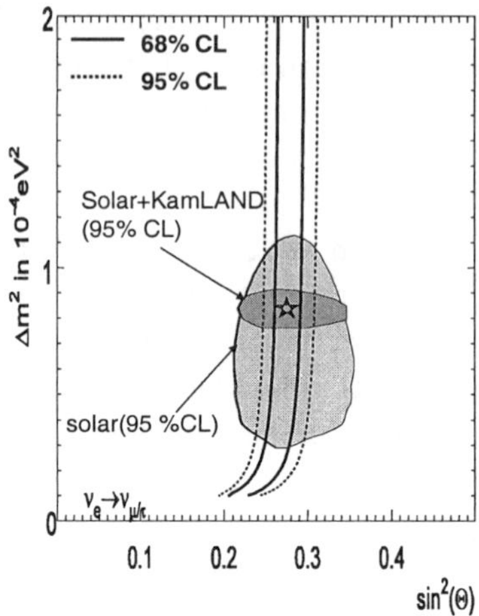

Figure 4.   Projected sensitivity for $\sin^2 2\Theta_{sol}$ derived from $\nu - e$–scattering of solar pp neutrinos in a 10 ton Xe detector after five years of data taking assuming a statistical error and standard solar model uncertainty of 1%.

Fig. 3 displays first results from the 100 kg prototype detector. It shows the efficient self shielding at low energies as expected from Monte Carlo simulations. More difficult is to shield the external 2.6 MeV gammas from the ^{208}Tl decay (thorium chain) at energy depositions close to $Q_{\beta\beta}$ of the ^{136}Xe double beta decay. Even a 20 ton xenon detector appears to be too small to reduce the external 2.6 MeV gammas coming from the state-of-the-art low-radioactivity photo multiplier tubes.

A different approach for background reduction for ^{136}Xe double beta decay is pursued by the EXO collaboration. The daughter nuclei ^{136}Ba will be detected by laser fluorescence spectroscopy. This, in principle, should discriminate all backgrounds except the 2ν double beta decay channel. High energy resolution, however, is required to separate this background channel. Simultaneous readout of scintillaton photons and charge is pursued for the energy measurement.

The XENON[13] proposal focus on dark matter search using the prompt scintillation light together with the delayed light from electro-luminescence signal after extraction of the electrons in the gas phase. A similar technique is pursued in the ZEPLIN IV project[14]. Their intensity ratio allows to discriminate ionizing events from nuclear recoil events. A schematic drawing of the XENON detector setup is displayed in Fig. 5. The collaboration plans

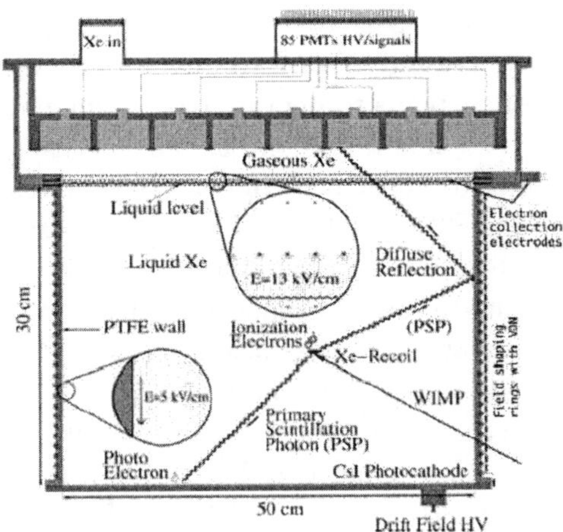

Figure 5. Schematic view of the two-phase XENON detector for dark matter search. Drifted electrons from the CsI photo cathode at the bottom of the detector together with the primary drifted charge provide a position information along the vertical axis.

to set up a farm of detectors each of them having approximately 100 kg fiducial mass.

4. Germanium projects

High-purity germanium diodes are used both in double beta decay and in dark matter experiments. The reason are the following: the double beta decay isotope ^{76}Ge, the high intrinsic purity now achievable with zone refinement techniques, and the detector performance at liquid nitrogen temperature and at 20 mK. Signals at liquid nitrogen temperatures are created by drifting electrons and holes after an ionizing event, while at mK temperatures additionally the phonon signal is used. Germanium diodes enriched in ^{76}Ge are being used for double beta decay search, while natural isotopic germanium, as well as enriched in ^{73}Ge, are used in dark matter experiments.

Two projects are under preparation using enriched ^{76}Ge for double beta decay search: the GERDA[15] experiment proposed at LNGS (Italy) and MAJORANA[16] experiment in USA/Canada. The first goal is to scrutinize the current claim for neutrinoless double beta decay, and if refuted, to improve the sensitivity continuously to reach ultimatly a sensitivity of 10 meV for the effective neutrino mass. To achieve this, the mass needs to be enlarged up to about one ton and the background reduced to 10^{-3} cts/(kev · kg · y) and below, i.e. an improvement in background reduction of at least two orders of magnitude with respect to the current experiments. The first step is to reduce the background from external sources. Here the GERDA and MAJORANA experiments pursue complementary strategies: the first plans to operate bare germanium diodes in liquid nitrogen serving as a cooling liquid and as high-purity passive shield. The latter project intends to construct copper cryostats strongly reduced in uranium, thorium and ^{60}Co made out of electro-formed copper produced underground. For the reduction of internal backgrounds, which are understood to come primarily from cosmogenic ^{60}Co and ^{68}Ge, both projects plan to implement similar techniques based on detector segmentation, pulse shape analysis and anti-coincidence methods.

The GERDA experiment will proceed in several phases. Phase I encompasses the operation of almost 20 kg of existing enriched germanium detectors, used in the past in the Heidelberg-Moscow and the IGEX experiments. Within one year of measurement, the sensitivity should allow a statistically unambiguous statement concerning neutrinoless double beta decay with a lifetime around $1.2 \cdot 10^{25}$ y, as claimed by the authors of ref[1]. Both projects are cooperating on various experimental aspects and consider to merge at a later stage to operate a common experiment with a target mass at the one ton scale.

Dark matter experiments using germanium detectors at tens of mK temperature (EDELWEISS[17], CDMS[18]) provide currently the best limits for direct dark matter searches. These experiments measure simultaneously the charge and heat signal of an energy deposition. This allows an event-by-event discrimination of ionizing and nuclear recoil events. The current limitation of sensitivity is related to fast neutrons penetrating the shielding system and inducing nuclear recoils similar to the expected dark matter signal. Ongoing improvements include the reduction of neutron backgrounds and the enlargement of the target to several tens of kg.

5. Liquid Argon Germanium hybrid detectors: LArGe

Operating bare germanium diodes for neutrinoless double beta decay search in liquid argon instead of nitrogen provides the opportunity to measure simultaneously the charge signal inside the diodes and the scintillation light of the liquid argon. Background events coming from nuclear decays deposit typically only a part of their energy inside of the diodes. Thus measuring the energy deposition in the liquid argon provides a strong suppression method. First experiments have shown the feasibility of the method [19] and showed suppression factors as expected by Monte Carlo calculations. This technique is pursued as an R&D projects by the GERDA and MAJORANA collaborations commonly. A prototype detector, the LARGE-TB is under construction for installation at LNGS. Fig. 6 displays the detector system including the expected background index related to external radiation from ^{208}Tl.

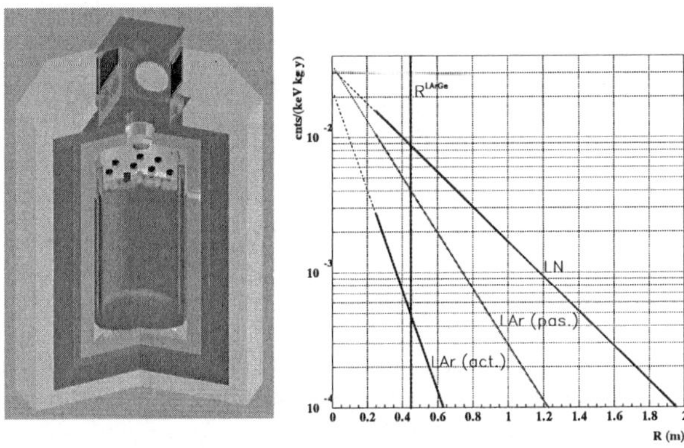

Figure 6. Left: schematic view of the LARGE-TB operating bare germanium detectors in liquid argon for double beta decay search. Right: Background index as a function of radius of cryogenic vessel for external 2.6 MeV gammas from ^{208}Tl (bulk activity: $20\mu Bq\ ^{232}$Th/kg) for liquid nitrogen and liquid argon with and w/o scintillation readout. R^{LArGe}: radius of LARGE-TB.

Fig. 7 shows the potential of background suppression of intrinsic ^{60}Co decays. According to MC calculations, a factor of 100 reduction at $Q_{\beta\beta}$ appears possible.

6. Outlook

Future progress in the field of low-energy particle physics and astrophysics will be driven by new detectors which are currently under development or construction. Larger target masses, lower background rates and event specific discrimination techniques will be employed. Detectors using metal loaded as well as unloaded high-purity organic liquid scintillators, liquefied noble gas as well as high-purity germanium detectors will most likely take a central role in future experiments searching for neutrinoless double beta decay, dark matter, Θ_{13} and spectroscopy of low-energy solar neutrinos.

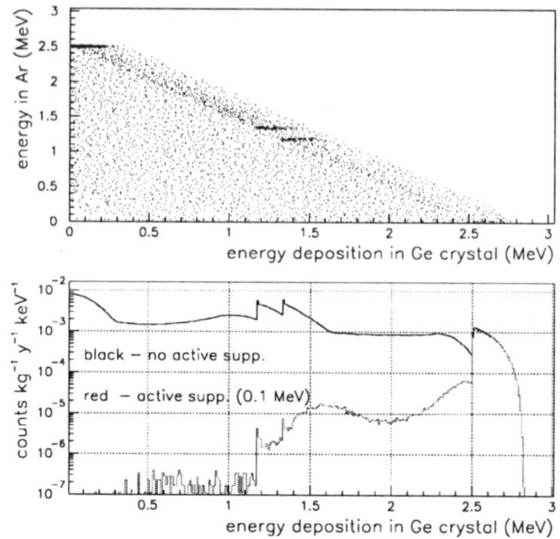

Figure 7.   Monte Carlo results for the background suppression of ^{60}Co internal to a 2 kg germanium diode. An initial exposure to cosmic rays of 10 days has been assumed corresponding to a specific ^{60}Co activity of 0.17 μBq/kg. A threshold of 100 keV for the energy deposition in the liquid argon has been assumed as for vetoing. A suppression factor of $\sim$ 100 can in principle be achieved.

References

1. H.V.Klapdor-Kleingrothaus *et. al.*, *NIM* **A522** (2004) 371.
2. A. McDonald *et. al.*, *Rev. Sci. Inst.* **75(2)** (2003) 293.
3. G. Alimonti *et. al.* (BOREXINO collab.) *Astropart. Phys.* **16** (2002) 205.
4. K. Inoue, *NOW 2004 workshop*, http://www.ba.infn.it/ now2004/ (2004)
5. G. Alimonti *et al.*, (BOREXINO collab. *Phys. Lett.* **B** 422 (1998) 349.
6. K. Anderson *et al.*, www.hep.anl.gov/minos/reactor13/white.html
7. F. Adrellier *et al.* (LoI DOUBLE-CHOOZ collab.), hep-ex/0405032.
8. C. Buck, Dissertation, Univ. Heidelberg, May 2004
9. D. McKinsey *et. al.*, LoI to SNOlab, (2004).
10. R. Brunetti *et. al.*, (WARP collab.) unpublished, (2004)
11. Techn. and applic. of xenon detect., Y. Suzuki *et. al.* *World Scientific* (2002)
12. M. Nakahata, *NOW 2004 workshop*, http://www.ba.infn.it/ now2004/ (2004)
13. E. Aprile *et. al.*,astro-ph/0407575 (2004)
14. N. Smith, DARK 2004, (2004)
15. I. Abt *et. al.*, LoI, hep-ex/0404039 (2004); I. Abt *et. al.* (GERDA) collab., proposal P38/04 to the LNGS (unpublished) (2004)
16. MAJORANA collab., nucl-ex/0311013, (2003)
17. O.Martineau *et. al.*, *NIM* **A530** (2004) 426; astro-ph/0310657
18. CDMS Collaboration, astro-ph/0405033 (2004)
19. P. Peiffer *et. al.*, XXIst Int. Neutr. Conf. 2004, *Nucl. Phys. B, proc. suppl.*

NEUTRINO SEESAW AND CP VIOLATION FROM DYNAMICAL ELECTROWEAK SYMMETRY BREAKING*

THOMAS APPELQUIST

Department of Physics
Yale University
New Haven, CT 06520-8120 USA
E-mail: thomas.appelquist@yale.edu

1. Introduction

In this talk, I will describe two features of a recently proposed class of extended technicolor (ETC) models [1-5]. The first is the possible existence of a seesaw mechanism for neutrino masses, and the second is CP violation as manifested in the quark electric (and chromo-electric) dipole moments.

The class of ETC models is based on the gauge group $SU(5)_{ETC}$ which commutes with the standard-model (SM) gauge group. It breaks sequentially to a residual exact $SU(2)_{TC}$ technicolor gauge symmetry, naturally producing a hierarchy of charged lepton and quark masses. Thus, $SU(5)_{ETC} \to SU(4)_{ETC}$ at a scale Λ_1, with the first-generation standard-model fermions separating from the other four components of the original 5 or $\bar{5}$ representations; then $SU(4)_{ETC} \to SU(3)_{ETC} \to SU(2)_{TC}$ at still lower scales with the second- and third-generation fermions separating in the same way, leaving the technifermions.

The models of Ref. [5] exhibit charged-current flavor mixing, intra-family mass splittings without excessive contributions to the difference $\rho - 1$ where $\rho = m_W^2/(m_Z^2 \cos^2 \theta_W)$, a dynamical origin of CP-violating phases in the quark and lepton sectors, and a potential see-saw mechanism for light neutrinos without the presence of a grand unified scale [2]. The choice of $SU(2)$

---

*This work is supported by U.S. Department of Energy grant DE-FG02-92ER-40704.

for the technicolor group (i) minimizes the TC contributions to the electroweak S parameter, (ii) with a standard-model family of technifermions in the fundamental representation of SU(2)$_{TC}$, can yield an approximate infrared fixed point [7] and associated walking behavior, and (iii) makes possible the mechanism for light neutrinos.

The sequential breaking of the SU(5)$_{ETC}$ to SU(2)$_{TC}$ is driven by the condensation of standard-model-singlet fermions which are part of the models. At the scale Λ_{TC}, technifermion condensates break the electroweak symmetry. The models do not yet yield fully realistic fermion masses and mixings, and they have a small number of unacceptable Nambu-Goldstone bosons arising from spontaneously broken $U(1)$ global symmetries. Additional interactions at energies not far above Λ_1 must be invoked to give them sufficiently large masses. Nevertheless, the models share interesting generic features that are worth studying in their own right. The bilinear fermion condensates forming at each stage of ETC breaking have generically nonzero phases, providing a natural, dynamical source of CP violation. Below the electroweak symmetry breaking scale, the effective theory consists of the standard-model interactions, mass terms for the quarks, charged leptons, and neutrinos, and a tower of higher-dimension operators generated by the underlying ETC theory.

2. Structure of the Models

The full gauge group of the class of models is $G = \mathrm{SU}(5)_{ETC} \times \mathrm{SU}(2)_{HC} \times G_{SM}$. An additional gauge interaction, $SU(2)_{HC}$ (HC = hypercolor), is introduced along with $SU(5)_{ETC}$ and G_{SM}. Both the $SU(2)_{HC}$ and $SU(5)_{ETC}$ interactions become strong, triggering the sequential breaking pattern. The fermion content of one representative model[2] of the class is listed below, where the numbers indicate the representations under SU(5)$_{ETC}$ × SU(2)$_{HC}$ × SU(3)$_c$ × SU(2)$_L$ and the subscript gives the weak hypercharge:

$$(5, 1, 3, 2)_{1/3, L} \ , \qquad (5, 1, 3, 1)_{4/3, R} \ , \qquad (5, 1, 3, 1)_{-2/3, R}$$

$$(5, 1, 1, 2)_{-1, L} \ , \qquad (5, 1, 1, 1)_{-2, R} \ , \qquad (\overline{10}, 1, 1, 1)_{0, R} \ , \qquad (10, 2, 1, 1)_{0, R}(1)$$

Thus the fermions include quarks and techniquarks in the representations $(5, 1, 3, 2)_{1/3, L}$, $(5, 1, 3, 1)_{4/3, R}$, and $(5, 1, 3, 1)_{-2/3, R}$, left-handed charged leptons and neutrinos and technileptons in $(5, 1, 1, 2)_{-1, L}$, and right-handed charged leptons and technileptons in $(5, 1, 1, 1)_{-2, R}$, together with

SM-singlet fermions $\psi_{ij,R}$ and $\zeta_R^{ij,\alpha}$ transforming as $(\overline{10},1,1,1)_{0,R}$ and $(10,2,1,1)_{0,R}$. There are also fermions $\omega_{p,R}^\alpha$, $p = 1,2$ transforming as $(1,2,1,1)_{0,R}$. The lepton number assigned to $\psi_{ij,R}$ is taken to be $L = 1$ in order that Dirac terms $\bar{n}_{i,L}\psi_{jk,R}$ conserve lepton number. The full model is a chiral gauge theory, while the $SU(2)_{HC}$ and $SU(2)_{TC}$ sub-sectors are vectorial.

To analyze the stages of symmetry breaking, plausible preferred condensation channels are identified using a generalized most-attractive-channel (GMAC) approach. As the energy decreases from high values down to $E \sim \Lambda_1 \sim 10^3$ TeV, the coupling α_{ETC} become sufficiently large to produce condensation in the attractive channel $(\overline{10},1,1,1)_{0,R} \times (\overline{10},1,1,1)_{0,R} \to (5,1,1,1)_0$, breaking $SU(5)_{ETC} \to SU(4)_{ETC}$. With respect to the unbroken $SU(4)_{ETC}$, we have $(\overline{10},1,1,1)_{0,R} = (\bar{4},1,1,1)_{0,R} + (\bar{6},1,1,1)_{0,R}$; we denote the $(\bar{4},1,1,1)_{0,R}$ as $\alpha_{1i,R} \equiv \psi_{1i,R}$ for $2 \leq i \leq 5$ and the $(\bar{6},1,1,1)_{0,R}$ as $\xi_{ij,R} \equiv \psi_{ij,R}$ for $2 \leq i,j \leq 5$. The associated condensate is then

$$\langle \epsilon^{1ijk\ell}\xi_{ij,R}^T C\xi_{k\ell,R}\rangle = 8\langle \xi_{23,R}^T C\xi_{45,R} - \xi_{24,R}^T C\xi_{35,R} + \xi_{25,R}^T C\xi_{34,R}\rangle \ . \tag{2}$$

This condensate and the resultant dynamical Majorana mass terms for the six components of ξ in eq. (2) produce a violation of total lepton number by $|\Delta L| = 2$ units.

At lower scales, depending on relative strengths of couplings, different symmetry-breaking sequences can occur. In one sequence, leading to a neutrino seesaw, the $SU(4)_{ETC}$ interaction produces a condensation at a scale $\Lambda_{BHC} \leq \Lambda_1$ (BHC = broken HC), in the channel $(6,2,1,1)_{0,R} \times (6,2,1,1)_{0,R} \to (1,3,1,1)_0$ with condensate $\langle \epsilon_{1ijk\ell}\zeta_R^{ij,1}{}^T C\zeta_R^{k\ell,2}\rangle + (1 \leftrightarrow 2)$. This breaks $SU(2)_{HC} \to U(1)_{HC}$ and gives dynamical masses $\sim \Lambda_{BHC}$ to the twelve $\zeta_R^{ij,\alpha}$ fields involved. Let $\alpha = 1,2$ correspond to $Q_{HC} = \pm 1$ under the $U(1)_{HC}$. This gives dynamical masses $\sim \Lambda_{BHC}$ to the twelve $\zeta_R^{ij,\alpha}$ fields involved. At the lower scale Λ_{23} there is condensation in the $4 \times 4 \to 6$ channel with condensate $\langle \epsilon_{\alpha\beta}\zeta_R^{12,\alpha}{}^T C\zeta_R^{13,\beta}\rangle$, which then breaks $SU(4)_{ETC} \to SU(2)_{ETC}$ and is $U(1)_{HC}$-invariant. The $U(1)_{HC}$ interaction naturally also forms the condensates $\langle \zeta_R^{12,\alpha}{}^T C\omega_{p,R}^\beta\rangle$ and $\langle \zeta_R^{13,\alpha}{}^T C\omega_{p,R}^\beta\rangle$ with $p = 1,2$, $\alpha \neq \beta$. Finally, technifermion condensation occurs at Λ_{TC}.

3. Neutrino Seesaw

The neutrino masses and mixing are calculated by diagonalizing the full mass matrix of neutrino-like states. The nonzero entries in this matrix arise in two different ways: (i) directly, as dynamical masses associated with various condensates, and (ii) via loop diagrams involving dynamical mass insertions on internal fermion lines. In the effective low energy theory at scales below Λ_{TC}, the light neutrinos include the three electroweak-doublet neutrinos $\nu_{\ell,L}$ with $\ell = e, \mu, \tau$, arising from a 5 of $SU(5)_{ETC}$, and two electroweak-singlet neutrinos, $\alpha_{1j,R}$, $j = 2,3$ arising from a $\overline{10}$ of $SU(5)_{ETC}$. The fact that these transform as different representations leads to a strong suppression of Dirac neutrino mass terms, as in Ref. [1], because the relevant entries require mixing of ETC gauge bosons on internal lines of these loop diagrams. The Dirac neutrino mass matrix is denoted b_{ij}, with $i = 1,2,3$ and $j = 2,3$. The dynamically generated technineutrino mass terms rely on the fact that we use $SU(2)_{TC}$ so that 2 and $\bar{2}$ representations are equivalent. For the given symmetry breaking sequence we find

$$|b_{23}| = |b_{32}| \propto \frac{\Lambda_{TC}^4}{\Lambda_{23}^3} \tag{3}$$

with comparable values for b_{22} and b_{33}, and much smaller values for b_{1j}, $j = 2,3$. The relevant Majorana mass matrix r_{ij} is defined by the operator product $\alpha_{1i,R}^T C r_{ij} \alpha_{1j,R}$, $i,j = 2,3$. We find $r_{22} = r_{33} = 0$ and $r_{23} \propto \Lambda_{BHC}^n \Lambda_{23}^n / \Lambda_1^{2n-1}$ where n has the respective values 2 and 3 in a theory with walking all the way up to Λ_1 or just up to Λ_{23}. The Majorana mass r_{23} also requires ETC gauge boson mixing and is suppressed relative to the largest ETC scale.

The electroweak-doublet neutrinos are, to a very good approximation, linear combinations of three mass eigenstates having a normal hierarchy, with

$$m(\nu_j) \propto \frac{(b_{23} \pm b_{22})^2}{r_{23}} \tag{4}$$

where $j = 3,2$ correspond to the $+, -$. Hence, for the case of a theory with walking up to Λ_1, we have $m(\nu_3) \propto (1+y)^2 \Lambda_{TC}^8 \Lambda_1^3 / (\Lambda_{23}^8 \Lambda_{BHC}^2)$, where $y = b_{22}/b_{23}$. For fixed Λ_1 and Λ_{TC}, one can envision models with scales Λ_{BHC} and Λ_{23} that yield $m(\nu_3) \simeq 0.005$ eV, in agreement with the value extracted from the experimental measurement of $|\Delta m_{atm}^2|$ assuming a hierarchical neutrino mass spectrum. The prediction $m(\nu_2)/m(\nu_3) = (b_{23} - b_{22})^2/(b_{23} + b_{22})^2$ can account for the ratio of $m(\nu_2)/m(\nu_3) \sim 0.2$ extracted from the data on $|\Delta m_{atm}^2|$ and Δm_{sol}^2. The model naturally yields a large ν_μ, ν_τ mixing angle θ_{23}, although the θ_{12} mixing angle appears to be too small. Also, the breaking sequence invoked here does not lead to realistic quark and charged lepton masses.

4. Quark mass matrices

The quark dipole operators of interest here are related to the dimension-3 mass terms of the up-type and down-type quarks, given in general by

$$\mathcal{L}_m = -\bar{f}_{L,j} M_{jk}^{(f)} f_{R,k} + h.c. \tag{5}$$

where f label the ETC eigenstates of the $Q = 2/3$ and $Q = -1/3$ quarks, respectively and the indices j, k label generation number. The mass matrices $M^{(f)}$ can be brought to real, positive diagonal form $M^{(q)}$ by the bi-unitary transformation

$$U_L^{(f)} M^{(f)} U_R^{(f)\,-1} = M^{(q)} . \tag{6}$$

The interaction eigenstates f are mapped to mass eigenstates q via

$$f_\chi = U_\chi^{(f)\,-1} q_\chi , \qquad \chi = L, R \tag{7}$$

where $q = (u, c, t)$ and $q = (d, s, b)$ for $Q = 2/3$ and $Q = -1/3$ respectively. In this way, the Cabibbo-Kobayashi-Maskawa (CKM) quark mixing matrix entering the charged current interactions is generated:

$$V = U_L^{(u)} U_L^{(d)\,\dagger} . \tag{8}$$

Using the conventions of [17], we write

$$U_\chi^{(f)} = P_\alpha^{(f)\chi} R_{23}(\theta_{23}^{(f)\chi}) P_\delta^{(f)\chi\,*} R_{13}(\theta_{13}^{(f)\chi}) P_\delta^{(f)\chi} R(\theta_{12}^{(f)\chi}) P_\beta^{(f)\chi} \tag{9}$$

where $R_{mn}(\theta_{mn}^{(f)\chi})$ is the rotation through $\theta_{mn}^{(f)\chi}$ in the mn subspace, $P_\alpha^{(f)\chi}$ and $P_\beta^{(f)\chi}$ are given by

$$P_a^{(f)\chi} = \mathrm{diag}(e^{ia_1^{(f)\chi}}, e^{ia_2^{(f)\chi}}, e^{ia_3^{(f)\chi}}), \qquad a = \alpha, \beta \tag{10}$$

and $P_\delta^{(f)\chi} = \mathrm{diag}(e^{i\delta^{(f)\chi}}, 1, 1)$. The mixing angles are typically small if the off-diagonal $M_{jk}^{(f)}$'s are more suppressed than the diagonal ones.

In ETC models, the off-diagonal entries of the quark mass matrices $M^{(f)}$ arise via mixing among the ETC gauge bosons. In the model of Ref. [5], employing a relatively conjugate ETC representation for the down-type quarks, this is true also of the diagonal elements [10]. We note that in this model, $M^{(u)}$ is hermitian, so that $U_L^{(u)} = U_R^{(u)}$, while $M^{(d)}$ is a more general complex matrix.

As for the phases, a complete theory should allow the computation of all the observable ones [9]. Here, I will describe bounds on combinations of mixing angles and phases. The mixing angles are then bounded with the assumption that the phases are generically of order unity. An important future study will be to see whether this is naturally the case.

5. Electromagnetic and Color Dipole Moment Matrices

The magnetic and electric dipole-moment matrices $D^{(f)}$ of the quarks appear in the dimension-5 operators

$$\mathcal{L}_{DM} = \frac{1}{2}\bar{f}_L D^{(f)}\sigma_{\mu\nu}f_R F_{em}^{\mu\nu} + h.c. \tag{11}$$

Similarly, the color (chromo-) magnetic and electric dipole-moment matrices $D_c^{(f)}$ enter the operators

$$\mathcal{L}_{CDM} = \frac{1}{2}\bar{f}_L T_a D_c^{(f)}\sigma_{\mu\nu}f_R G_a^{\mu\nu} + h.c., \tag{12}$$

where T_a and $G_a^{\mu\nu}$ denote a generator, and the field-strength tensor, for color $SU(3)_c$. (I include the color $SU(3)_c$ coupling g_s in the definition of $D_c^{(f)}$, just as it is usual to include the electromagnetic coupling e in the definition of the electric dipole moment.)

In the class of ETC models being considered, the ETC gauge bosons do not carry SM quantum numbers. Hence, in the respective diagrams that produce the dipole moment matrices and color dipole moment matrices, the photon and gluon couple only to the virtual (techni)fermions. Therefore,

$$D_c^{(f)} = \frac{g_s}{eQ_f}D^{(f)} \tag{13}$$

where $g_s = g_s(\mu)$ is evaluated at the appropriate scale μ.

Transforming to the mass-eigenstate basis,

$$\bar{f}_L D^{(f)}\sigma_{\mu\nu}f_R F_{em}^{\mu\nu} + h.c. = \bar{q}_L D^{(q)}\sigma_{\mu\nu}q_R F_{em}^{\mu\nu} + h.c., \tag{14}$$

where

$$D^{(q)} = U_L^{(f)} D^{(f)} U_R^{(f)\,-1} . \tag{15}$$

Analogously, $D_c^{(q)} = U_L^{(f)} D_c^{(f)} U_R^{(f)\,-1}$. Both $D^{(q)}$ and $D_c^{(q)}$ are independent of $P_\beta^{(f)\chi}$, $\chi = L, R$.

Decomposing $D^{(q)}$ into hermitian and anti-hermitian parts, $D^{(q)} = D_H^{(q)} + D_{AH}^{(q)}$, where $D_{H,AH}^{(q)} = (1/2)(D^{(q)} \pm D^{(q)\,\dagger})$, the dipole operator takes the form $(1/2)[\bar{q}D_H^{(q)}\sigma_{\mu\nu}q + \bar{q}D_{AH}^{(q)}\sigma_{\mu\nu}\gamma_5 q]F_{em}^{\mu\nu}$. Then the EDM of q_j is

$$d_{q_j} = -iD_{AH,jj}^{(q)} \ . \tag{16}$$

Defining $D_{c,AH}^{(q)}$ analogously, the chromo-EDM of q_j is

$$d_{c,q_j} = -iD_{c,AH,jj}^{(q)} \ . \tag{17}$$

In Refs. [5,6], an estimate of the mass matrix $M^{(f)}$ and dipole matrix $\tilde{D}^{(f)}$ from an underlying ETC theory was described, and it was noted how they are related in the presence of the mixing of ETC interaction eigenstates to form mass eigenstates of the fermions and gauge bosons. An important result from that analysis is the relation

$$D_{jk}^{(f)} \simeq \frac{eM_{jk}^{(f)}}{\Lambda_{jk}^2} \tag{18}$$

where each Λ_{jk} is a dimensionful parameter of order the scale above which the $(j \leftrightarrow k)$ ETC propagator becomes soft. This structure reflects the fact that the leading dipole contribution contains two additional inverse factors of the ETC scale(s) relative to the mass, and that the corresponding integral is again sensitive to physics at the ETC scales. Λ_{jk} is no greater than $\min(\Lambda_j, \Lambda_k)$, and can be less. The fact that its (j, k) dependence is nontrivial implies that $D_{jk}^{(f)}$ is not, in general, $\propto M_{jk}^{(f)}$. It is therefore not diagonalized by the transformation that diagonalizes $M^{(f)}$; this transformation yields, instead, a non-diagonal and complex form for the dipole matrix $D^{(q)}$ of Eq. (15). Thus mixing has an important effect on quark dipole moments in ETC models.

For numerical estimates of the dipole matrix, I take the ETC breaking scales to be $\Lambda_1 \simeq 10^3$ TeV, $\Lambda_2 \simeq 10^2$ TeV, and $\Lambda_3 \simeq 4$ TeV. Since the ETC interactions are strong at these scales, there is resultant uncertainty in the calculations. I focus on the contribution to each element of the dipole matrix in $D^{(q)}$ in the mass-diagonal basis of third-family physics arising at the lowest ETC scale Λ_3. Then

$$D_{jk}^{(q)} \simeq \frac{eQ_q \, m_{q3} F_{jk,3}^{(f)}}{\Lambda_3^2} \tag{19}$$

where $m_{q3} = m_t, m_b$ for the u, d sectors respectively, and where $F_{jk,3}^{(f)}$ is a dimensionless function of the parameters in $U_L^{(f)}$ and $U_R^{(f)}$, of $O(1)$ for generic values of these parameters. Terms involving exchange of heavier ETC vector bosons with masses Λ_1 and Λ_2 are also present but are suppressed by the propagator mass ratios Λ_3^2/Λ_j^2, $j = 1, 2$. Part of this propagator suppression may be compensated for by the property that these other terms involve fewer small mixing angle factors, and hence they are not necessarily negligible; however, the Λ_3-scale terms should provide a rough measure of the overall ETC contributions.

The experimental constraints will demand small mixing angles, so I record here the small-angle form of the function $F_{jk,3}^{(f)}$. Using Eq. (15) and Eq. (9),

$$F_{33,3}^{(f)} \simeq 1 + \ldots \, , \tag{20}$$

$$F_{23,3}^{(f)} \simeq e^{i(\alpha_2^{(f)L} - \alpha_3^{(f)L})} \theta_{23}^{(f)L} + \ldots \, , \tag{21}$$

$$F_{32,3}^{(f)} \simeq e^{i(-\alpha_2^{(f)R} + \alpha_3^{(f)R})} \theta_{23}^{(f)R} + \ldots \, , \tag{22}$$

and, for $j, k \neq 3$,

$$F_{jk,3}^{(f)} \simeq \eta_{jk} e^{i[(\alpha_j^{(f)L} - \alpha_3^{(f)L}) - (\alpha_k^{(f)R} - \alpha_3^{(f)R})]} \theta_{j3}^{(f)L} \theta_{k3}^{(f)R} + \ldots \, , \tag{23}$$

where each expression is accurate up to a real coefficient of order unity, η_{jk} can contain $\delta^{(f)\chi}$ phases, and ... denote higher order terms.

6. The Strong CP Problem

Before considering the phenomenology of the dimension-5 operators, with their CP violating phases, I will discuss briefly the strong CP problem within the class of ETC models being considered. Can these models lead to the necessary condition

$$|\bar{\theta}| \leq 10^{-10} \tag{24}$$

where

$$\bar{\theta} = \theta - [arg(det(M^{(u)})) + arg(det(M^{(d)}))] \, , \tag{25}$$

with θ appearing via the topological term

$$\frac{\theta g_s^2}{32\pi^2} G_a{}_{\mu\nu} \tilde{G}_a^{\mu\nu} \tag{26}$$

in the QCD Lagrangian?

The quark mass matrices $M^{(f)}$ $(f = u, d)$ in the effective theory below Λ_{TC} are generated by integrating out short-distance physics at scales ranging from Λ_{TC} to the highest ETC scale Λ_1. Above Λ_1, all fermions are massless. Some of the global chiral symmetries are anomalous, and hence are broken by instantons. The $F_{\mu\nu}\tilde{F}^{\mu\nu}$ terms associated with each (non-abelian) gauge interaction may be rotated away by chiral transformations through the relevant global anomalies. In particular, this renders $\theta = 0$ for $SU(3)_c$ in the underlying theory. In the effective low energy theory, then, we have $\bar{\theta} = -arg(det(M^{(u)})) - arg(det(M^{(d)}))$. If $\bar{\theta} \neq 0$, the rotation (2) to the real diagonal mass basis will, of course, regenerate the topological term through the anomaly. In the models of Refs. [1-5], $M^{(u)}$ is hermitian, so $\bar{\theta}$ resides in $M^{(d)}$.

More generally, the condition $|arg(det(M^{(u)})) + arg(det(M^{(d)}))| \leq 10^{-10}$ can be rewritten by letting

$$U_\chi^{(f)} = e^{i\phi_\chi^{(f)}}\mathcal{U}_\chi^{(f)} \in U(3) \tag{27}$$

where $\mathcal{U}_\chi^{(f)} \in SU(3)$ and $\chi = L, R$. Then $det(U_\chi^{(f)}) = e^{i\phi_\chi^{(f)}}$, and, from Eq. (6),

$$det(M^{(f)}) = e^{i(-\phi_L^{(f)}+\phi_R^{(f)})}det(M_{diag.}^{(f)}) . \tag{28}$$

Hence, the necessary condition reads

$$|\sum_{f=u,d}(-\phi_L^{(f)} + \phi_R^{(f)})| \leq 10^{-10} . \tag{29}$$

Other phases, entering the CKM matrix or the dipole operators, enter through the unimodular matrices $\mathcal{U}_\chi^{(f)}$, and are, in this sense, distinct from the strong CP phase.

In the notation of Eq. (9),

$$det(U_\chi^{(f)}) = exp[i\sum_j(\alpha_j^{(f)\chi} + \beta_j^{(f)\chi})] , \quad \chi = L, R \tag{30}$$

and, in terms of these quantities, Eq. (29) reads

$$\left|\sum_{f=u,d}\sum_j\left[-(\alpha_j^{(f)L} + \beta_j^{(f)L}) + (\alpha_j^{(f)R} + \beta_j^{(f)R})\right]\right| \leq 10^{-10} .$$

(independent of $\delta^{(f)\chi}$). So in terms of these parameters, only one linear combination (a sum over flavors j) of the CP-violating phases $\alpha_j^{(f)\chi}$ and $\beta_j^{(f)\chi}$ is tightly constrained.

Whether a resolution of the strong CP problem will emerge in the class of models considered here is not yet clear [9]. These models include certain

Nambu-Goldstone bosons, one of which can be associated with a Peccei-Quinn dynamical relaxation of $\bar{\theta}$ to zero. But, as I noted earlier, these Nambu-Goldstone bosons must be given large masses by new interactions, eliminating this approach to solving the problem. Whatever the resolution of the strong CP problem turns out to be, an important point is that the relevant phase combination, entering in Eq. (29) or equivalently Eq. (31), involves a sum over the generational phases (labelled by j), whereas other CP-violating phase combinations which contribute to the quantities considered here (cf. Eqs. (21)-(23)) involve differences of generational phases (and $\delta^{(f)x}$).

7. Electric and Chromoelectric Dipole Moments

The diagonal electric and chromo-electric dipole moments for the up and down quarks arising from ETC interactions provide tight constraints on certain combinations of phase differences and mixing angles. These moments derive from the diagonal elements of $D_{jk}^{(q)}$ (Eq. 19). Recall that this expression is correct to leading order in ETC scales, depending explicitly on the inverse square of only the lowest ETC scale Λ_3. Thus,

$$d_u = \frac{e\,Q_u}{g_s}d_{c,u} \simeq \frac{e\,Q_u\,m_t\,Im(F_{11,3}^{(u)})}{\Lambda_3^2} \tag{31}$$

$$d_d = \frac{e\,Q_d}{g_s}d_{c,d} \simeq \frac{e\,Q_d\,m_b\,Im(F_{11,3}^{(d)})}{\Lambda_3^2}\ , \tag{32}$$

with the small-angle expressions for the $F_{11,3}^{(f)}$ given by Eq. (23) and involving flavor-differences of phases.

CP-violating electric dipole moments such as those of the neutron and certain atoms like ^{199}Hg receive contributions from the quark EDM's and the quark color EDM's. The latter lead to CP-violation in the hadronic wave function (the CEDM enters as a correction to t-channel gluon exchange between the bound quarks). There are also contributions from the CP-violating triple-gluon operator $c_{abc}G_a\,_{\lambda\mu}\tilde{G}_b^{\mu\nu}G_c^\lambda\,_\nu$ [21], and loop-induced CP-violating $W^+W^-\gamma$ vertices [22]. I focus here on the quark-EDM and CEDM contributions. Because the direct SM contribution to these quantities is many orders of magnitude smaller than the expected ETC contribution, I neglect the SM contribution. And because the computation of the hadronic matrix elements, involving only first-generation quarks, is rather uncertain, I neglect the RG running of the ETC contribution.

8. Experimental Bounds

The current experimental upper bound on the neutron electric dipole moment d_n is $|d_n| < 6.3 \times 10^{-26}$ e-cm [23]. In setting constraints, I assume that there are no accidental cancellations between different contributions to the experimentally observable EDM's. For an estimate of the hadronic matrix element of the quark EDM operators, $\langle n|\bar{f}\sigma_{\mu\nu}\gamma_5 f|n\rangle$, $f = u, d$, various methods yield roughly similar results, which are comparable to the static quark model relation $d_n = (1/3)(4d_d - d_u)$. Using these estimates, I infer from the above limit on $|d_n|$ that $|Im(F_{11,3}^{(u)})| \leq 1 \times 10^{-6}$ and $|Im(F_{11,3}^{(d)})| \leq 3 \times 10^{-5}$.

The same quantities enter into the quark color EDM's and can, in principle, be bounded from their contributions to d_n. Since QCD is nonperturbative at the low energies relevant here, there are uncertainties in the proportionality factors connecting the CEDM's $d_{c,f}$, $f = u, d$, to d_n (e.g., [24]), and there is the related question of what value to use for the color gauge coupling g_s in Eq. (13) at such low energies. In general, from these CEDM's one obtains limits that are comparable to the ones above on $|Im(F_{11,3}^{(f)})|$, $f = u, d$.

The most stringent limits on these quantities are obtained from upper bounds on EDM's of atoms, in particular from ^{199}Hg. Experimentally, $|d_{^{199}Hg}| < 2.1 \times 10^{-28}$ e-cm [25], which is about a factor of 50 smaller than the upper limit on $|d_n|$. From this I obtain the bounds

$$|Im(F_{11,3}^{(u)})| \leq 0.3 \times 10^{-7} \tag{33}$$

$$|Im(F_{11,3}^{(d)})| \leq 0.6 \times 10^{-6} . \tag{34}$$

Note that in the class of models Refs. [1-5], $Im(F_{11,3}^{(u)})$ (23) vanishes identically. This is because $M^{(u)}$ is hermitian in these models and therefore $U_L^{(u)} = U_R^{(u)}$.

From Eq. (23), we see that Eqs. (33) and (34) constrain a product of $\theta_{13}^{(f)L}\theta_{13}^{(f)R}$ times the imaginary part of a phase factor, for $f = u, d$. If the phase differences are of order unity, then

$$|\theta_{13}^{(d)L}\theta_{13}^{(d)R}| \leq 0.6 \times 10^{-6} , \tag{35}$$

with a tighter bound on $|\theta_{13}^{(u)L}\theta_{13}^{(u)R}|$ if $M^{(u)}$ is not hermitian. The bound (35) is comparable to that on the corresponding product of charged-lepton mixing angles coming from the current limit on the electron EDM [6]. It may be satisfied with $|\theta_{13}^{(d)L}| \simeq |\theta_{13}^{(d)R}| \leq 0.0008$. The corresponding angle θ_{13} in the CKM matrix V has the measured value $\simeq 0.004$, only a factor of five larger. Furthermore, the CKM θ_{13} contains terms proportional to products such as $\theta_{12}^{(u)L}\theta_{23}^{(d)L}$ which could be the dominant contribution and naturally be of order 0.004. Hence, although the bound (35) and its analogue for $f = u$ do imply quite small values for the indicated products of rotation angles, they can plausibly be satisfied in ETC models that successfully predict the CKM matrix.

Finally, an s-quark (chromo-) EDM can also contribute to the ^{199}Hg EDM. This contribution is more difficult to compute since there are no valence s quarks in the nucleon. It can be roughly estimated, leading to a bound on $|Im(F_{22,3}^{(d)})|$. This bound is much weaker than the above bound on $|Im(F_{11,3}^{(d)})|$, but it could have important implications for the mixing angles $\theta_{23}^{(d)L}$ and $\theta_{23}^{(d)R}$.

To summarize, quark EDM's and chromo-EDM's correspond to the diagonal elements of the matrices $D^{(q)}$ and $D_c^{(q)}$. The bounds on these quantities, in particular from the measured limit on the EDM of ^{199}Hg, lead to tight constraints on mixing angles and/or CP-violating phase differences in the individual sub-sectors $q = u, d$. These derive from the contributions to the elements of $D^{(q)}$ and $D_c^{(q)}$ from physics at the lowest ETC scales Λ_3 (e.g. Eq. (19)). If the phase differences are of order unity, then the down-type mixing angles are bounded as in Eq. (35), with an even tighter bound in the up sector if $M^{(u)}$ is not hermitian. The bound (35) is comparable to that on the corresponding product of charged-lepton mixing angles coming from the current limit on the electron EDM [6], and not difficult to satisfy in the class of theories being considered.

References

1. T. Appelquist, J. Terning, *Phys. Rev. D* **50** (1994) 2116.
2. T. Appelquist, R. Shrock, *Phys. Lett.* **B548** (2002) 204.
3. T. Appelquist and R. Shrock, in M. Harada, Y. Kikukawa, K. Yamawaki, *Strong Coupling Gauge Theories and Effective Field Theories* (World Scientific, Singapore, 2003), p. 266.
4. T. Appelquist, R. Shrock, *Phys. Rev. Lett.* **90** (2003) 201801.
5. T. Appelquist, M. Piai, and R. Shrock, *Phys. Rev. D* **69** (2004) 015002.
6. T. Appelquist, M. Piai, and R. Shrock, *Phys. Rev. B*, in press (hep-ph/0401114).
7. B. Holdom, *Phys. Lett. B* **150** (1985) 301; K Yamawaki, M. Bando, K. Matumoto, *Phys. Rev. Lett.* **56** (1986) 1335; T. Appelquist, D. Karabali, L.C.R. Wijewardhana, *Phys. Rev. Lett.* **57** (1986) 957; T. Appelquist and L.C.R. Wijewardhana, *Phys. Rev. D* **35** (1987) 774.

8. T. Appelquist, N. Christensen, M. Piai, and R. Shrock, to appear.

9. For recent discussions of CP violation in the ETC context, see K. Lane, hep-ph/0106328, hep-ph/0202255.

10. In this model, the ETC mixing arises from the coupling to certain condensing SM-singlet fermions, and takes the form of two-, three-, and four-point ETC gauge-boson vertices.

11. See, e.g., the reviews A. Buras, hep-ph/9806471; M. Neubert, hep-ph/9809377; G. Isidori, hep-ph/0401079 and references therein to the extensive literature.

12. A recent experimental review is M. Nakao, in *Proc. Lepton-Photon Symposium* 2003, hep-ex/0312041.

13. F. Parodi, in *Proc. Fifth International Conference on Hyperons, Charm, and Beauty Hadrons*, ed. C. Kalman et al., *Nucl. Phys. B* (Proc. Suppl.) **115** (2003) 212; M. Battaglia et al., hep-ph/0304132; M. Ciuchini et al., hep-ph/0307195; http://cern.ch/ckm-workshop. With $\bar{\eta}$ and $\bar{\rho}$ defined via the rephasing-invariant ratio $-V_{ud}V_{ub}^*/(V_{cd}V_{cb}^*) = \bar{\rho} + i\bar{\eta}$, these fits give $\bar{\rho} = 0.18 \pm 0.05$ and $\bar{\eta} = 0.34 \pm 0.03$.

14. K. Abe et al. (Belle Collab.), hep-ex/0308038.

15. B. Aubert et al. (BABAR Collab.), hep-ex/0403035.

16. J. Soares, *Nucl. Phys. B* **367** (1991) 575; A. Kagan, *Phys. Rev. D* **51** (1995) 6196; A. Kagan and M. Neubert, *Phys. Rev. D* **58** (1998) 094012.

17. See http://pdg.lbl.gov (limits are 90 % CL).

18. See, e.g., S. Bertolini, M. Fabbrichesi and J. O. Eeg, *Rev. Mod. Phys.* **72**, 65 (2000).

19. Some reviews of EDM's are W. Bernreuther and M. Suzuki, *Rev. Mod. Phys.* **71** (1991) 1463; S. Barr, *Int. J. Mod. Phys. A* **8** (1993) 209; I. Khriplovich and S. Lamoreaux, *CP Violation without Strangeness* (Springer, New York, 1997).

20. Some recent reviews of experimental searches for neutron and atomic EDM's are J. Pendlebury and E. Hinds, *Nucl. Instr. Meth. A* **440** (2000) 471; E. N. Fortson, P. Sandars, and S. Barr, *Phys. Today* (A.I.P., Jun. 2003); Y. Semertzidis, hep-ex/0401016.

21. S. Weinberg, *Phys. Rev. Lett.* **63** (1998) 2333.

22. T. Appelquist, G.-H. Wu, *Phys. Rev. D* **51** (1995) 240.

23. P. Harris et al., *Phys. Rev. Lett.* **82** (1999) 904; see also S. Lamoreaux and R. Golub, *Phys. Rev. D* **61** (2000) 051301.

24. T. Falk, K. Olive, M. Pospelov, R. Roiban, *Nucl. Phys. B* **560** (1999) 3.

25. M. Romalis, W. Griffith, J. Jacobs, and E. N. Fortson, *Phys. Rev. Lett.* **86** (2001) 2505.

ALTERNATIVES TO THE SEESAW MECHANISM

A. YU. SMIRNOV

International Centre for Theoretical Physics,
Strada Costiera 11,
34014 Trieste, Italy
E-mail: smirnov@ictp.trieste.it

The observed pattern of lepton mixing does not give an evidence of the seesaw. It is easier to disprove seesaw showing that one of the alternatives gives dominant contribution to the neutrino mass. We consider alternative mechanisms based on (i) small (tree level) effective couplings, (ii) small VEV, (iii) radiative generation of masses, (iv) protection by SUSY breaking scale or by μ-term, (v) small overlap of wave functions of the left and right handed neutrino components in extra dimensions. Seesaw can be the mechanism of suppression of the Dirac mass terms and not dominant mechanism of the neutrino mass giving just a sub-leading contribution.

1. What is wrong with the Seesaw?

Needless to say, the seesaw [1,2,3] is the most appealing mechanism of small neutrino mass generation. We admire its simplicity, elegance and naturalness. We (at least some of us) admire, and doubt, and the reasons for doubt can be summarized in the following way.

1). *No clear evidence of the seesaw* is seen in the observed pattern of neutrino masses and lepton mixing.

Such an evidence would be obtained if, *e.g.*, the "HDM + solar SMA MSW " scenario advocated in 90ties is realized. *A priory* this scenario implied nearly quadratic neutrino mass hierarchy: $m_\nu \sim m_u^2$ and small (similar to quark) mixing. The heaviest neutrino was in the eV-range providing the hot component of the dark matter (HDM) in the Universe with substantial contribution to the energy density balance. If this is realized we would agree that the seesaw works, the right handed neutrino masses are at the intermediate mass scale ($10^{10} - 10^{12}$ GeV) and there are similar structures of the Dirac matrices of neutrinos, charged leptons and quarks.

Instead, the bi-large lepton mixing and weak (or none) neutrino mass hierarchy have been found. Generically, the seesaw does not reproduce the observed pattern. For this one needs (i) some tuning of parameters; (ii) particular structures of the RH neutrino mass matrix (very strong hierarchy, off-diagonal elements dominance, *etc..*),. or the Dirac mass matrix which differs from the matrices of the charged leptons and quarks.

Though the data do not exclude the seesaw: both the large mixing and weak mass hierarchy can be reproduced by seesaw [4].

2). *Even more doubts* in the seesaw will appear if (i) light sterile neutrinos are found, *e.g.*, if MiniBOONE confirms the LSND result; or (ii) neutrino mass spectrum turns out to be quasi-degenerate. Again, both these features can be accommodated in the seesaw.

In fact, even without introduction of new fields one can supply the seesaw with an additional symmetry which leads to the three light active neutrinos and one light right handed (sterile) neutrino [5].

As far as the degenerate spectrum is concerned, one can use the seesaw type-II with, *e.g.*, SO(3) flavor symmetry[6]. Also various symmetries can lead to the degenerate spectrum in the context of the seesaw type-I. The double seesaw [7] may reproduce the degenerate spectrum. Indeed, let us consider (in addition to the RH neutrinos) three SO(10) singlets S and the mass matrix of the form

$$m = \begin{pmatrix} 0 & m_D & 0 \\ m_D & 0 & M_D \\ 0 & M_D & M \end{pmatrix}, \tag{1}$$

where $m_D \ll M_D \ll M$. It leads to the light neutrino mass matrix

$$m = m_D M_D^{-1} M M_D^{-1T} m_D^T. \tag{2}$$

Assume that

$$M_D = A m_D, \quad M = M_0 I, \tag{3}$$

where I is the unit matrix and A is a constant, that is, the heavy and light Dirac mass matrices are proportional each other (the Yukawa couplings of S follow the family structure). Then the mass matrix becomes

$$m_\nu = M_0 A^{-2} I. \tag{4}$$

Small deviation from this structure gives small split of masses and mixing.

3). *No way to prove...* What are signatures of the seesaw? One can mention the neutrinoless double beta decay, leptogenesis, flavor changing decays.

Indeed, discovery of the $\beta\beta_{0\nu}$ decay will be in favor of seesaw. However, this is neither necessary (due to possible cancellations), nor sufficient: the positive result of the $\beta\beta_{0\nu}$ searches is not the prove that the seesaw is the main mechanism of neutrino mass generation.

Leptogenesis: it is difficult to establish either.

In the SUSY context the seesaw leads to lepton violating decays, additional contributions to EDM, *etc.*. However, those provide indirect tests which rely on a number of assumptions (see general discussion in [8]).

True signature of the seesaw is detection of the RH Majorana neutrinos, measurements of their masses and couplings with the W-bosons. Particular versions of the low scale seesaw can be tested, in principle, in high energy accelerator experiments.

Apparently it is easier to disprove the seesaw as the dominant mechanism of the neutrino mass generation.

4). *Unpredictable neutrinos:* As it was already in the past, neutrinos may not follow our prejudices about simplicity, elegance and naturalness...

5). *String theory* tells us that

(i) Majorana neutrinos are not particularly favored;

(ii) Appearance of the 126-plets in SO(10) is problematic;

(iii) Dirac masses can be very small:

- some selection rules may exist which lead to small masses;

- values of the Yukawa couplings can be in huge range as the "Landscape paradigm" admits;

- singular structures of the Yukawa matrices may appear.

6). *In a more general context...* One can put the seesaw in some general context and then argue *pro and contra* the context itself.

An example: the seesaw with $M_R = (10^8 - 10^{15})$ GeV in SUSY or SUSY GUT. It can lead to leptogenesis at temperatures $T > 10^8$ GeV. Inflation should occur at higher temperatures, but in this case the gravitino problem appears. Furthermore, SUSY GUT's have problems with proton decay, FCNC, *etc.*. Another example is the consistent anomaly mediation [9].

2. Why alternatives?

Leading or sub-leading. The effective operator [10]

$$\frac{\lambda_{ij}}{M}(l_i H)^T (l_j H), \qquad i,j, = e, \mu, \tau, \tag{5}$$

where l_i and H are the leptons and Higgs doublets correspondingly, generates the Majorana neutrino mass $m_{ij} = \lambda_{ij}\langle H\rangle^2/M$. For $M = M_{Pl}$ and $\lambda_{ij} \sim 1$ it gives $m_{ij} \sim 10^{-5}$ eV. Such a small contribution is still relevant for phenomenology [11]. Sub-dominant structures of the neutrino mass matrix can be generated by the Planck scale interactions [12]. So, the neutrino mass matrix can obtain substantial contributions from new physics at all possible scales from the EW to the Planck scale and from various mechanisms. We can write the following "superformula" for neutrino masses:

$$m_\nu = \sum m_{seesaw} + m_{triplet} + \sum m_{radiative} + m_{SUSY} + m_{Planck} + ..., \quad (6)$$

where in order the terms correspond to contributions from 1). the seesaw realized at different energy scales, 2). the Higgs triplets, 3). one, two, *etc.* loops effects, 4). SUSY contributions, 5). the Planck scale physics, *etc.*.

One can imagine two possibilities: (i). The seesaw gives leading contribution, whereas other mechanisms produce sub-leading effects. (ii). The seesaw may turn out to be the sub-leading mechanism.

Questions to alternatives. There are two questions to any alternative to the seesaw:

- Where are the RH neutrinos, ν_R?
- If ν_R exist, why the Dirac mass terms effects are small or absent?

There are two possible answers to the second question:

1). The Dirac masses are forbidden by symmetry with immediate objection that this is unnatural - why neutrino but not other fermion masses are suppressed?

2). Dirac mass contributions are suppressed by couplings with the heavy degrees of freedom. Here again one can consider two possibilities:

(i). Introduction of large Majorana mass of the RH neutrinos. In this way we come back to the seesaw. So, the seesaw can be the mechanism of suppression of the Dirac mass term and not the main contribution to neutrino mass;

(ii). Introduction of another (large) Dirac mass terms formed by ν_R and new singlet N. In the basis (ν, ν_R, N) consider the mass matrix

$$m = \begin{pmatrix} 0 & m_D & 0 \\ m_D & 0 & M_D \\ 0 & M_D & 0 \end{pmatrix}, \quad (7)$$

which leads to one strictly massless neutrino [13]. For $m_D \ll M$ the admixture of the heavy lepton is negligible. We will refer to this possibility as to the *multi-singlet* mechanism of the suppression.

General context. A general context for consideration of neutrino masses can be formulated in the following way. Beyond the Standard Model there are three RH neutrinos, ν_{Rj}, and also a number of other SM singlets, S_i. The Yukawa couplings of these singlets with the active neutrinos,

$$h_{kj}\bar{l}_k\nu_{Rj}H + f_{ik}\bar{l}_k S_i H, \tag{8}$$

are small due to symmetry, or their contributions to neutrino masses are suppressed by the seesaw or by "multi-singlet" mechanism.

Prove or Disprove. It is easier to test alternatives to the seesaw than the seesaw itself. In fact, a number of alternatives is related to new physics at the electroweak scale which can be tested at high energy accelerators. Therefore the way to proceed is to exclude alternatives. Though it may not be possible to exclude all of them, still more confidence in the seesaw will be obtained. Or it may happen that validity of the alternative will be proven. In this case the seesaw still can play a role of the sub-leading mechanism and we will put bounds on its contribution.

So, we need to consider the alternatives because this is probably the most efficient way to disprove or prove the seesaw.

3. Classifying alternatives

There are different ways to classify the alternatives. One can use

- diagrams (tree level, one loop, two loops, *etc.*) [14];
- possible field operators which lead to the neutrino masses [15];
- physical context.

Let us consider the latter possibility. The first step in classification is the nature of the neutrino mass terms: Dirac or Majorana and their gauge symmetry properties. In the Majorana case (weak isotriplet) the mass can be generated by the effective operator which includes interaction with two Higgs doublets or with Higgs triplet.

In turn, smallness of the mass can be due to (i) small VEV; (ii) small (effective) coupling which may appear at tree level or radiatively; (iii) small overlap of wave functions of the left and right handed neutrino components in extra dimensions.

Small Yukawa couplings. The observed neutrino masses can be reproduced if $h_{ij} \sim 10^{-13}$ in (8). For usual Dirac type Yukawa couplings similar to the quark or charged lepton couplings these values look very unnatural and require some explanation.

One can consider the following scenario: the usual Yukawa couplings for the ν_L and ν_R are not small (of the same size as quark and lepton couplings). However the corresponding masses are strongly suppressed by the seesaw or multi-singlet mechanisms.

Neutrino masses which we observe in the oscillation experiments are formed by ν_L and new singlets, S, (see second term in (8)) which have no analogy in the quark sector. These singlets may have some particular symmetry properties or/and come from the hidden sector of theory. As a consequence, their couplings, f_{ij}, can be small.

Clearly, scenario with small Dirac couplings will be excluded if the neutrinoless double beta decay is discovered and it will be shown that the decay is due to light Majorana neutrinos.

Small effective couplings. Non-renormalizable operators

$$a_{ij}\bar{l}_i S_j H \frac{S}{M} \tag{9}$$

can generate small effective Yukawa couplings

$$h_{ij} = a_{ij}\frac{\langle S \rangle}{M} \tag{10}$$

for $a_{ij} \sim O(1)$, if $\langle S \rangle / M \sim 10^{-13}$. (Renormalizable coupling can be suppressed by symmetry). One can consider the SUSY or GUT scales for M, if $\langle S \rangle$ is at the electroweak scale, or take $m_{3/2}/M_{Pl}$. Another possibility is to assume a small VEV of S [16].

Hierarchy $\langle S \rangle / M$ can be substantially reduced if the effective coupling appears in higher order non-renormalizable interactions:

$$h_{ij} = a_{ij}\frac{\Pi_{k=1...n}\langle S_k \rangle}{M^n}. \tag{11}$$

Higgs triplet mechanism. The Majorana neutrino mass can be generated at tree level by coupling with the Higgs triplet [17,18,19,20] $\Delta \equiv (\Delta^{++}, \Delta^+, \Delta^0)$: $g_{\alpha\beta}l_\alpha^T l_\beta \Delta$.

The electroweak precision measurements give $\langle\Delta\rangle/\langle H\rangle < 0.03$. To avoid appearance of the triplet Majoron [18] the coupling $\mu\Delta HH$ with the Higgs doublets should be introduced. If $g_{\alpha\beta} \sim 1$, then $\langle\Delta^0\rangle \sim 1$ eV.

Various scenarios depend on the triplet mass M_Δ. If $M_\Delta, \mu \gg \langle H\rangle$, the induced VEV appears $\langle\Delta^0\rangle \sim \langle H\rangle^2\mu/M_\Delta^2$ [17] and we arrive at the seesaw type-II. If in contrast, $M_\Delta \sim \langle H\rangle$ and $\mu \ll \langle H\rangle$, we find $\langle\Delta^0\rangle \sim \mu$. The pseudo-Majoron mass $\sim \mu\langle H\rangle^2/\langle\Delta^0\rangle$ can be made large enough to avoid the experimental bounds, in particular, from measured Z^0 width [19,20].

One can consider the effective coupling of neutrinos with triplet which arises from the non-renormalizable interactions:

$$g_{\alpha\beta}\frac{S}{M}l_\alpha^T l_\beta\Delta, \tag{12}$$

where the singlet S acquires VEV. This allows us to increase the required VEV of Δ. Another possibility apears in models with the triplet and two Higgs doublets [20].

4. Mechanisms never die

Zee mechanism. There is no RH neutrinos, instead new scalar bosons are introduced: the charged singlet of SU(2), η^+, and second Higgs doublet H_2. Their couplings

$$l^T\hat{f}i\sigma_2 l\eta^+ + \sum_{i=1,2}\bar{l}\hat{f}_i eH_i, \tag{13}$$

where $\hat{f}_i$ is the matrix of the Yukawa couplings of Higgs H_i (i = 1,2), generate neutrino masses in one loop [21] (fig. 1 a)

$$m_\nu = A[(\hat{f}\hat{m}^2 + \hat{m}^2\hat{f}^T) - v(\cos\beta)^{-1}(\hat{f}\hat{m}\hat{f}_2 + \hat{f}_2^T\hat{m}\hat{f}^T)]. \tag{14}$$

Here $A = \sin 2\theta_Z \ln(M_2/M_1)/(8\pi^2 v\tan\beta))$, $\hat{m} = diag(m_e, m_\mu, m_\tau)$, $\tan\beta \equiv v_1/v_2$, $v^2 \equiv v_1^2 + v_2^2$. In the minimal version only one Higgs doublet couples to leptons, $\hat{f}_2 = 0$, and consequently, only the first term in (14) contributes to the mass. The neutrino mass matrix has zero diagonal elements and therefore experimentally excluded [22]: it can not reconcile two large mixings, one small mixing and hierarchy of Δm^2.

There are several ways to revive the Zee-model [22].

1). Introduction of the non-zero couplings of both Higgs doublets with leptons (non-zero second term in (14)) leads to non-zero diagonal mass terms of the mass matrix and a possibility to describe all experimental results. The model predicts decays $\tau \to \mu\mu\mu$, $\mu \to eee$, $\tau \to \mu\mu e$ due to the

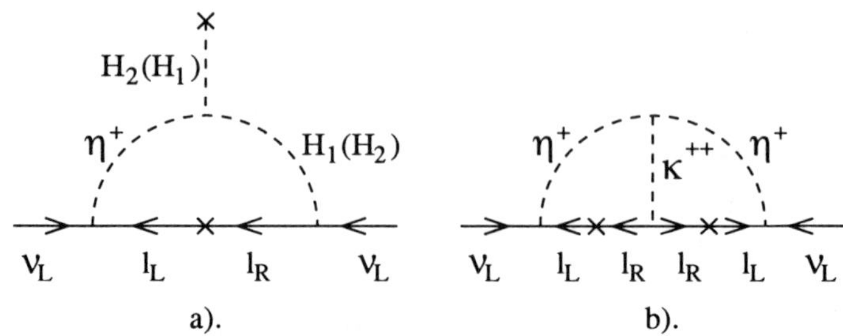

Figure 1. Diagrams of the radiative neutrino mass generation. a). Zee mechanism. b). Babu-Zee mechanism. Flavor indices are omitted.

Higgs exchange with the branching ratios being 2 - 3 orders of magnitude below the present experimental bounds.

2). Some other mechanism can give additional contributions to the neutrino mass matrix, in particular, to the diagonal terms: Higgs triplet, scalar singlet, two loop contribution [22].

3). Additional contributions to the mass matrix appear if new leptons, in particular, sterile neutrinos, exist.

The model is testable in the precision electroweak measurements, searches for charged Higgses and rare decays.

Still the problem exists with explanation of smallness of the couplings: the neutrino data require inverse hierarchy of $f_{\alpha\beta}$ and $f_{\alpha\beta} \sim 10^{-4}$.

Zee-Babu mechanism. There is no RH neutrinos. New scalar bosons, singlets of SU(2), η^+ and k^{++}, are introduced with the following couplings

$$l^T \hat{f} l \eta^+ + l_R^T \hat{h} l_R k^{++}. \tag{15}$$

Here $\hat{f}$ and $\hat{h}$ are the matrices of Yukawa couplings in the flavor basis. The Majorana neutrino masses are generated in two loops [21,23] (fig. 1b):

$$m_\nu \sim 8\mu \hat{f} \hat{m}_l \hat{h} \hat{m}_l \hat{f} I, \tag{16}$$

where $\hat{m}_l \equiv diag(m_e, m_\mu, m_\tau)$.

The main features of the model (see [24,25] for recent discussion) are: one massless neutrino; inverted hierarchy of the couplings in the flavor basis; values of the couplings: $f, h \sim 0.1$. The model is testable: new charged scalar bosons exist at the electroweak scale, the decay rates for $\mu \to e\gamma$, and $\tau \to 3\mu$ are within a reach of the forthcoming experiments.

R-parity violating SUSY. Terms of the superpotential

$$W = -\mu_\alpha l_\alpha H_u - 0.5\lambda_{\alpha\beta m} + \lambda'_{\alpha nm} l_\alpha Q_n d^c_m + h_{mn} H_u Q_n u^c_m \qquad (17)$$

violate the lepton number. Here $\alpha = 0, 1, 2, 3$, $l_0 \equiv H_d$, $m = 1, 2, 3$. No RH neutrinos are introduced.

The bi-linear terms in (17) [26] give the dominant tree level contribution to neutrino masses. In the basis where sneutrinos have zero VEV's, $\langle \tilde{\nu} \rangle = 0$, the masses are produced via mixing with Higgsinos by the diagram (fig. 2a):

$$m_{ij} = \mu_i \mu_j \frac{\cos^2 \beta}{m_\chi}. \qquad (18)$$

In the basis with $\mu_m = 0$, the neutrino masses are generated by the electroweak seesaw: light neutrinos are mixed with wino (neutralino) after sneutrinos get VEV's (fig. 2b): $m_{ij} = A\langle \tilde{\nu}_i \rangle \langle \tilde{\nu}_j \rangle$. Here $A = h_b^2/(16\pi^2 m_{\tilde{W}}^2)$.

Only one neutrino acquires mass at this tree level (in assumption of universality of the soft symmetry breaking terms) [26,27,28,29,30]. Moreover, mixing is determined by the ratios of the mass parameters: μ_i/μ_j.

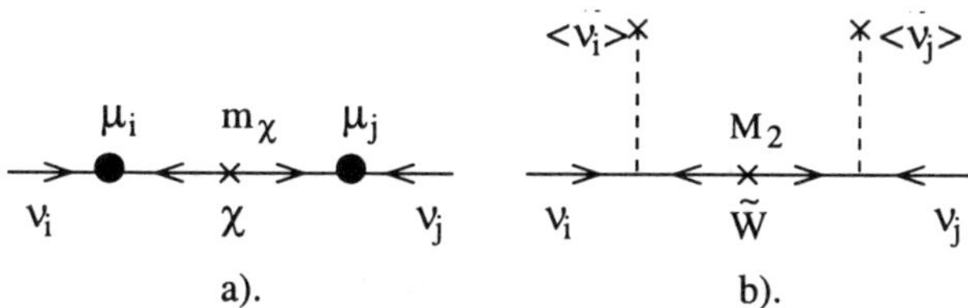

Figure 2. Diagrams for the neutrino mass generation at tree level in the model with the R-parity violation. a) In the basis where $\langle \tilde{\nu} \rangle = 0$. b). In the basis where $\mu_m = 0 (m = 1, 2, 3)$.

The trilinear RpV couplings in (17) and soft symmetry breaking terms (characterized by the mass parameters B_i) generate one loop contributions (fig. 3). As a result, natural neutrino mass hierarchy appears: at tree level one (largest) mass as well as one large mixing are generated; loop contributions produce other small masses and small mixings.

Correct scale of the neutrino masses requires $\mu_m \sim 10^{-4}$ GeV, which in turn, implies further structuring - explanation of the hierarchy $\mu_m \ll m_0$.

In the original version (the universal SUSY breaking at high scale), generically only one large mixing can be obtained. Explanation of neutrino data with two large mixings requires violation of universality of the soft symmetry breaking terms. Both the Higgs-lepton $(\mu - \mu_i)$ and flavor universality should be broken[29,30], that is, $B_i \neq B_j$ at the high scale.

230

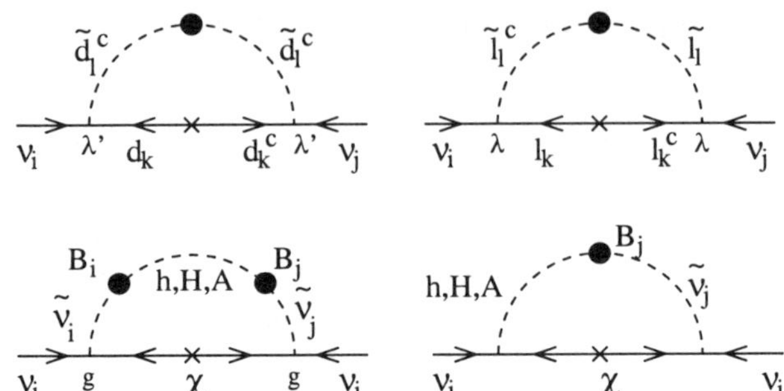

Figure 3. One loop diagrams for the neutrino mass generation in the model with R-parity violation.

The RpV models have very rich phenomenology: new physics at colliders, relatively fast flavor violating decays, new neutrino interactions *etc.*.

5. Old and New

SUSY violation and neutrino mass [31,32,33,34,35,36,37]. It was observed long time ago [31] that

$$m = \frac{m_{3/2}v_{EW}}{M_{Pl}} \sim 10^{-4} \text{ eV}, \tag{19}$$

where $m_{3/2} \sim 1$ TeV is the gravitino mass and $v_{EW} \equiv \langle H \rangle$ is the electroweak VEV. The value (19) is close to the scale of observed neutrino masses and certainly is interesting from the phenomenology point of view. It can be generated by the Yukawa interaction

$$\lambda \bar{l}SH, \qquad \lambda = \frac{m_{3/2}}{M_{Pl}}. \tag{20}$$

It can mix active neutrinos with singlets of SM, *e.g.*, form usual mass term, or mix neutrinos with modulino [31].

The interaction (20) may follow from non-renormalizable term in the superpotential or from the Kähler potential similarly to appearance of the μ - term in the Giudice-Masiero mechanism:

$$K = \frac{1}{M_{Pl}}P(S, z, z^*)\bar{l}H + h.c., \tag{21}$$

where z are the Wilson lines [31,33].

The mass (19) is too small to explain observations, but there are various ways to enhance it. In general, the mass can be written as

$$m = \frac{\alpha\eta m_{3/2}v_{EW}}{M_{Pl}}, \tag{22}$$

where η describes the renormalization group effect and α is an additional numerical factor.

1). The gravitino mass can be larger: the value $m_{3/2} \sim 10^2$ TeV brings the mass to the correct range 10^{-2} eV. Such a scale for $m_{3/2}$ appears, *e.g.*, in the model of "consistent anomaly mediation" [9].

2). One can take M_{GUT} instead of M_{Pl} which leads to $\alpha \sim 10^2$.

3). Large factor α may appear as a consequence of particular mechanism of mass generation. For instance, the terms in the Kähler potential

$$K = \frac{1}{M} P(S, \sigma, \sigma^*) \bar{l} H N + h.c. \tag{23}$$

may have the cut-off parameter $M = 10^{17}$ GeV - below the Planck mass. Here σ are the fields of the hidden sector. Then the dominant contribution to the neutrino mass is given by [34] $v_{EW} F_\sigma / M^2$, where $F_\sigma = \sqrt{3} M_{Pl} m_{3/2}$. It leads to a correct range

$$m = \frac{\eta m_{3/2} v_{EW}}{M} \frac{\sqrt{3} M_{Pl}}{M} \sim 0.05 \ \text{ eV}. \tag{24}$$

Variation on the theme. Small Dirac type Yukawa couplings can appear from the superpotential, whereas the Majorana masses follow from the Kähler potential [36]:

$$W = g\frac{X}{M} lNH, \quad K = h\frac{Y^*}{M} NN. \tag{25}$$

Here $M = M_{Pl}/\sqrt{8\pi}$, X and Y are the fields of hidden sector, with VEV's $\langle X_A \rangle = m_I$ and $\langle Y_F \rangle = m_I^2$ at the intermediate mass scale $m_I = \sqrt{m_{3/2} M_{Pl}}$. Then $m_D = g v_{EW} \sqrt{m_{3/2}/M}$, $m_N = h m_{3/2}$ are small, and the TeV-scale seesaw gives $m_\nu \sim v_{EW}^2 / M g^2 / h \sim 10^{-3}$ eV.

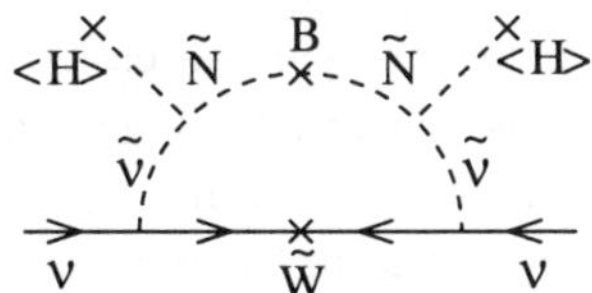

Figure 4. Diagram for the neutrino mass generation in the model[36].

If in addition the term in the Kähler potential

$$K = ... + h\frac{Y^* Y X^*}{M^3} NN \tag{26}$$

232

is introduced [36], the main contribution, ~ 0.05 eV, follows from the one loop diagram shown in fig. 4.

μ-term mixing. Small Dirac mass of neutrino can be related to (protected by) a small value of the μ - parameter in the context of SUSY models and not to the EW scale [38] or $m_{3/2}$. Such a possibility is realized in the SU(5) GUT with R-symmetry [38]. The following R-charges for the matter and Higgs fields are prescribed (second number in the bracket):

$$\bar{F}(\bar{5},1), \ H(10,1), \ N(1,-1), \ \ H(5,0), \ \bar{H}(5,0), \ H'(5,2), \ \bar{H}'(5,0). \quad (27)$$

Notice that the R-charge of the RH neutrinos differs from the charges of other matter fields. Also new Higgs multiplets $(H', \bar{H}')$ are introduced and one of them has non-zero R-charge. The superpotential includes

$$W = f\bar{F}H'N + M_H H'\bar{H}' + ... \quad (28)$$

The μ-term as well as the Majorana mass terms for the RH neutrinos are forbidden by the R-symmetry.

SUSY breaking leads to the R-symmetry breaking and can generate the following operators at the TeV scale:

$$W_R = \mu H\bar{H} + \mu' H\bar{H}'. \quad (29)$$

The last terms in (28) and (29) mix H and H' with the angle $\sim \mu'/M_H$. Consequently, the first term in (28) generates small Yukawa coupling: $(f\mu'/M_H)\bar{F}HN$. As a result, neutrinos acquire the Dirac masses $f\mu'v_{EW}/M_H$.

6. Extra dimensions and Extra possibilities

Theories with extra space dimensions provide qualitatively new mechanism of generation of the small *Dirac* neutrino mass. There are different scenarios, however their common feature can be called the overlap suppression: the overlap of wave functions of the left, $\nu_L(y)$, and right , $\nu_R(y)$ handed components in extra dimensions (coordinate y). The suppression occurs due to different localizations of the $\nu_L(y)$ and $\nu_R(y)$. The effective Yukawa coupling is proportional to the overlap. One can introduce also suppression of overlap of the neutrino and Higgs fields. Let us consider realizations of the overlap mechanism in different extra dimensional scenarios.

... in large flat extra dimensions. The setup is the $3D$ spatial brane in $(3+\delta)D$ bulk [39]. Extra dimensions have large radii $R_i \gg 1/M_{Pl}$ which allows one to reduce the fundamental scale of theory down to $M^* \sim 10-100$ TeV [39]. The left handed neutrino is localized on the brane, whereas the right handed component (being a singlet of the gauge group) propagates in the bulk (see fig. 5a).

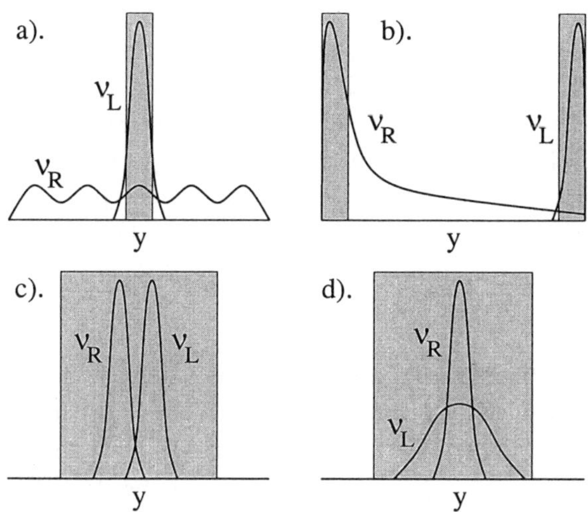

Figure 5. The overlap mechanism of small Dirac neutrino mass generation in models with extra spatial dimensions. a). Large flat extra dimensions. b). Warped extra dimensions. c - d). Models of "fat" branes.

Let us consider one extra dimension of radius R. Since the RH component is not localized, we find from the normalization condition that its wave function has typical value $\nu_R(y) \sim 1/\sqrt{R}$. The width of the brane is of the order $d \sim 1/M^*$, so the amplitude of probability to find the RH neutrino on the brane equals

$$d^{1/2}\nu_R \sim \frac{1}{\sqrt{M^*R}}. \tag{30}$$

Since the LH neutrino is localized on the brane, this amplitude describes the overlap of the wave functions. For δ extra dimensions we get for the overlap factor $1/\sqrt{M^{*\delta}V_\delta}$, where V_δ is the volume of extra dimensions. If λ is the Yukawa coupling for neutrinos in the $(4+\delta)D$ theory, the effective coupling in 4D will be suppressed by this overlap factor. Consequently,

$$m_D = \lambda v_{EW} \frac{1}{\sqrt{M^{*\delta}V_\delta}} = \lambda v_{EW} \frac{M^*}{M_{Pl}}, \tag{31}$$

where in the second equality the relation $M_{Pl}^2 = M^{*2+\delta}V_\delta$ have been used. For $M^* \sim 100$ TeV and $\lambda \sim 1$ we obtain $m_D \sim 10^{-2}$ eV.

...in warped extra dimensions. The setting is one extra dimension compactified on the S^1/Z_2 orbifold, and non-factorizable metric. The coordinate in the extra dimension is parameterized by $r_c\phi$, where r_c is the radius of extra dimension and the angle ϕ changes from 0 to π. Two branes are localized in different points of extra dimension: the "hidden" brane at $\phi = 0$ and the observable one at $\phi = \pi$ [40]. The wave function of the RH neutrino $\nu_R(\phi)$ is centered on the hidden brane, whereas the LH one - on the visible brane (see fig. 5 b). Due to warp geometry $\nu_R(\phi)$ exponentially decreases from the hidden to the observable brane. On the observable brane it is given by

$$\nu^R(\pi) \sim \epsilon^{\nu-1/2}, \quad \epsilon = e^{-kr_c\pi} = \frac{v_{EW}}{M_{Pl}}. \tag{32}$$

Here M_{Pl} is the Planck scale, $k \sim M_{Pl}$ is the curvature parameter. In (32) $\nu \equiv m/k$ and $m \sim M_{Pl}$ is the Dirac mass in 5D. Essentially $\nu_R(\pi)$ gives the overlap factor and the Dirac mass on the visible brane equals

$$m_D = \lambda\nu^R(\pi)v_{EW} \sim M\left(\frac{V_{EW}}{M}\right)^{\nu+1/2}. \tag{33}$$

For $\nu = 1.1 - 1.6$ we obtain the mass in the required range.

Different realization when ν_R is on the TeV brane, whereas ν_L is on the Planck brane, has been suggested in [41].

...on the fat brane. The LH and RH neutrino wave functions can be localized differently on the same "fat" brane [42]. There are various possibilities to suppress the overlap:

1). localize ν_L and ν_R in different places of the brane (fig. 5 c);

2). arrange parameters in such a way that $e.g.$ the RH neutrino is localized in the narrow region of the fat brane whereas the LH neutrino wave function is distributed in whole the brane (fig.5d) [43].

7. Conclusion

1. Comprehensive tests of the alternatives to the seesaw is probably the only way to uncover mechanism of neutrino mass generation. A number of alternatives are related to interesting ideas and concepts of physics beyond the SM and therefore their tests provide (often unique) tests of this new physics.

The alternatives include the radiative mass generation, small effective coupling, small VEV, mechanisms related to SUSY breaking or generation of μ term, the overlap mechanism in extra dimensions, *etc.*.

2. It is easy to test alternatives (at least some of them related to the EW scale physics) than the seesaw itself. And it may happen that one of the alternatives gives the main contribution to the neutrino mass, thus excluding the seesaw as the dominant mechanism.

3. It is possible that some alternatives produce relevant sub-leading contributions to the neutrino mass matrix being responsible for physics "beyond the seesaw".

References

1. P. Minkowski, Phys. Lett. B **67**, 421 (1977), for earlier work see H. Fritzsch, M. Gell-Mann and P. Minkowski, Phys. Lett. B **59**, 256 (1975), H. Fritzsch and P. Minkowski, Phys. Lett. B **62**, 72 (1976).
2. M. Gell-Mann, P. Ramond and R. Slansky, in *Supergravity*, eds P. van Niewenhuizen and D. Z. Freedman (North Holland, Amsterdam 1980); P. Ramond, *Sanibel talk*, retroprinted as hep-ph/9809459; T. Yanagida, in *Proc. of Workshop on Unified Theory and Baryon number in the Universe*, eds. O. Sawada and A. Sugamoto, KEK, Tsukuba, (1979); S. L. Glashow, in *Quarks and Leptons*, Cargèse lectures, eds. M. Lévy, (Plenum, 1980, New York) p. 707; R. N. Mohapatra and G. Senjanović, *Phys. Rev. Lett.* **44**, 912 (1980).
3. R. N. Mohapatra and G. Senjanović, *Phys. Rev.* D **23**, 165 (1981), C. Wetterich, *Nucl. Phys.* B **187**, 343 (1981).
4. A. Yu. Smirnov, *Phys. Rev.* D **48** 3264 (1993); M. Tanimoto, *Phys. Lett.* B **345**, 477 (1995); T.K. Kuo, Guo-Hong Wu, S. W. Mansour, *Phys. Rev.* D **61**, 111301 (2000); G. Altarelli F. Feruglio and I. Masina, *Phys. Lett.* B **472**, 382 (2000); S. Lavignac, I. Masina, C. A. Savoy, *Nucl. Phys.* B **633**, 139 (2002); A. Datta, Fu-Sin Ling and P. Ramond, Nucl.Phys. B **671**, 383 (2003), M. Bando, *et al.*, Phys. Lett. B **580**, 229 (2004).
5. B. Brahmachari, S. Choubey, R. N. Mohapatra, Phys. Lett. B **536**, 94 (2002).
6. D. Caldwell, R.Mohapatra, Phys. Rev.D **50**, 3477 (1994); S. T. Petcov and A. Yu. Smirnov, Phys. Lett. B **322**, 109 (1994); B. Bamert, C. Burgess, Phys. Lett. B**329**, 289 (1994).
7. R.N. Mohapatra, Phys. Rev. Lett. **56**, 561 (1986).
8. S. Davidson and A. Ibarra, JHEP **0109** 013 (2001), talk at this conference.
9. H. Murayama, hep-ph/0410140.
10. S. Weinberg, *Phys. Rev. Lett.* **43**, 1566 (1979).
11. R. Barbieri, J. Ellis and M. K. Gaillard, *Phys. Lett.* B **90**, 249 (1980); E. Kh. Akhmedov, Z. G. Berezhiani, G. Senjanović, *Phys. Rev. Lett.* **69**, 3013 (1992).
12. F. Vissani, M. Narayan and V. Berezinsky, Phys. Lett. B **571**, 209 (2003).
13. R.N. Mohapatra, J. W. F. Valle, Phys. Rev. D**34**, 1642 (1986).
14. E. Ma, Phys. Rev. Lett. **81**, 1171 (1998).

15. K. S. Babu and C. N. Leung, Nucl. Phys. B **619** 667, (2001).

16. Z. Chacko, et al., hep-ph/0405067.

17. M. Magg, C. Wetterich, Phys. Lett. **94B**, 61 (1980); J. Schechter and J. W. F. Valle Phys. Rev. D **22**, 2227 (1980); G. Lazarides, Q. Shafi, C. Wetterich, Nucl. Phys. B **181**, 287 (1981); R. N. Mohapatra, G. Senjanovic, Phys. Rev. D **23**, 165 (1981).

18. G. B. Gelmini and M. Roncadelli, Phys. Lett. B **99**, 411 (1981).

19. E. Ma, U. Sarkar, Phys. Rev. Lett., **80**, 5716 (1998).

20. E. Ma, Phys. Rev. D **66**, 037301 (2002). P. H. Frampton, M. C. Oh, T. Yoshikawa, Phys. Rev. D **66**, 033007 (2002).

21. A. Zee, *Phys. Lett.* B **93**, 389 (1980), *ibidem* **161**, 141 (1985).

22. For the latest discussion see, *e.g.*, P. H. Frampton, M. C. Oh, T. Yoshikawa, Phys. Rev. D **65** 073014, (2002); T. Kitabayashi and M. Yasue, Int. J. Mod. Phys. A **17**, 2519 (2002); Xiao-Gang He, Eur.Phys.J.C **34**, 371 (2004)

23. K. S. Babu, *Phys. Lett.* B **203**, 132 (1988).

24. K.S. Babu and C. Macesanu, Phys. Rev. D **67**, 073010 (2003).

25. I. Aizawa, *et al.*, Phys. Rev. D **70**, 015011 (2004).

26. L. J. Hall and M. Suzuki, *Nucl. Phys.* B **231**, 419 (1984), A. Joshipura, M. Nowakowski, *Phys. Rev.* D **51**, 2421 (1995); A. Yu. Smirnov, F. Vissani, *Nucl. Phys.* B **460**, 37 (1996); R. Hempfling, *Nucl. Phys.* B **478**, 3 (1996).

27. A. S. Joshipura, R. D. Vaidya, S. K. Vempati, Nucl. Phys. B **639** 290 (2002).

28. See review and references therein R. Barbier et. al., hep-ph/0406039.

29. Y. Grossman and S. Rakshit, Phys. Rev. D **69**, 093002 (2004).

30. M. A. Diaz, et. al., Phys. Rev. D **68** 013009 (2003).

31. K. Benakli and A. Yu. Smirnov, Phys. Rev. Lett. **79** 4314, (1997).

32. N. Arkani-Hamed, *et al.*, Phys. Rev. D **64**, 115011, (2001); hep-ph/0007001.

33. F. Borzumati, Y. Nomura, Phys. Rev. D **64**, 053005 (2001); F. Borzumati, *et al.*, hep-ph/0012118.

34. S. Abel, A. Dedes and K. Tamvakis, hep-ph/0402287.

35. R. Arnowitt, B. Dutta, B. Hu, Nucl. Phys. B **682**, 347 (2004).

36. J. March-Russell, S. M. West Phys. Lett. B **593**, 181 (2004).

37. J.A. Casas, J. R. Espinosa, I. Navarro Phys. Rev. Lett. **89**, 161801 (2002).

38. R. Kitano, Phys. Lett. B **539**, 102 (2002).

39. N. Arkani-Hamed, S. Dimopoulos, G. R. Dvali and J. March-Russell, *Phys. Rev.* D **65**, 02432 (2002); K. R. Dienes, E. Dudas and T. Ghergetta, *Nucl. Phys.* B **557**, 25 (1999).

40. Y. Grossman and M. Neubert, *Phys. Lett.* B **474**, 361 (2000).

41. T. Gherghetta, Phys. Rev. Lett. **92**, 161601 (2004).

42. N. Arkani-Hamed, M. Schmaltz, Phys. Rev. D**61**, 033005 (2000).

43. P.Q. Hung, Phys. Rev. D **67**, 095011 (2003).

SEARCHING FOR THE ABSOLUTE NEUTRINO MASS SCALE

CH. WEINHEIMER

Helmholtz-Institut für Strahlen- und Kernphysik
Nussallee 14-16
Rheinische Friedrich-Wilhelms-Universität Bonn, D-53359 Bonn, Germany
present address: Institut für Kernphysik, Universität Münster, Germany
E-mail: weinheimer@hiskp.uni-bonn.de

The discovery of neutrino oscillation proved recently that neutrinos have non-vanishing masses in contrast to their description within the Standard Model of particle physics. However, the neutrino mass scale, which is very important for particle physics as well as for cosmology and astrophysics, cannot be resolved by oscillation experiments. Two methods of laboratory experiments allow to search for the neutrino mass scale, the search for the neutrinoless double beta decay and the direct way by investigating a β decay spectrum near its endpoint. In the former case one group has reported evidence for a positive signal. New experiments have just started and will check this claim. The direct mass experiments at Mainz and Troitsk using tritium and the experiments at Gran Sasso using ^{187}Re have given upper limits on the neutrino mass scale of about 2 eV/c^2 in the case of tritium β decay experiments. The new Karlsruhe Tritium Neutrino Experiment (KATRIN) will enhance the sensitivity on the neutrino mass by another order of magnitude down to 0.2 eV/c^2 by using a very strong windowless gaseous tritium source and a huge ultra-high resolution electrostatic spectrometer.

1. Introduction

The recent discovery of neutrino oscillation by experiment with atmospheric, solar, reactor and accelerator neutrinos [1] proved that neutrino mix and that they have non-zero masses in contrast to their description in the Standard Model of particle physics. Unfortunately, these oscillation experiments are sensitive to the differences of squared neutrino mass states, but not directly to the neutrino masses. Theories beyond the Standard Model need to explain the smallness of neutrino masses in comparison with the much heavier charged fermions. Prominent ways to fulfill this are the Seesaw type I mechanism using heavy Majorana neutrinos which result in hierarchical pattern of neutrino masses. Seesaw type II models usually produce a scenario of quasi-degenerate neutrino masses with the help of a Higgs triplet. Then the masses are of at least a few 0.1 eV/c^2 with small mass differences between each other to explain the oscillations. In the latter case, due to huge abundance of relic neutrinos left over by the big bang in the universe neutrinos would make up not the major, but a significant contribution to the dark matter. Therefore the open question of the value of the neutrino mass is not only crucial for particle physics to decide between different theories but also very important for astrophysics and cosmology.

Vice versa, information on the absolute scale of the neutrino mass can be obtained from astrophysical observations like the power spectrum of the matter and the energy distribution in the universe at different scales. Usually these analysis's need the combination of Cosmic Microwave Background data (*e.g.* from the WMAP experiment), the distribution of the galaxies in our universe the so-called "Large Scale Structure" and information from the so-called "Lyman α-Forest" or x-ray clusters to describe the distribution at large, medium and small scales, respectively. In most cases they give upper limits on the mass of the neutrinos on the order of several 0.1 eV/c^2 [2], in some cases non-zero neutrino masses are found [3] illustrating the dependence on the assumptions and the data used to obtain the cosmological limits. One should not forget that these models describe 95 % of the matter and energy distribution of the universe by yet non-understood quantities like the cosmological constant and the Cold Dark Matter. The limits on the neutrino mass rely on the existence of the yet not observed relic neutrinos [4].

There are two laboratory methods to access the neutrino mass scale [5]:

- search for the neutrinoless double β decay
- investigation of the shape of a β decay spectrum near its endpoint.

This paper is organized as following: Section 2 describes the status of the search for neutrinoless double β decay. In section 3 the recent direct neutrino mass experiments investigating the β decay spectra of tritium and ^{185}Re are presented. Section 4 describes the new KATRIN experiment. The conclusions are given in section 5.

2. Neutrinoless Double β Decay

The process of a double β decay without emission of neutrinos requires the neutrino to be Majorana particles. Due to the left-handed structure of charged current weak interactions the matrix element of this transition is directly proportional to the neutrino mass (in the absence of right-handed weak charged currents or the exchange of other new particles). The signature of a neutrinoless double β decay in the case of a $\beta^-\beta^-$ decay ($\beta^+\beta^+$ decay) is the emission of two decay electrons (positrons) which carry the whole available energy Q. There are more than 10 isotopes which are candidates for undergoing double β decay. Whereas double β decay with neutrinos have been observed in about 10 cases there are – with one exception – only lower limits for the neutrinoless double β decay process.

The most sensitive experiment is the Heidelberg-Moscow experiment using 5 low background, high purity and high resolution Germanium detectors in the Gran Sasso underground lab. The double β isotope ^{76}Ge has been enriched from natural abundance to 86 % . The subgroup which has claimed evidence for having observed neutrinoless double β decay has recently presented new data and a re-analysis of the old data [6] showing a line at the position expected for neutrinoless double β decay. They state to see neutrinoless double β decay at the level of 4 σ with an effective neutrino mass of 0.1 eV/c^2 $< m_{ee} <$ 0.9 eV/c^2 at 99 % confidence level. Unfortunately, not all other lines in the vicinity of the signal line are fully understood. Clearly this result requires further checks.

Two new experiments have been started within the last years which might reach the sensitivity to check the claim. The Neutrino Ettore Majorana Observatory (NEMO 3) at the Frejus tunnel is using foils with ^{82}Se, ^{100}Mo, ^{116}Cd and ^{130}Te inbetween tracking chambers in a magnetic field in order to identify both electrons from neutrinoless double β^- decay emitted from the thin foils and to measure the electron energies. Presently NEMO3 has observed a very nice signal from double β decay with 2 neutrinos of the isotope ^{100}Mo but is not giving results on neutrinoless double β decay yet. An alternative way is to run large crystal of TeO$_2$ as cryogenic bolometers. The CUORICINO collaboration is operating a total mass of 41 kg of these detectors in the Gran Sasso underground lab. Their recent result on neutrinoless double β decay is an upper limit on the effective neutrino mass of $m_{ee} <$ 0.37 − 1.9 eV/c^{2a} at 90 % confidence level.

Both running experiments have a realistic chance to confirm the claim but – also because of the large uncertainties of the nuclear matrix elements of about a factor of 2 – not the guarantee to really exclude it. Several next generation double β decay experiments have been proposed and R&D work has been started. They all want to use very large masses in the 1 t region

---

[a]The exact value of the limit depends on the nuclear matrix element.

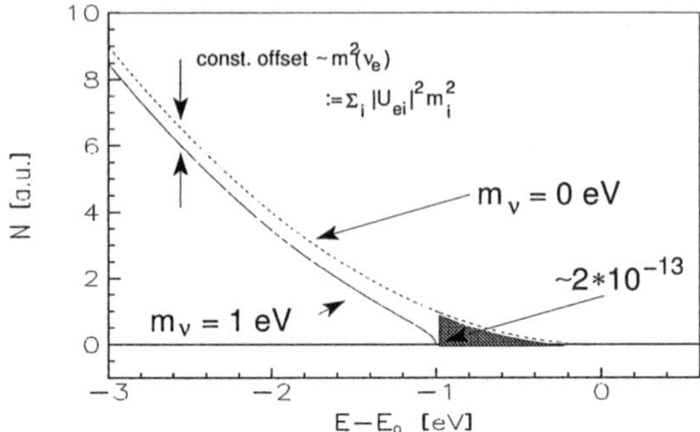

Figure 1. Expanded tritium β spectrumaround tritium endpoint E_0 for $m(\nu_e) = 0$ (dashed line) and for a arbitrarily chosen neutrino mass of 1 eV/c^2 (solid line) The gray shaded area corresponds to a fraction of $2 \cdot 10^{-13}$ of all tritium β decays. The offset between the two curves explains what the "$m(\nu_e)$" is: the average over all neutrino mass states with their contribution according to the neutrino mixing matrix U.

with enrichment and new methods to suppress background.

3. Direct Neutrino Mass Experiments

The most sensitive way to determine the neutrino mass scale without further assumptions is to measure the shape of a β spectrum near its endpoint. Here tritium is the standard isotope due to its low endpoint of 18.6 keV, its rather short half-life of 12.3 y, its super-allowed shape of the β spectrum, and its simple electronic structure. Tritium β decay experiments have been performed in search for the neutrino mass for more than 50 years.

The mass of the electron neutrino is determined by investigating the shape of the electron energy spectrum (β spectrum) (figure 1). Unfortunately the region of sensitivity for the neutrino mass is only the very upper end of the β spectrum; this is due to the necessity that the neutrino have to reveal its mass by becoming non-relativistic From fig. 1 it is clearly visible that the main requirement for such an experiment is to cope with the vanishing count rate near the endpoint by providing the strongest possible signal rate at lowest background rate. Additionally, to be really sensitive to the shape of the β spectrum a high energy resolution on the order of eV is required.

A major break-through in this field was achieved in the nineties by a new type of spectrometer, the so-called MAC-E-Filter (Magnetic Adiabatic Collimation followed by an Electrostatic Filter), which was developed in-

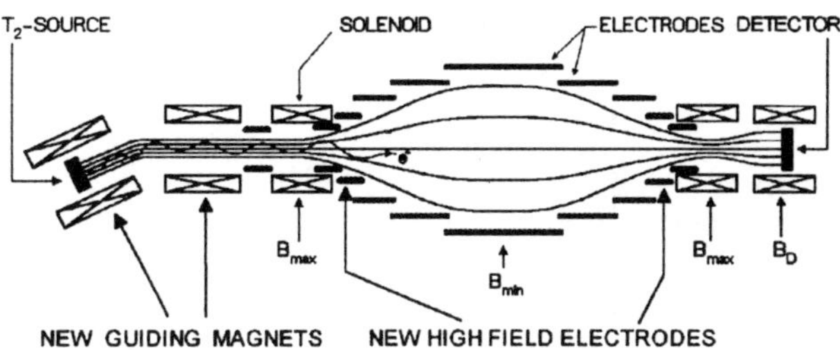

Figure 2. The upgraded Mainz setup shown schematically. The outer diameter amounts to 1 m, the distance from source to detector is 6 m.

dependently in Mainz, Germany and at Troitsk, Russia [7]. This integrating spectrometer provides high luminosity and low background combined with a large energy resolution.

The two recent tritium β decay experiments at Mainz and at Troitsk use similar MAC-E-Filters with an energy resolution of 4.8 eV (3.5 eV) at Mainz (Troitsk). The spectrometers differ slightly in size: The diameter and length of the Mainz (Troitsk) spectrometer are 1 m (1.5 m) and 4 m (7 m). The major differences between the two setups are the tritium sources: Mainz uses a thin film of molecular tritium quench-condensed on a cold graphite substrate as tritium source, whereas Troitsk has chosen a windowless gaseous molecular tritium source. After the upgrade of the Mainz experiment in 1995-1997 both experiments run with similar signal and similar background rates.

3.1. *The Mainz Neutrino Mass Experiment*

Fig. 2 shows the Mainz setup after its upgrade in 1995-1997, which included the installation of a new tilted pair of superconducting solenoids between the tritium source and spectrometer and the use of a new cryostat providing temperatures of the tritium film below 2 K. The first measure eliminated source correlated background and allowed the source strength to be increased significantly. The second measure avoids the roughening transition of the homogeneously condensed tritium films with time [8]. The upgrade of the Mainz setup was completed by the application of HF pulses on one of the electrodes in between measurements every 20 s, and a full automation of the apparatus and remote control. This former improvement lowers and stabilizes the background, the latter one allows long–term measurements.

Figure 3 shows the endpoint region of the Mainz 1998, 1999 and 2001 data in comparison with the former Mainz 1994 data. An improvement of the signal-to-background ratio by a factor 10 by the upgrade of the Mainz experiment as well as a significant enhancement of the statistical quality of the data by longterm measurements are clearly visible. The main systematic uncertainties of the Mainz experiment are the inelastic scattering of β electrons within the tritium film, the excitation of neighbor molecules due to the β decay, and the self-charging of the tritium film by its radioactivity. As a result of detailed investigations in Mainz [9,10,11] – mostly by dedicated experiments – the systematic corrections became much better understood and their uncertainties were reduced significantly.

The high-statistics Mainz data from 1998-2001 allowed the first determination of the amplitude of the so-called neighbor excitation [b] to occur in $(5 \pm 1.6 \pm 2.2)$ % of all β decays [11] in good agreement with the theoretical expectation [12].

The most sensitive analysis on the neutrino mass, in which only the last 70 eV of the β spectrum below the endpoint are used, resulted in the following for Mainz 1998, 1999 and 2001 data [11]

$$m^2(\nu_e) = (-0.7 \pm 2.2 \pm 2.1)\ \mathrm{eV}^2/\mathrm{c}^4 \tag{1}$$

which corresponds to an upper limit of

$$m(\nu_e) < 2.3\ \mathrm{eV}/\mathrm{c}^2 \quad (95\ \%\ \mathrm{C.L.}) \tag{2}$$

This is the lowest upper limit of the neutrino mass obtained thus far.

3.2. *The Troitsk Neutrino Mass Experiment*

The windowless gaseous tritium source of the Troitsk experiment [13] is essentially a tube of 5 cm diameter filled with T_2 resulting in a column density of 10^{17} molecules/cm^2. The source is connected to the ultrahigh vacuum of the spectrometer by a series a differential pumping stations.

From its first data taking in 1994 the Troitsk experiment reports an anomalous excess in the experimental β spectrum as a sharp step of the count rate at a varying position of a few eV below the endpoint of the β spectrum E_0 [13], which seems to be an experimental artefact appearing with varying intensity at the Troitsk setup. The Troitsk experiment is correcting for this anomaly by fitting an additional line to the β spectrum run-by-run.

---

[b]The sudden change of the nuclear charge during β decay can result in the excitation of neighboring molecules.

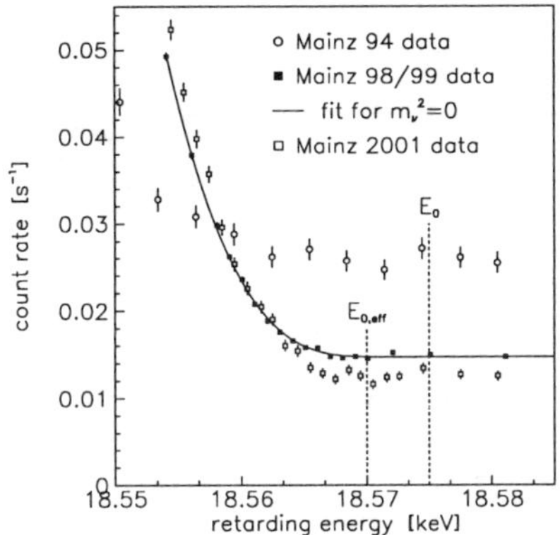

Figure 3. Averaged count rate of the 1998/1999 data filled squares) with fit (line) and of the 2001 data (open squares) in comparison with previous Mainz data from 1994 (open circles) as a function of the retarding energy near the endpoint E_0 and effective endpoint $E_{0,eff}$ (accounting for the width of response function of the setup and the mean rotation-vibration excitation energy of the electronic ground state of the $^3\mathrm{HeT}^+$ daughter molecule).

Combining the 2001 results with the previous ones since 1994 gives [14]

$$m^2(\nu_e) = (-2.3 \pm 2.5 \pm 2.0) \ \mathrm{eV^2/c^4} \qquad (3)$$

from which the Troitsk group deduce an upper limit

$$m(\nu_e) < 2.05 \ \mathrm{eV/c^2} \quad (95 \ \% \ \mathrm{C.L.}) \qquad (4)$$

This value does not contain the systematic uncertainty which is needed to account for describing the timely-varying anomalous excess count rate at Troitsk run-by-run with an additional line.

3.3. Rhenium β decay experiments

Due to the complicated electronic structure of ^{187}Re and its β decay the advantage of the 7 times lower endpoint energy E_0 of ^{187}Re with respect to tritium can only be exploited if the β spectrometer measures the entire released energy, except that of the neutrino. This situation can be realized by using a cryogenic bolometer as the β spectrometer, which at the same time contains the β emitter ^{187}Re.

One disadvantage connected to this method is the fact that one measures always the entire β spectrum. Even for the case of the very low endpoint energy of ^{187}Re, the relative fraction of events in the last eV below E_0 is of order 10^{-10} only (compare to figure 1). Considering the long time constant

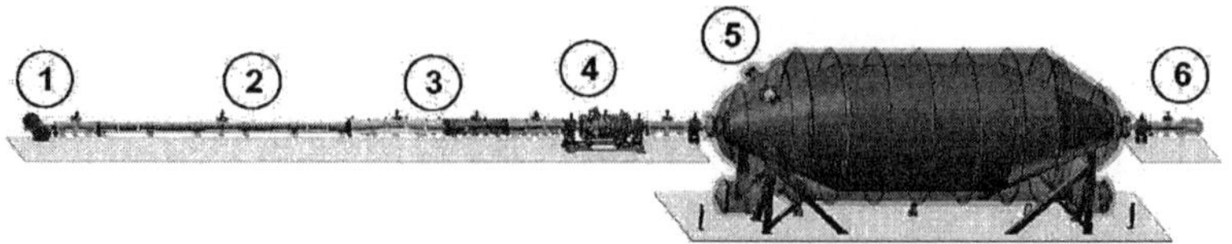

Figure 4. Schematic view of the KATRIN experiment with the rear monitoring and calibration system (1), the windowless gaseous tritium source (WGTS) (2), the differential and cryopumping electron transport section (3), the pre spectrometer (4), the main spectrometer (5) and the electron detector array (6). The main spectrometer has a length of 24 m, a diameter of 10 m, the overall length over the experimental setup amounts to about 70 m. Not shown is the monitor spectrometer.

of the signal of a cryogenic bolometer (typically several hundred μs) only large arrays of cryogenic bolometers can deliver the signal rate needed.

Two groups are working on ^{187}Re β decay experiments at Milan (Mi-Beta) and Genoa (MANU2) using $AgReO_4$ and metallic rhenium, respectively. Although cryogenic bolometers with an energy resolution of 5 eV have been produced with other absorbers, this has yet not been reached for rhenium. The lowest neutrino mass limit of $m(\nu_e) < 15$ ev comes from MiBeta [15]. Further improvements in the energy resolution and the number of crystals are envisaged aiming for a sensitivity of a few eV/c^2.

4. The KATRIN Experiment

The very important tasks presented in the introduction – to distinguish hierarchical from quasi-degenerate neutrino mass scenarios and to check the cosmological relevance of neutrino dark matter for the evolution of the universe – require the improvement of the direct neutrino mass search by one order of magnitude at least.

The KATRIN collaboration has taken this challenge and has proposed to build an ultra-sensitive tritium β decay experiment based on the successful MAC-E-Filter spectrometer technique and a very strong Windowless Gaseous Tritium Source (WGTS) [16] at the Forschungszentrum Karlsruhe, Germany. Figure 4 shows a schematic view of the proposed experimental configuration.

The windowless gaseous tritium source (WGTS) allows for the measurement of the endpoint region of the tritium β decay and consequently the determination of the neutrino mass with a maximum of signal strength combined with a minimum of systematic uncertainties from the tritium source.

The WGTS consists of a 10 m long cylindrical tube of 90 mm diameter filled with molecular tritium gas of high isotopic purity (> 95 %). The tritium gas will be continuously injected by a capillary at the middle and pumped out by a series of differential turbo molecular pump stations at the end giving rise to a density profile over the source length of nearly triangular shape with a total column density of $5 \cdot 10^{17}/\text{cm}^2$ providing a count rate about a factor 100 larger than in Mainz and Troitsk. The β electrons are leaving the WGTS directly to both ends following the magnetic field lines at a magnetic field of 3.6 T whereas the pumped-out tritium gas is then purified and re-circulated. Of special importance is the control of the column density on the 1 per mill level by regulating the pressure in the tritium supply buffer vessel and the temperature of the WGTS tube. To allow a very stable and low WGTS temperature the WGTS tube is placed inside a LNe cryostat.

The electron transport system adiabatically guides β decay electrons from the tritium source to the spectrometer by a system of superconducting solenoids at a magnetic field of 5.6 T. At the same time it is eliminating any tritium flow towards the spectrometer by a differential pumping system consisting of 1 m long tubes inside the magnets alternated by pump ports with turbo molecular pumps yielding a tritium reduction factor of about a factor of 10^8. In the second part the surfaces of the liquid helium cold vacuum tube act as a cryo-trapping section to suppress the tritium partial pressure further to an insignificant level. To reduce the molecular beaming effect, the direct line-of-sight is prohibited by 20 degree bents between each pair of superconducting magnets of 1 m length.

Between the tritium source and the main spectrometer a pre-spectrometer of MAC-E-Filter type will be installed. It acts as an electron pre-filter running at a retarding energy about 100 eV below the endpoint of the β spectrum to reject all β electrons except the very high energetic ones in the region of interest close to the endpoint E_0. This minimizes the chances to cause background by ionization of residual gas in the main spectrometer.

New ideas of background suppression by a nearly massless wire grid and new technical solutions, *e.g.* putting the retarding high voltage directly to the vacuum vessel and thereby using the entire vessel as a retarding electrode, have been developed for the KATRIN main spectrometer. These new ideas are applied also to the KATRIN pre spectrometer, which is already being set up at the Forschungszentrum Karlsruhe, and checked therein in 2005.

A key component of the new experiment will be the large electrostatic main spectrometer with a diameter of 10 m and an overall length of about 24 m. This high-resolution MAC-E-Filter will allow to scan the tritium β decay endpoint at a resolution of $\Delta E = 0.93$ eV, which is – at a much higher luminosity – a factor of 4-5 better than the MAC-E-Filters in Mainz and Troitsk.

To suppress background from the spectrometer stringent vacuum conditions have to be fulfilled. Special selection of materials as well as surface cleaning and out-baking at 350 °C will allow to reach a residual gas pressure of better than 10^{-11} mbar. Also the main spectrometer will be equipped with a large system of a nearly massless wire electrode system at a slightly more negative potential than the vessel to repel secondary electrons created by cosmic rays or environmental radioactivity from the vessel wall. This new method of strong background reduction has been developed and successfully tested at the Mainz spectrometer.

The detector requires high efficiency for electrons at $E_0 = 18.6$ keV and low γ background. A high energy resolution of $\Delta E < 600$ eV for 18.6 keV electrons should suppress background events at different energies. The present concept of the detector is based on a large array of about 1000 PIN photodiodes surrounded by low-level passive shielding and an active veto counter to reduce background.

After publishing the Letter of Intent the KATRIN collaboration has done significant work to increase the sensitivity of the experiment. The major improvements of the setup are the design of a tritium re-circulating and purification system providing a nearly maximum tritium purity of > 95 %, the increase of the diameter of the windowless tritium source from 75 mm to 90 mm and correspondingly of the diameter of the main spectrometer from 7 m to 10 m. Together with the investigation of the systematics and the plans for dedicated experiments to reduce the uncertainties of the former and combined with an optimized distribution of measurement points around the endpoint of the tritium β spectrum the detailed simulations of the KATRIN experiment yield the following: A sensitivity of 0.20 eV/c^2 will be achieved with the KATRIN experiment after 3 years or pure data taking. To this value statistical and systematic uncertainties contribute about equally. This value of 0.20 eV/c^2 corresponds to an upper limit with 90 % C.L. in the case that no neutrino mass will be observed. To the contrary, a non-zero neutrino mass of 0.30 eV/c^2 would be detected with 3 σ significance, a mass of 0.35 eV/c^2 even with 5 σ.

The main systematic uncertainties comprise the inelastic scattering within the tritium source and the stability of the retarding voltage of the main spectrometer. The former will be determined and repeatedly monitored with the help of a high-precision electron gun injecting electrons from the rear system. For the latter, a dedicated high-precision high voltage divider is being developed with the support of the Physikalisch Technische Bundesanstalt at Braunschweig, Germany. For redundancy, the retarding high voltage of the main spectrometer is applied in parallel to a third spectrometer, the monitor spectrometer[c], which continuously measures a sharp electron line. Different well-defined sources are in preparation, *e.g.* a condensed ^{83m}Kr conversion electron source or a cobalt photoelectron source irradiated by γs from a ^{241}Am source. Further systematic uncertainties are the electrical potential distribution within the WGTS, which will be checked by running the WGTS in a second "high temperature regime" of 120-150 K with the conversion electron emitter ^{83m}Kr added to the gaseous molecular tritium and the source contamination by other hydrogen isotopes than tritium, which will be monitored with the help of laser Raman spectroscopy.

The design of the experiment is nearly finished. The ordering process of the major components has been started. The first hardware components are already being set up, *e.g.* the pre spectrometer with its detector. Many dedicated test experiments are being performed at different places to investigate the inner tritium loop, cryo-trapping, vacuum, new background reduction methods, calibration sources, detector and data acquisition, etc. The full setup of the KATRIN experiment will be finished and data taking will start in 2008.

5. Conclusions

Neutrino oscillation experiments have pointed to new physics beyond the Standard Model by proving that neutrinos have non-zero neutrino masses. The next goal is to determine the absolute scale of the neutrino mass due to its high importance for particle physics, astrophysics and cosmology.

The two ways to address the absolute neutrino mass scale with laboratory experiments are the search for neutrinoless double β decay and the direct neutrino mass search by investigating the endpoint region of β decay spectra. Both methods yield complementary information. Neutrinoless double β decay is sensitive – for Majorana neutrinos only – to a coherent

---

[c]The Mainz spectrometer will be modified for this purpose into a high-resolution spectrometer with $\Delta E \approx 1$ eV.

sum of all neutrino mass states $m(\nu_i)$ contributing to the electron neutrino $m_{ee} = \sum_i |U_{ei}^2 \cdot m(\nu_i)|$ whereas the direct neutrino mass is sensitive to the incoherent sum $m^2(\nu_e) = \sum_i |U_{ei}|^2 \cdot m^2(\nu_i)$. In the former case complex phases of the neutrino Majorana mixing matrix U can lead to a partial cancellation. Also the large uncertainties of the nuclear matrix element and the possibility that the exchange of other more exotic particles adds to the neutrinoless double β decay signal disfavors double β decay for a precise neutrino mass determination. On the other hand the high sensitivity of the next generation of double β decay experiments and their unique possibility to prove the Majorana nature of neutrinos underlines the very high importance of neutrinoless double β decay experiments.

The investigation of the endpoint spectrum of the tritium β decay is still the most sensitive direct method. The tritium β decay experiments at Mainz and Troitsk have been finished yielding upper limits of about $2\ \mathrm{eV}/\mathrm{c}^2$. The new KATRIN experiment will enhance the sensitivity further by one order of magnitude down to $0.2\ \mathrm{eV}/\mathrm{c}^2$.

Acknowledgments

The work by the author for the KATRIN experiment is supported by the German Bundesministerium für Bildung und Forschung and within the virtual institute VIDMAN by the Helmholtz Gemeinschaft.

References

1. K. Heeger, *these proceedings*
2. S. Hannestad, *these proceedings*
3. S.W. Allen *et al.*, arXiv:astro-ph/0303076
4. J.F. Beacom *et al.*, arXiv:astro-ph/0404585
5. Ch. Weinheimer, ch, 2 of "Massive Neutrinos", ed. G. Altarelli and K. Winter, *Springer Tracts in Modern Physics*, Springer, 2003, p25-52
6. K.V. Klapdor-Kleingrothaus *et al.*, Phys. Lett. B **586**, 198 (2004)
7. A. Picard *et al.*, Nucl. Instr. Meth. B **63**,345 (1992)
8. L. Fleischmann *et al.*, Eur. Phys. J. B **16**, 521 (2000)
9. V.N. Aseev *et al.*, Eur. Phys. J. D **10**, 39 (2000)
10. B. Bornschein *et al.*, J. Low Temp. Phys. 131, (2003)
11. C. Kraus *et al.*, Eur. Phys. J. C **33** s805 (2003)
12. W. Kolos *et al.*, Phys. Rev. A **37**, 2297 (1988)
13. V.M. Lobashev *et al.*, Phys. Lett. B **460** (1999) 227
14. V.M. Lobashev , Nucl. Phys. A **719**, 153c (2003)
15. M. Sisti *et al.*, Nucl. Instr. Meth. A **520** (2004) 125
16. A. Osipowicz *et al.*, arXiv:hep-ex/0109033

PARAMETRIZATIONS OF THE SEESAW
or
CAN THE SEESAW BE TESTED?

SACHA DAVIDSON *

Dept of Physics, University of Durham, Durham, DH1 3LE, England
E-mail: sacha.davidson@durham.ac.uk

This proceedings contains a review, followed by a more speculative discussion. I review different coordinate choices on the 21-dimensional parameter space of the seesaw, and which of these 21 quantities are observable. In MSUGRA, there is a 1-1 correspondance between the parameters, and the interactions of light (s)particles. However, not all of the 21 can be extracted from data, so the answer to the title question is "no". How to parametrise the remaining unknowns is confusing— different choices seem to give contradictory results (for instance, to the question "does the Baryon Asymmetry depend on the CHOOZ angle?"). I speculate on possible resolutions of the puzzle.

1. Introduction

The seesaw mechanism [1] is a theoretically elegant way to get the small neutrino masses we observe. It predicts that the light neutrino masses are majorana, which could be verified in neutrinoless double β decay experiments. In the absence of Supersymmetry, it predicts that lepton flavour violation (LFV), and CP violation are suppressed by powers of the neutrino mass, making the rates very low outside the neutrino sector. On the other hand, if spartners were discovered, for instance at the LHC, observable CP and flavour violation can be imprinted by the seesaw into the slepton mass matrices. Experimentally verifying these predictions would increase our confidence in the seesaw. Measuring something different—for instance majorana masses, no SUSY, and large neutrino magnetic moments—would indicate that there is other new physics, or more new physics in the lepton sector than just the seesaw (see *e.g.* Smirnov, in this volume). The aim

---

*work supported by a PPARC Advanced Fellowship

here is to ask if we can *test* the seesaw, or as discussed below, a particular implementation of the seesaw mechanism.

This proceedings is written from a bottom-up phenomenological perspective. I want to make as few assumptions as possible about the theory at scales above m_W, so I assume the particle content is the Standard Model (SM), or the MSSM with universal soft masses, plus three ν_R, and allow all possible renormalisable interactions. This gives the Lagrangian (in the SM case)

$$ L = Y_e \bar{e}_R H_d \cdot \ell_L + Y_\nu \bar{\nu}_R H_u \cdot \ell_L + \frac{M}{2} \bar{\nu^c}_R \nu_R + h.c. \tag{1} $$

where ℓ are the lepton doublets, $\bar{\nu^c}_R = (C\nu_R^*)^\dagger \gamma_0$, and generation indices are suppressed. The index order on the Yukawa matrices is right-left.

To test this implementation of the seesaw mechanism, we need to

(1) extract the unknown parameters of eqn (1) from data
(2) predict an additional observable calculated from those parameters
(3) verify the prediction

These proceedings discuss the first step. If it could be accomplished successfully, we could calculate the baryon asymmetry produced in various leptogenesis[2,3,4] mechanisms (see Hambye and Raidal in this volume), which would be a fabulous cross-check of particle physics and cosmology.

There are many other versions of the seesaw (2 ν_R, type II with scalar triplets, with extra singlets...), which are motivated from various theoretical perspectives (see T Hambye in this volume). The model used here contains three ν_R because there are three generations, and only three ν_R because it is useful to know how well the simple model works before adding complications.

I want to test the seesaw *mechanism*, rather than a particular model, so GUT models, textures, and theoretical considerations of "naturalness" are avoided (insofar as possible). The seesaw mechanism can accomodate *any* neutrino masses and mixing angles (And almost any sneutrino mass matrix[6]). Particular models may prefer certain ranges for observables, so data can provide hints about the theory that gives the Lagrangian of eqn (1). This is discussed elsewhere in this volume (G Ross and P Ramond). However, if these theoretical expectations are not fulfilled, it is difficult to know if the model was wrong, or if there is more new physics in addition to the seesaw mechanism.

2. Parametrisations

Twenty-one parameters are required[5] to fully determine the Lagrangian of eqn (1). Three of the possible ways these can be chosen are discussed here.

The usual **"top-down"** description of the theory is as follows. At energy scales $\Lambda \gtrsim M$, where the ν_R are propagating degrees of freedom, one can always choose the ν_R basis where the mass matrix M is diagonal, with positive and real eigenvalues: $M = D_M$. Similarly, one can choose the ℓ basis such that the charged lepton Yukawa Y_e is diagonal on its LH indices: $Y_e^\dagger Y_e = D_{Y_e}^2$. The remaining neutrino Yukawa matrix Y_ν is an arbitrary complex matrix, from which three phases can be removed by phase redefinitions on the ℓ_i. It is therefore described by 9 moduli and 6 phases, giving in total 21 real parameters for the seesaw. See [5] for a more elegant counting, in particular of the phases.

To relate various parametrisations of the seesaw, it is useful to diagonalise Y_ν, which can be done with independent unitary transformations on the left and right:

$$Y_\nu = V_R^\dagger D_{Y_\nu} V_L \tag{2}$$

So in the top-down approach, the lepton sector can be described by the nine eigenvalues of D_M, D_{Y_ν} and D_{Y_e}, and the six angles and six phases of V_L and V_R. Notice that in this parametrisation, the inputs are masses and coupling constants of the propagating particles at energies Λ, so it makes "physical" sense.

The effective mass matrix m of the light neutrinos can be calculated, in the D_{Y_e} basis (charged lepton mass eigenstate basis):

$$m \equiv \kappa \langle H_u \rangle^2 = Y_\nu^T D_M^{-1} Y_\nu \langle H_u \rangle^2 = V_L^T D_{Y_\nu} V_R^* D_M^{-1} V_R^\dagger D_{Y_\nu} V_L \langle H_u \rangle^2 \tag{3}$$

κ is introduced to avoid the Higgs vev $\langle H_u \rangle$ cluttering up formulae. The leptonic mixing matrix U is extracted by diagonalising κ:

$$\kappa = U^* D_\kappa U^\dagger \tag{4}$$

where $D_\kappa = diag\{\kappa_1, \kappa_2, \kappa_3\}$, and U is parametrised as

$$U = \hat{U} \cdot \text{diag}(1, e^{i\alpha}, e^{i\beta}) \quad . \tag{5}$$

α and β are "Majorana" phases, and $\hat{U}$ has the form of the CKM matrix

$$\hat{U} = \begin{bmatrix} c_{13}c_{12} & c_{13}s_{12} & s_{13}e^{-i\delta} \\ -c_{23}s_{12} - s_{23}s_{13}c_{12}e^{i\delta} & c_{23}c_{12} - s_{23}s_{13}s_{12}e^{i\delta} & s_{23}c_{13} \\ s_{23}s_{12} - c_{23}s_{13}c_{12}e^{i\delta} & -s_{23}c_{12} - c_{23}s_{13}s_{12}e^{i\delta} & c_{23}c_{13} \end{bmatrix} \quad . \tag{6}$$

Alternatively, the (type I) seesaw Lagrangian of eqn (1) can be described with inputs from the left-handed sector [6]. This is refered to as a **"bottom-up"** parametrisation, because the left-handed (SU(2) doublet) particles have masses $\lesssim$ the weak scale. D_{Y_e}, U and D_κ, can be taken as a subset of the inputs. To identify the remainder, imagine sitting in the ℓ basis where κ is diagonal, so as to emphasize the parallel between this parametrisation and the previous one (this is similar to the ν_R basis being chosen to diagonalise M). If one knows $Y_\nu^\dagger Y_\nu \equiv W_L D_{Y_\nu}^2 W_L^\dagger$ in the D_κ basis, then the ν_R masses and mixing angles can be calculated:

$$M^{-1} = D_Y^{-1} W_L^* D_\kappa W_L^\dagger D_Y^{-1} = V_R^* D_M^{-1} V_R^\dagger \tag{7}$$

In this parametrisation, there are three possible basis choices for the ℓ vector space: the charged lepton mass eigenstate basis (D_{Y_e}), the neutrino mass eigenstate basis (D_κ), and the basis where the Y_ν is diagonal. The first two choices are physical, that is, U rotates between these two bases. D_{Y_e},D_κ and U contain the 12 possibly measurable parameters of the SM seesaw. The remaining 9 parameters can be taken to be D_{Y_ν} and V_L (or $W = V_L U$). In SUSY one can hope to extract these parameters from the slepton mass matrix.

The **Casas-Ibarra** [7] parametrisation is very convenient for calculations. It uses D_M, D_κ and D_{Y_e} as inputs, and the transformations U and R, which go between the bases where these matrices are diagonal. U is the usual leptonic mixing matrix. The matrix $R = D_M^{-1/2} Y_\nu D_\kappa^{-1/2}$, is a complex *orthogonal* matrix, which transforms between the D_M and D_κ bases. (Since M and κ are respectively in the RH and LH neutrino vector spaces, it is unsurprising that the transformation matrix is not unitary.) R can be written as $R = \text{diag}\{\pm 1, \pm 1, \pm 1\}\hat{R}$ where the ± 1 are related to the CP parities of the N_i, and $\hat{R}$ is an orthogonal matrix with complex angles:

$$\hat{R} = \begin{bmatrix} c_{13}c_{12} & c_{13}s_{12} & s_{13} \\ -c_{23}s_{12} - s_{23}s_{13}c_{12} & c_{23}c_{12} - s_{23}s_{13}s_{12} & s_{23}c_{13} \\ s_{23}s_{12} - c_{23}s_{13}c_{12} & -s_{23}c_{12} - c_{23}s_{13}s_{12} & c_{23}c_{13} \end{bmatrix} . \tag{8}$$

The aim of this proceedings is to reconstruct the RH seesaw parameters from the LH ones, many of which are accessible at low energy. However,as discussed in the following section, reconstruction is impossible. We can at best try to establish relations between observables, which turns out to be quite confusing. R will be helpful in discussing these puzzles.

In summary, the lepton sector of the SM + seesaw can be parametrised with D_{Y_e}, the real eigenvalues of two more matrices, and the transformations among the bases where the matrices are diagonal. The matrices-to-be-diagonalised can be chosen in various ways:

(1) " top-down"—input the ν_R sector: D_M, $D_{Y_\nu Y_\nu^\dagger}$, and V_R and V_L.
(2) " bottom-up"—input the ν_L sector: D_κ, $D_{Y_\nu^\dagger Y_\nu}$, and V_L and U.
(3) "intermediate"—the Casas-Ibarra parametrization: D_M, D_κ, and U and a complex orthogonal matrix R.

3. (Supersymmetric) reconstruction?

If the matrices D_κ, D_{Y_e}, $D_{Y_\nu^\dagger Y_\nu}$, V_L and U were known, it would be possible to reconstruct the masses and mixing angles of the ν_R. Can the elements of these matrices be determined [6]?

We know the masses of the charged leptons, so we know D_{Y_e} (modulo $\tan\beta$ in SUSY models).

We know two mass differences in the neutrino sector. If the light neutrinos are degenerate, measuring the overall scale of their masses is possible and would determine D_κ. However, if the mass pattern is hierarchical or inverse hierarchical, we would know only κ_3 and κ_2. See the contribution of K Heeger, for present and future accuracy on D_κ, and U.

In the mixing matrix U, we currently know two angles. We hope to measure the third, and also the "Dirac" phase δ. But the "majorana" phases appear only in slow lepton number changing processes, so at the moment do not seem experimentally accessible [8].

The remaining parameters to be determined are the eigenvalues of Y_ν, and the matrix V_L. In supersymmetric models Y_ν contributes via loops to the slepton mass matrix. Consider a model, such as gravity-[9] or anomaly-mediated [10] SUSY breaking, where the soft masses are universal at a scale $\Lambda > M_3$. In renormalisation group running between Λ and m_W, the slepton mass matrix will acquire flavour off-diagonal terms, due to loops involving the ν_R (see Masiero in this volume). Using the leading log approximation for the RG running, the sneutrino mass matrix, in the D_{Y_e} basis, is:

$$\left[m_{\tilde{\nu}}^2\right]_{ij} \simeq (\text{diag part}) - \frac{3m_0^2 + A_0^2}{8\pi^2}(\mathbf{Y}_\nu^\dagger)_{ik}(\mathbf{Y}_\nu)_{kj} \log\frac{\Lambda}{M_k} \qquad (9)$$

where m_0 and A_0 are the universal soft parameters at scale Λ.

It is tantalising that the seesaw contribution to flavour violation in the sleptons is potentially observable, and depends on the heavy neutrino masses in a different way than κ. *If* we could determine $\left[m_{\tilde{\nu}}^2\right]$ exactly (the three masses, three mixing angles, and three phases), and *if* we take seriously the assumption of universal soft masses, then we could reconstruct the renormalisable interactions of the high-scale seesaw—that is, the ν_R masses and Yukawa couplings—from the mass matrices and mixing angles of weak-scale particles.

Unfortunately, neither of these conditions is likely to be fulfilled. Firstly, not all the parameters of $\left[m_{\tilde{\nu}}^2\right]$ can be measured with the required accuracy. The diagonal elements of the second term of eqn (9) shift $m_{\tilde{\nu}}^2$ by of order $y_i^2\%$, so a large Y_ν eigenvalue ~ 1 could have a measurable effect. However, if Y_ν has a hierarchy similar to the quark Yukawas, the effects of the first and second generation y_i are (undetectably) small.

The flavour-changing elements of eqn (9) could be seen at colliders [11], and induce rare decays, such as $\mu \to e\gamma$ [12]. A very optimistic experimental sensitivity of order $BR(\tau \to \ell\gamma) \sim 10^{-9}$ (the current limit is $\sim 10^{-7}$), could probe $|[V_L]_{3\ell}[V_L]^*_{3\tau}y_3^2| \gtrsim 10^{-(1\div 2)}$. $\mu - e$ flavour violation is more encouraging: there are plans to reach $BR(\mu \to e\gamma) \sim 10^{-13}$, which would be sensitive to $|[V_L]_{3e}[V_L]^*_{3\mu}y_3^2| \gtrsim 10^{-(3\div 4)}$. However, to extract a "measurement" of either of the $|[V_L]_{3\ell}|$ from rare decays would require knowing all the masses and mixing angles for the other SUSY particles contributing to the decay.

For hierarchical Y_ν eigenvalues, eqn (9) implies that the three off-diagonal elements of $\left[m_{\tilde{\nu}}^2\right]$, are determined by two matrix elements of V_L. So one angle of V_L is unknown, and there should be some correlation between $\left[m_{\tilde{\nu}}^2\right]_{\tau\mu}$, $\left[m_{\tilde{\nu}}^2\right]_{\tau e}$, and $\left[m_{\tilde{\nu}}^2\right]_{\mu e}$. Notice, however, that this is a prediction of hierarchical Y_ν. In the bottom-up parametrisation, the slepton mass matrix determines V_L and D_{Y_ν}, rather than the seesaw making predictions for $\left[m_{\tilde{\nu}}^2\right]$.

Now we come to the three phases of V_L. To extract all of these is quite hypothetical; it would require three independent measurements of CP violation in the sleptons. Two possibilities at colliders are charged lepton asymmetries in slepton decays [13], and sneutino-anti-sneutrino oscillations [14]. The slepton phases also contribute to CP violating observables in the leptons, in conjunction with phases from other SUSY particles. This is discussed in this volume by Hisano.

The second objection to extracting seesaw parameters from eqn (9), is that we do not know that soft masses are universal. It is a reasonable assumption in top-down analyses, because we know that flavour violation mediated by sparticles must be suppressed. But I know of no way to distinguish contributions to $\left[m_{\tilde{\nu}}^2\right]$ that come from the RG running with the seesaw, from those that come from non-universal soft masses, threshold effects, other particles with flavour off-diagonal couplings, etc... So measuring $[m_\nu^2]$ exactly could be used to set an upper bound on the see-saw contributions (if one makes the reasonable assumption that there are no cancellations among different contributions), but would not determine them.

It is also possible-in-principle to reconstruct the **non-SUSY** seesaw: Broncano *et al.* [15] observed that the 21 parameters can be extracted from the coefficients of dimension 5 and 6 operators in the Standard Model. However, the coefficients of the dimension 6, lepton number conserving operators are suppressed by two powers of the ν_R mass, so are (unobservably) small.

In summary, the parameters of the type I seesaw cannot be extracted from data. This should hardly be surprising—we do not usually expect to reconstruct high-scale theories (*e.g.* which GUT, and how does it break?) from weak-scale observations. So why do we even ask if it is possible in the seesaw? I am aware of two peculiarities, which make the seesaw "reconstructable in principle": the ν_R only have interactions with light particles (via Y_ν), and the effective operators induced at low energy are experimentally accessible (in principle!) for all flavour indices. To see why these features are significant, compare to proton decay— which I assume to be mediated by a "triplet higgsino" dressed with a squark loop. However accurately we mesure every available proton decay channel, we cannot determine the mass and couplings of the triplet higgsino, because we must always sum over squark flavours in the loop, and we only measure proton decay with first generation quarks in the initial state (unlike the three generation ν and $\tilde{\nu}$ mass matrices).

4. Independence, orthogonality and relations when we cannot reconstruct

The 21 parameters of the seesaw cannot be determined from observation, but some sort of partial reconstruction, using the available data, could be possible. This turns out to be much more confusing than one would anticipate. To identify the problem, imagine calculating the baryon asymmetry as a function of parameters separated into three categories: those we know now, those we hope to know, and those we will never know. It then seems straightforward to study how the asymmetry depends on, for instance, θ_{13}. But in practise it is anything but transparent (see eqn 11): the asymmetry is independent of U in the Casas-Ibarra parametrisation, but does depends on U in the bottom-up version. That is, the choice of parametrisation for the unmeasurables, changes the dependence of one observable (the baryon asymmetry) on another (θ_{13}). It would be better to ask "is ϵ_1 *sensitive* to θ_{13}?"—this has a unique and useful answer, as discussed in the next section.

The aim of this section is to explore how different coordinates on seesaw parameter space depend on each other, and what we mean by "depends on" and "independent" . I start by reviewing some contradictory statements which can be derived using various parametrisations. Then I present a toy model using parametrisations of the plane, where these same contradictions arise, and where the resolution is obvious. Lastly I suggest how the analogy of the plane could be related to the seesaw.

It has been claimed in various papers that ϵ_1, the CP asymmetry of thermal leptogenesis, is independent of the leptonic mixing matrix U. This seems intuitively reasonable, because leptogenesis involves the ν_R, and is independent of Y_e. In the limit of hierarchical ν_R:

$$\epsilon_1 \simeq -\frac{3M_1}{8\pi[Y_\nu Y_\nu^\dagger]_{11}}\Im\{Y_\nu\kappa^*Y^T\} = -\frac{3M_1}{8\pi}\frac{\Im\{R_{1j}^2\kappa_j^2\}}{|R_{1k}|^2\kappa_k} \tag{10}$$

$$\propto \Im\left\{V_L U D_\kappa^3 U^T V_L^T D_{Y_\nu}^{-2} V_L^* U^* D_\kappa U^\dagger V_L^\dagger D_{Y_\nu}^{-2}\right\} \tag{11}$$

where the second equality of (10) is in the Casas-Ibarra parametrisation. To translate eqn (10) into bottom-up coordinates, requires calculating the mass and eigenvector (first colomn of V_R) of ν_{R1}, which gives short analytic formulae in some limits. However, for hierarchical ν_R (the limit in which eqn (10) is valid), ϵ_1 is proportional to a Jarlskog invariant[16], which gives eqn (11). We see that U does not appear in the expression for ϵ_1 in Casas-Ibarra, but does appear in the bottom-up parametrisation. So it is unclear whether ϵ_1 depends on U—what do we mean by "depend"? If a mathematical definition can be constructed, then there should be a unique answer.

We can draw an analogy between coordinate choices on a manifold, and parametrisation choices for the seesaw. Different coordinate choices on the plane, all used with the same [a] metric $\delta_{\alpha\beta}$, give confusing results that ressemble the puzzle about whether ϵ depends on θ_{13}. If we use the appropriate metrics, results are independent of the coordinate system, which can be chosen for calculational convenience. There is no metric given on "seesaw parameter space", but this analogy suggests that inventing one would resolve the confusion.

---

[a]Of course, we know that this is wrong; the metric should change with the coordinate system.

Consider two choices of coordinates on the upper half plane:

(1) the cartesian (y, z) with $y > 0$ and metric $g_{\alpha\beta} = I$.
(2) $R = \sqrt{y^2 + z^2}$ and $Z = z$, with $R > Z$ and metric

$$g_{AB} = \frac{R^2}{R^2 - Z^2} \begin{bmatrix} 1 & -Z/R \\ -Z/R & 1 \end{bmatrix} \tag{12}$$

These are equally good coordinate choices for the same flat 2-d surface. The seesaw analogy we want to address is: does R "depend" on Z?

In any coordinate system, the coordinates vary independently. So by definition

$$\frac{\partial R}{\partial Z} = 0 \tag{13}$$

which could be taken to mean that "R is independent of Z". A more intuitive quantity is the total derivative, or by analogy with general relativity, the change of R, treated as a scalar function, along the curve of varying Z:

$$\left(\frac{\partial R}{\partial y} \ \frac{\partial R}{\partial z} \right) \begin{bmatrix} 1 & 0 \\ 0 & 1 \end{bmatrix} \left(\begin{matrix} \frac{\partial Z}{\partial y} \\ \frac{\partial Z}{\partial z} \end{matrix} \right) = \frac{Z}{R} \tag{14}$$

which is the expected answer. Notice that we need to know how to transform to cartesian coordinates (equivalently, the metric on R, Z space) for this calculation.

To summarise, ϵ and θ_{13} are functions of seesaw parameter space, and can be defined such that

$$\frac{\partial \epsilon}{\partial \theta_{13}} = 0 \tag{15}$$

by a suitable choice of parameters. However, a better measure of whether ϵ depends on θ_{13} would be something like eqn (14). To evaluate this, we need a metric on seesaw parameter space [b].

How to choose this metric? The top-down parametrisation is the most natural, so in two generations, the obvious choice is to take $\{D_{Y_e}, V_L, D_{Y_\nu}, V_R, D_M\}$ as cartesian coordinates. With this metric, it is straightforward to show that ϵ does vary with the angle of the matrix U. (This is simple, because in 2 generations it is easy to calculate the angle of W_L in terms of RH parameters.) However, in three generations, "distance"

---

[b] When doing seesaw parameter space scans, one must choose the distribution of input points in parameter space. This number density ("measure on parameter space") is motivated by some theoretical model for the origin of seesaw parameters, so is not intrinsic to the seesaw. Therefore it is not related to this "metric".

on the unitary transformations should be invariant under reparametrizations ($e.g. V_R = U_{12}U_{13}U_{23}$ or $= U_{23}U_{13}U_{12}$), suggesting a metric similar to the one for polar coordinates.

It *is* clear, from this section, that the "dependance" of one seesaw observable on another in *not* clear. For example, the coordinates on seesaw parameter space can be chosen such that either ϵ_1 is a function of the MNS matrix, or it is not. This confusion can be resolved by inventing a notion of "orthogonality" for coordinates, that is, a metric on parameter space. However, the metric seems an esoteric solution, and how to find the correct one is not obvious.

5. Rethink: what happens in the Standard Model?

In the Standard Model, the Lagrangian parameters can be reconstructed from data—in fact, there are many more measurements than parameters, so the SM is tested at part-per-mil accuracy. But some parameters are better determined than others, so the difference with respect to the seesaw is just the size of the error bars.

In the SM, the key is the *sensitivity* of data to a parameter. For instance, to determine m_t from electroweak data, one should choose an observable with large m_t^2 corrections, and a parametrisation (*eg*, definition of s_W^2), where these are easy to identify. If the parameters other than m_t are sufficiently well determined, a range for m_t can be extracted [c]. This is self-evident; the data allows a model to occupy a subset (often a multi-dimensional ellipse) in parameter space.

We say an observable Ob is sensitive to a parameter P, if measuring Ob constrains P to sit in a certain range. Conversely, Ob is insensitive to P, if measuring Ob is consistent with any value of P (possibly because one ajusts other unknowns to compensate for variations in P).

So returning to the seesaw, one could conclude that "does ϵ_1 depend on θ_{13}?" is the wrong question. If instead, one asks "is ϵ_1 *sensitive* to θ_{13}?", then the answer at present is clearly no. It is easy to see, in the parametrisation using R, that any value of θ_{13} is consistent with the observed baryon a symmetry.

---

[c]In reality this is a crude approx to doing a combined fit.

6. Summary

The seesaw generates small neutrino masses, by introducing heavy majorana ν_Rs, which share a Yukawa coupling with the lepton doublets of the Standard Model. It is theoretically possible to establish a 1-1 correspondance between observables (in the quantum mechanical sense), and the 21 parameters of the seesaw (type I, 3 generations). This correspondance, and two other parametrisations of the seesaw, are discussed in section 2. Unfortunately, this peculiarity of the seesaw does not mean the parameters can be extracted from data; some of the "observables" are not realistically measurable, and others cannot be determined accurately enough (see section 3). This makes the seesaw mechanism difficult to test, according to the definition of test outlined in the introduction.

This is sad because the dream test of the seesaw would be to extract its parameters from data, calculate the baryon asymmetry produced in leptogenesis—and get the right answer. More realistically, we can ask "is the baryon asymmetry *sensitive* to any of the seesaw's measurable parameters?" For instance, does generating the baryon asymmetry by a specific leptogenesis scenario imply that θ_{13}, or the phase δ, should occupy restricted ranges? Again, the answer sadly seems to be "no". More generally, one could study which observables are sensitive to which parameters, *e.g.* would $BR(\mu \to e\gamma) \neq 0$ restrict the majorana phases of MNS [d]? Most studies to date have looked at whether an observable O "depends" on a parameter P—which is not such a useful question, because the answer depends on the parametrisation. Section 4 attempts to construct a parametrisation-independent definition of "depend", not very successfully. So it is better to ask if O is *sensitive* to P, which does have a unique answer, as discussed in section 5.

Acknowledgements

I wish the seesaw many happy returns, and thank the organisers for a very enjoyable birthday celebration, with many interesting speakers and participants. In particular, I thank S Vempati for clarifying questions, S Lavignac for discussions and a careful reading of the manuscript, and S Petcov for many interesting discussions about seesaw parametrisations.

---

[d]In the Casas-Ibarra parametrisation, $BR(\mu \to e\gamma)$ depends on theses phases [18]

References

1. P. Minkowski, Phys. Lett. B **67** (1977) 421; M. Gell-Mann, P. Ramond and R. Slansky, *Proceedings of the Supergravity Stony Brook Workshop*, New York 1979, eds. P. Van Nieuwenhuizen and D. Freedman; T. Yanagida, *Proceedinds of the Workshop on Unified Theories and Baryon Number in the Universe*, Tsukuba, Japan 1979, ed.s A. Sawada and A. Sugamoto; R. N. Mohapatra, G. Senjanovic, *Phys.Rev.Lett.* **44** (1980)912.
2. M. Fukugita and T. Yanagida, Phys. Lett. B **174** (1986) 45.
3. *see e.g.* W. Buchmüller and M. Plümacher, Int. J. Mod. Phys. A **15** (2000) 5047 [arXiv:hep-ph/0007176], *or* G. F. Giudice, A. Notari, M. Raidal, A. Riotto and A. Strumia, Nucl. Phys. B **685** (2004) 89 [arXiv:hep-ph/0310123].
4. *a list of other possibilities can be found in, e.g.,* L. Boubekeur, S. Davidson, M. Peloso and L. Sorbo, Phys. Rev. D **67** (2003) 043515 [arXiv:hep-ph/0209256].
5. G. C. Branco, L. Lavoura and M. N. Rebelo, Phys. Lett. B **180** (1986) 264. A. Santamaria, Phys. Lett. B **305** (1993) 90 [arXiv:hep-ph/9302301].
6. S. Davidson and A. Ibarra, JHEP **0109** (2001) 013 [arXiv:hep-ph/0104076].
7. J. A. Casas and A. Ibarra, Nucl. Phys. B **618** (2001) 171 [arXiv:hep-ph/0103065].
8. V. Barger, S. L. Glashow, P. Langacker and D. Marfatia, Phys. Lett. B **540** (2002) 247 [arXiv:hep-ph/0205290].
9. F. Borzumati and A. Masiero, Phys. Rev. Lett. **57** (1986) 961.
10. M. Ibe, R. Kitano, H. Murayama and T. Yanagida, arXiv:hep-ph/0403198.
11. N. Arkani-Hamed, H. C. Cheng, J. L. Feng and L. J. Hall, Phys. Rev. Lett. **77** (1996) 1937 [arXiv:hep-ph/9603431].
12. *see, e.g.* [7], *or* J. Hisano and D. Nomura, Phys. Rev. D **59** (1999) 116005 [arXiv:hep-ph/9810479]. S. Lavignac, I. Masina and C. A. Savoy, Nucl. Phys. B **633** (2002) 139 [arXiv:hep-ph/0202086].
13. N. Arkani-Hamed, J. L. Feng, L. J. Hall and H. C. Cheng, Nucl. Phys. B **505** (1997) 3 [arXiv:hep-ph/9704205].
14. Y. Grossman and H. E. Haber, Phys. Rev. Lett. **78** (1997) 3438 [arXiv:hep-ph/9702421].
15. A. Broncano, M. B. Gavela and E. Jenkins, Phys. Lett. B **552** (2003) 177 [arXiv:hep-ph/0210271].
16. S. Davidson and R. Kitano, JHEP **0403** (2004) 020 [arXiv:hep-ph/0312007].
17. S. Davidson, JHEP **0303** (2003) 037 [arXiv:hep-ph/0302075].
18. S Petcov at ν'04.

Horizontal Gauge Symmetry and Masses of Neutrinos

Tsutomu YANAGIDA

Department of Physics, Tohoku University, Sendai 980

Recently several authors have studied a possible unification of electronic and muonic matter by adding the horizontal local-symmetry, $SU_F(2)$,[1]~[3] to the weak and electromagnetic $SU(2) \times U(1)$.[4] As a consequence of gauging the symmetry, the conservation of muon number is violated. The exchange of horizontal gauge bosons, S_μ^a, also induce the superweak type of CP-nonconservation.[1] From the data on CP-violation in $K_L^0 \to 2\pi$ decay, the effective coupling constant, G_S, of S_μ^a with leptons and quarks is determined as $G_S \sim 10^{-15}$ GeV^{-2} unless the accidental cancellation occurs.[1] The strength is enough weak to avoid unwanted flavour-changing transitions.

262

If there exist six leptons and six quarks, we extend the horizontal $SU_F(2)$ to $SU_F(3)$. The weak-SU(2) doublet-and singlet-fermions transform as triplets under the horizontal $SU_F(3)$. The triangle anomalies[5] appearing in the lepton sector can be removed by assuming right-handed neutrinos.[*]

The purpose of this short note is to point out the possibility that the spontaneous breakdown of the symmetry generates the masses of right-handed and left-handed neutrinos and each neutrino becomes a massive Majorana particle. The mass of each particle may be $m_\xi \sim 10^5$ GeV and $m_\zeta \sim 10 \sim 10^4$ eV, respectively.

The assignment for leptons is the following:

$$
SU(2) \quad \psi_L = \begin{pmatrix} \nu_e & \nu_\mu & \nu_\tau \\ e & \mu & \tau \end{pmatrix}_L , \qquad
\begin{matrix} \nu_R = [\nu_e \ \ \nu_\mu \ \ \nu_\tau]_R \ , \\ (3, \ \ 1, \ \ 0) \\ \ell_R = [e \ \ \ \mu \ \ \ \tau]_R \ . \\ (3, \ \ 1, \ \ -2) \end{matrix}
\tag{1}
$$
$$
(3, \quad 2, \quad -1)
$$
$$
\leftarrow SU_F(3) \rightarrow \qquad\qquad \leftarrow SU_F(3) \rightarrow
$$

Here the first two values in each parenthesis denote the representation dimensions of $SU_F(3) \times SU(2)$ and the last one the U(1) hypercharge.

In order to make the horizontal gauge bosons, S_μ^a ($a = 1 \sim 8$), heavy sufficiently we introduce a Higgs scalar, $\chi_{ij} = (6, 1, 0)$. Another Higgs scalar, $\phi^a = (8, 2, -1)$ is also assumed to break the symmetry $SU(2) \times U(1)$ surviving down to the electromagnetic one.

---

[*] It is possible to assign fermions as triplets of $SU_F(2)$. In this case the anomalies are not generated without right-handed neutrinos (see Ref.2)).

The general form of the neutrino's mass term is given by

$$\mathcal{L}_{mass}^{\nu} = \frac{1}{2}G_{\chi}^{\nu} \, \overline{(\nu_R^i)^c} \, \langle x_{ij} \rangle \, \nu_R^i + G_{\phi}^{\nu} \, \bar{\psi}_L \langle \phi^a \rangle \lambda^a \nu_R + h.c, \tag{2}$$

where $(\nu_R^i)^c$ denotes the charge-conjugated field of ν_R^i and $(\nu^1 \ \nu^2 \ \nu^3)_R$ corresponds $(\nu_e \ \nu_\mu \ \nu_\tau)_R$. Eq(1) represents a 6×6 mass matrix among six Majorana particles, $\xi^i = \nu_R^i + (\nu_R^i)^c$ and $\zeta^i = \nu_L^i + (\nu_L^i)^c$ $(i = 1 \sim 3)$. The masses of these neutrinos are roughly obtained as

$$m_\xi \sim G_\chi^\nu \langle \chi \rangle \ , \tag{3}$$

$$m_\zeta \sim \frac{G_\phi^\nu \langle \phi \rangle}{G_\chi^\nu \langle \chi \rangle} \cdot G_\phi^\nu \langle \phi \rangle \ , \tag{4}$$

where $\langle \chi \rangle$ and $\langle \phi \rangle$ are vacuum-expectation values averaged and $\langle \phi \rangle / \langle \chi \rangle \sim 10^{-5}$.[1] To estimate magnitudes of these masses we tentatively assume that all Yukawa coupling constants are same order, $G_\chi^\nu \sim G_\phi^\nu \sim G_\phi^\ell$, where G_ϕ^ℓ is the coupling constant for leptons, $\ell = (e \ \mu \ \tau)$. Then we find $m_\xi \sim 10^5$ GeV and $m_\zeta \sim 10 \sim 10^4$ eV. Neutrino oscillations are also expected, but the oscillation length depends on the details of the mass matrix.

Finally we stress that the present scheme of the symmetry breaking is a realistic one in the sense that the Higgs scalars ϕ^a and χ_{ij} can be considered as bound states of fermion-antifermion $(\bar{\nu}_R \psi_L + \bar{\ell}_R \psi_L + \cdots)$ and fermion-fermion $(\nu_R \nu_R)$, respectively.[6] It is, therefore, due to the large violation of the horizontal symmetry that right-handed neutrinos disappear at low energy regions.

References

1) T. Maehara and T. Yanagida, Prog. Theor. Phys. 60 (1978), 822; 61 (1979), No.5.

2) F. Wilzek and A. Zee, Phys. Rev. Letters 42 (1979), 421.

3) K. Akama, Y. Chikashige and T. Matsuki, INS-Report-288 (1977). H. Terazawa, Y. Chikashige and K. Akama, Phys. Rev. D15 (1977), 480.

4) S. Weinberg, Phys. Rev. Letters 19 (1967), 1264. A. Salam, in Elementary Particle Physics, edited by N. Svartholm (Stockholm, 1968), P367.

5) C. Bouchiat, J. Iliopoulos and Ph.Meyer, Phys. Letter 38B (1972), 519. D. Gross and R. Jackiew, Phys. Rev. D6 (1972), 477.

6) Y. Nambu and G. Jona-Lasinio, Phys. Rev. 122 (1961) 345; 124 (1961) 246.

THE FAMILY GROUP IN GRAND UNIFIED THEORIES*

P. RAMOND[†]

California Institute of Technology, Pasadena, California 91125

CALT-68-709
DoE RESEARCH AND
DEVELOPMENT REPORT

Invited talk at the Sanibel Symposia
February 1979

We review the known ways of incorporating and breaking symmetries in a renormalizable way. We summarize the various grand unified theories based on SU_5, SO_{10},
and E_6 as family enlargement groups. An SU_5 model with an SU_2 gauged family
group is presented as an illustration. In it, the e-family (i.e., e,u and d) is classically
massless and acquires calculable mass corrections. The family group is broken by
the same agent that does the superstrong breaking. Finally, we sketch a way of
unifying the family group with SU_5 into SU_8.

The most important task facing model builders is to find the symmetries of the action that describes the fundamental interactions. With the
realization that strong, weak and electromagnetic forces are most probably
described by Yang-Mills theories came the possibility that all these interactions, although very different at our scale, are but manifestations of the
same master Yang-Mills theory[1,2]. In this talk we shall assume: 1) the
validity of QCD[3], the gauge theory based on color SU_3 to describe strong
interactions, 2) the Glashow-Weinberg-Salam-Ward[4] $SU_2 \times U_1$ Yang-Mills
model as the correct theory for the weak and electromagnetic interactions,
3) that gravity can be neglected in first approximation as the scales of interest will still be several orders of magnitude away from those where quantum
gravity is thought to be important. This survey will be divided in three
parts:

---

*Work supported in part by the U.S. Department of Energy contract EY76-C-03-0068.
[†]Robert Andrews Millikan Senior Research Fellow.

I) A general theoretical discussion of the known allowed ways to build renormalizable theories with all types of symmetries, broken explicitly, spontaneously, or not at all.

II) A description of the zoo of elementary fermions and of the unification picture it suggests. The concept of families of elementary particles will be seen to emerge. The SU_5[2] model will be reviewed as well as the most likely family enlargement models based on SO_{10} [5,6] and E_6 [7].

III) Last, but not least, a presentation of models that postulate interactions between the families ruled by "family groups". An illustrative model of this type will be discussed where the e, u, and d masses are zero classically, but acquire calculable quantum corrections.

 I) <u>Allowed symmetries of renormalizable Lagrangians and their breaking.</u>

It is convenient to split up the Lagrangian into several parts

$$\mathcal{L} = \mathcal{L}_{\rm kin} + \mathcal{L}_{\rm int}^{\rm g} + \mathcal{L}_{\rm int}^{Y} + \mathcal{L}_{\rm int}^{\rm sc} ,$$

where $\mathcal{L}_{\rm kin}$ contains the kinetic terms of the particles which we take to have spins 0, 1/2, 1. The spin-1 kinetic part will always be of the Yang-Mills variety and thus will include self-interactions. $\mathcal{L}_{\rm int}^{\rm g}$ describes the interactions of the spin-1 gauge fields with the spin-0,-1/2 fields, and vanishes as the gauge coupling, g, tends to zero. $\mathcal{L}_{\rm int}^{Y}$ contains fermion mass terms (if any) and fermion-fermion-spinless boson interactions. Finally $\mathcal{L}_{\rm int}^{\rm sc}$ displays the self-interactions of the spinless bosons, including their mass terms, and is equal to minus the classical potential. Before discussing each term in detail, let us ask what possible symmetries $\mathcal{L}$ can have. These come in two categories - continuous and discrete. Continuous symmetries can be of the graded-Lie type. Lagrangians with this symmetry are called supersymmetric[8]. Local (i.e., space-time dependent) supersymmetry necessarily involves gravity leading to supergravity theories[9]. There are also continuous Lie symmetries which can be either local or global; the former case yields Yang-Mills gauge theories. It is thought that the Lagrangian which describes all the fundamental interactions, save perhaps gravity, is <u>locally</u> invariant under a yet to be discovered Lie group. The success of continuous Lie groups is linked to the existence of additively conserved quantum numbers. Finally, discrete symmetries, although a definite logical possibility, have not found wide usage because they give rise to multiplicatively conserved quantum numbers, even though they have distinct theoretical advantages[10].

The kinetic part of $\mathcal{L}$, $\mathcal{L}_{\rm kin}$, is the most symmetric as its symmetry is always greater (or equal if $\mathcal{L} = \mathcal{L}_{\rm kin}$) than that of $\mathcal{L}$. Given N Weyl spinors, the fermion kinetic term has a global $U(N)$ symmetry which includes chiral symmetry. Not all of this symmetry can be gauged for fear of introducing unrenormalizable anomalies[11]. The spinless boson kinetic term for M bosons displays an $SO(M)$ global symmetry which can all be gauged (except for $M = 6$). It can happen that the number of spin-0 fields is twice the number of two-component Weyl spinors. Then the symmetry is $0(M) \times$ global $N = 1$ supersymmetry. Thus $\mathcal{L}_{\mathrm{kin}}$ has an enormous symmetry which will be nibbled away by the other terms. So it is convenient to discuss these terms by their action on the symmetries of $\mathcal{L}_{\mathrm{kin}}$.

The gauge interaction terms, $\mathcal{L}^{\mathrm{g}}_{\mathrm{int}}$, of dimension-4, respects the chiral symmetry of $\mathcal{L}_{\mathrm{kin}}$ but will in general explicitly break the global supersymmetry of $\mathcal{L}_{\mathrm{kin}}$ (if any), which is then restored as the gauge coupling vanishes.

The Yukawa term, $\mathcal{L}^{Y}_{\mathrm{int}}$, explicitly breaks the chiral symmetries of $\mathcal{L}_{\mathrm{kin}}$ and in general the supersymmetry. This term is the most intriguing since the observed patterns of fermion masses mixing angles and CP violation[12], depend on it. Unlike gauge couplings, there is no known principle governing its form. The restrictions imposed on it by the known gauge couplings unfortunately does not suffice to make it predictive. This is where additional symmetries, such as the family symmetry will prove invaluable. Certain patterns can be inferred from experiment: the existence of low mass fermions indicates that chiral symmetry may not be too badly violated while the absence of low mass spinless particles suggests that supersymmetry is badly broken.

Finally, the most unprincipled part of $\mathcal{L}$ is that which describes the interactions among the spinless bosons. It can occur with dimensions -4, -3, -2, and -1. It can be used to achieve two types of symmetry breaking. a) Explicit breaking which is allowed for all except the gauge symmetries. Two cases arise – if the breaking term has dimension -4, quantum corrections will spread the breaking action into other parts of $\mathcal{L}$. This is the so-called "hard breaking". When the breaking is done by terms of lower dimensions, ("soft breaking") the quantum corrections to the relations implied by unbroken symmetry will be calculable[13]. b) Spontaneous breaking. In this case, the field equations are not affected, just the choice of ground state. Its importance lies in the Higgs mechanismref14 – the only known way to break symmetry without losing renormalizability. Spontaneous breaking can be used on all symmetries with one important proviso – massless Goldstone bosons[15] will appear if the symmetry is continuous and global. Even then, there is an ingenious evasion mechanism: match the global symmetry with a minor local symmetry and break both spontaneously, leaving the sum invariant. In this way the Goldstone danger is avoided, but one is left with a global symmetry. Such a mechanism is at work in the SU_5 model where due to the reducibility of the fermion representation, a global U_1 symmetry exists. It is broken spontaneously together with a U_1 from within the SU_5, leaving a linear combination unbroken, which is baryon number minus lepton number. Finally, let us mention that a $\mathcal{L}_{\mathrm{int}}^{\mathrm{sc}}$ made up of dimension -4 terms alone will give rise to spontaneous breaking of symmetry[16], via quantum effects. Discrete symmetries can be spontaneously broken without an ensuing Goldstone boson, and therein lies their theoretical attractiveness. This breakup of $\mathcal{L}$ makes it convenient to discuss pseudo-symmetries[17], i.e., symmetries respected by some parts of $\mathcal{L}$ but not by all. For instance when the symmetry or "$\mathcal{L}_{\mathrm{kin}}$" + $\mathcal{L}_{\mathrm{int}}^{\mathrm{sc}}$ is larger than that of $\mathcal{L}_{\mathrm{int}}^{\mathrm{g}}$ (where "$\mathcal{L}_{\mathrm{kin}}$" does not contain the vector self-interactions). The spontaneous breakdown via $\mathcal{L}_{\mathrm{int}}^{\mathrm{sc}}$ leads to more Goldstone bosons but those not eaten by the gauge fields acquire a mass due to the explicit breaking in $\mathcal{L}_{\mathrm{int}}^{g}$. These are the pseudo-Goldstone bosons. One might consider an analogous case with supersymmetry where $\mathcal{L} - \mathcal{L}_{\mathrm{int}}^{\mathrm{g}}$ is supersymmetric. Then one will obtain in this way, by turning on $\mathcal{L}_{\mathrm{int}}^{\mathrm{g}}$ "pseudo-Goldstinos". Lastly the soft explicit breaking occurs when $\mathcal{L} = \mathcal{L}_0 + \tilde{\mathcal{L}}$ with $\tilde{\mathcal{L}}$ breaking the symmetry of $\mathcal{L}$ with fermion and spinless mass terms and possibly by cubic scalar self-interaction terms.

II) <u>The Zoo of Elementary Fermions - $SU_5 \subset SO_{10} \subset E_6 \subset ...$</u>

According to SU_3^c (c is for color), fermions come in two genres – leptons with no color (color singlets) and quarks (antiquarks) which are color triplets (antitriplets). No fermions with other color assignments are known at present, but their absence at low mass may just be a "caprice" of the mass matrix and may not have any fundamental significance. The masses of the observed fermions come in a rough pattern which enables us to define the concept of a family. First we have very low mass fermions which under $SU_2 \times U_1 \times SU_3^c$ transform as

$$(\underset{\sim}{2},\ \underset{\sim}{1^c}) \qquad (\underset{\sim}{1},\ \underset{\sim}{\overline{3}^c}) \qquad (\underset{\sim}{1},\ \underset{\sim}{\overline{3}^c}) \qquad (\underset{\sim}{2},\ \underset{\sim}{3^c}) \qquad (\underset{\sim}{1},\ \underset{\sim}{1^c})$$

$$\begin{pmatrix} \nu_{eL} \\ e_L \end{pmatrix} \qquad \underset{\sim}{\overline{d}}_L \qquad \underset{\sim}{\overline{u}}_L \qquad \begin{pmatrix} \underset{\sim}{u}_L \\ \underset{\sim}{d}_L \end{pmatrix} \qquad \overline{e}_L$$

[Notation: e_L = 2cpt-left-handed electron field, $\overline{e}_L = \sigma_2 e_R^*$ = right-handed positron field; quark fields are underlined by a wiggle, to indicate color.] We call this array the electron family.

The remarkable thing is that this pattern is repeated at a slightly higher mass, yielding the muon family with the same quantum numbers

$$\begin{pmatrix} \nu_{\nu L} \\ \nu_L \end{pmatrix} \qquad \underset{\sim}{\overline{s}}_L \qquad \underset{\sim}{\overline{c}}_L \qquad \begin{pmatrix} \underset{\sim}{c}_L \\ \underset{\sim}{s}_L \end{pmatrix} \qquad \overline{\nu}_L \ .$$

Not shown here are the slight Cabibbo mixings of $\underset{\sim}{d}_L$ and $\underset{\sim}{s}_L$. This family incorporates the GIM mechanism[18]. As if it were not enough, it seems a third family is being discovered with a much higher central mass – the τ family:

$$\begin{pmatrix} \nu_{\tau L} \\ \tau_L \end{pmatrix} \qquad \underset{\sim}{\overline{b}}_L \qquad \underset{\sim}{\overline{t}}_L \qquad \begin{pmatrix} \underset{\sim}{t}_L \\ \underset{\sim}{b}_L \end{pmatrix} \qquad \overline{\tau}_L \ ,$$

where the yet undiscovered charge 2/3 t-quark is heavier than 7 GeV (in the sense that no $t\overline{t}$ state exists at a mass of up to 15 GeV). Theoretically, the discovery of the t-quark at a reasonable mass would validate this emerging family picture. Alternatively, its non-existence at a mass $\lesssim 30$ GeV would give credence to the "topless" models advocated by exceptional group enthusiasts[7]. To conclude this preliminary classification, we note that neutrinos are all light although their families' central mass increases. This could be due to the absence of right-handed partners so that the only way they can acquire mass is by developing a Majorana-type mass which has different quantum numbers.

Faced with these three families of fermions we now briefly review various attempts at defining the families themselves with unifying groups. Then we will approach the problem of interaction between the families.

Since QCD and QFD are described by Yang-Mills theories, it is natural to consider a larger Yang-Mills theory which contains these two, i.e., the larger group of local invariance, G, will include $SU_2 \times U_1 \times SU_3^c$ as a subgroup. Let us for instance consider the imbedding of SU_3^c with G such that at least one representation of G exists with at most 1^c, 3^c and/or $\overline{3}^c$, in order to represent the fermions. These have been listed[19], but only few are noteworthy: only in three cases do the quarks and leptons share weak interactions as a result of the group structure. In all other cases, the quark-lepton universality of the weak interactions must arise from the specifics of the breaking mechanism of G down to $SU_3^c \times U_1 \times SU_2$. The three cases of interest are

$$G = SU_n \supset SU_{n-3} \times U_1 \times SU_3^c \;,$$

with the fermions in the $(\underset{\sim}{nx}...\underset{\sim}{xn})_A$ of SU_n. Call it case I. Then we have case II.

$$G = SO_n \supset SO_{n-6} \times U_1 \times SU_3^c \;,$$

with the fermions in the spinor representation of SO_n. For both cases, the electric charge ratio between quarks and leptons is arranged by hand through the use of the U_1 factor in the flavor group. Case III includes the exceptional groups. The relevant ones are

$$G = E_6 \supset SU_3 \times SU_3 \times SU_3^c \;,$$

with the fermions in the (complex) $\underset{\sim}{27}$, and

$$G = E_7 \supset SU_6 \times SU_3^c \;,$$

with the fermions in the (pseudoreal) $\underset{\sim}{56}$. In both cases, the factor of 3 between lepton and quark charges arises as a result of the group structure. In this sense, these are the most natural simple Lie groups.

There is in the literature an example of each of the above three cases. The most studied[20] and apparently successful is the SU_5 model of Georgi and Glashow, which is an example of case I. The imbedding

$$SU_5 \supset SU_2 \times U_1 \times SU_3^c$$

is defined by the fundamental of SU_5,

$$\underset{\sim}{5} = (\underset{\sim}{2},\ \underset{\sim}{1}^c) + (\underset{\sim}{1}, \underset{\sim}{3}^c) \;,$$

so that each family is described by a $\underset{\sim}{\overline{5}}$ and a $\underset{\sim}{10}$ (see Table I)

$$\underset{\sim}{10} = (\underset{\sim}{5} \times \underset{\sim}{5})_A = (\underset{\sim}{1},\underset{\sim}{\overline{3}}^c) + (\underset{\sim}{2},\underset{\sim}{3}^c) + (\underset{\sim}{1},\underset{\sim}{1}^c) \;.$$

This pattern is repeated thrice, once for each family. The spin-1 bosons belong to the adjoint representation, which is

$$24 = (1, 8^c) + (3, 1^c) + (1, 1^c) + (2, 3^c) + (2, \overline{3}^c) \, ,$$

corresponding to the gluon octet, the vectors of the GWSW model, and six others which play the dual role of causing lepton-quark and quark-quark transitions. These cause proton decay in second order in the SU_5 coupling constant. The fermion mass matrix consists of two parts, $\overline{5} \times 10 = 5 + \overline{45}$ which gives mass to the charged leptons and charge-1/3 quark within each family, and $(10 \times 10)_S = \overline{5} + 50$ which gives a mass to the charge 2/3 quarks. The minimal Higgs structure is a 24 to break SU_5 down to $SU_2 \times U_1 \times SU_3^c$ and a 5 to break $SU_2 \times U_1$ down to U_1^γ. Some consequences of this model are (at some scale)

$$\frac{m_d}{m_e} = \frac{m_s}{m_\mu} = \frac{m_b}{m_\tau} = 1 \, , \qquad \sin^2\theta_{\mathrm{w}} = \frac{3}{8} \, ,$$

where θ_{w} is the Weinberg angle. The theory is asymptotically free so that perturbation theory can be used reliably over large scales. By matching the strong and weak coupling constants at our scale as a boundary condition, one finds[21] that SU_5 symmetries are valid at very large masses of $0(10^{14}$ GeV$)$ [22] (i.e., very short distances). Then one finds the proton decays with a rate of 10^{-32} per year for all modes. Similarly one finds[20] renormalized values for $\sin^2\theta_w \sim 0.20 - .21$, $\frac{m_b}{m_\tau} \sim 2-3$, $\frac{m_s}{m_\mu} \sim 4-5$. While these results are spectacular, the ratio $\frac{m_d}{m_e}$ comes out all wrong, presumably because we are dealing with very light particles. As a further consequence of this theory with two 5-Higgs fields[23] one can explain the observed over-abundance of matter over antimatter in the universe, while starting from symmetric boundary conditions. Lastly, we re-emphasize an important aspect of the SU_5 theory: the neutrinos are, as in the GWSW model, strictly massless because they are forbidden from acquiring a Majorana mass by the exact conservation of baryon number minus lepton number, $B - L$. This law comes about because of the reducibility of the fermion representations which allows for a conserved quantum number which is 1 for the 10, -3 for the $\overline{5}$ of fermions and -2 for the 5 of Higgs. When the 5 of Higgs acquires a vacuum expectation value, this U_1 as well as the U_1 within SU_5 which has value 1 for the $(1, 3^c)$ and -3/2 for the $(2,1)$ of 5, are broken, but as the 5 only has one non-zero entry, a linear combination is preserved: $B - L$ for the fermions. In a more unified theory where the fermion family not be reducible, this conservation law will not exist. Then the neutrinos will be free to acquire Majorana masses.

The next group in this family description is SO_{10}[5,6] which falls in case II. Unlike SU_5 it is automatically free of anomalies[5]. The imbedding is given by

$$SO_{10} \supset SU_5 \times U_1 \, ,$$

with the fermions appearing in the (complex) spinor representation (see Table II)

$$\underset{\sim}{16} = \underset{\sim}{\overline{5}} + \underset{\sim}{10} + \underset{\sim}{1} \, .$$

Here, there is an extra neutral lepton helicity for each family. It can act as a right-handed neutrino, which means that the neutrinos, in this theory, are not automatically massless. To see this, consider the fermion mass matrix

$$\underset{\sim}{16} \times \underset{\sim}{16} = (\underset{\sim}{10} + \underset{\sim}{126})_S + \underset{\sim}{120}_A \, .$$

The Higgs structure is very rich. With just $\underset{\sim}{10}$'s of Higgs, the neutrino mass occurs in the same way as that of the charge 2/3 quark. Special measures have to be taken to insure the low mass of ν_L (these are the true leptons!). One way[24] is to use the Majorana mass of this extra right-handed neutrino, which appears in the $\underset{\sim}{126}$. Call this extra lepton helicity $\overline{\nu}_L$. Then the neutral lepton mass matrix is

$$\begin{pmatrix} \nu_L^T & \overline{\nu}_L^T \end{pmatrix} \begin{pmatrix} 0 & a \\ a & A \end{pmatrix} \begin{pmatrix} \nu_L \\ \overline{\nu}_L \end{pmatrix} \, ,$$

with a proportional to $m_{2/3}$. Take $a << A$. Then ν_L is approximately massless and very slightly mixed. This is, however, slightly disturbed by radiative corrections. Note that a high scale is introduced in the mass matrix. Another way[6] is to invent yet another neutral lepton helicity which will act as the true Dirac partner of $\overline{\nu}_L$, which we label Y_L. Then there are three neutral lepton helicities per family giving a mass matrix (per family),

$$\begin{pmatrix} \nu_L^T & \overline{\nu}_L^T & Y_L^T \end{pmatrix} \begin{pmatrix} 0 & a & 0 \\ a & 0 & A \\ 0 & A & 0 \end{pmatrix} \begin{pmatrix} \nu_L \\ \overline{\nu}_L \\ Y_L \end{pmatrix} \, ,$$

with again $a << A$. The ν_L is slightly mixed but strictly massless. This extra Diract mass introduces a 16 Higgs which is ten used to break SO_{10} down to SU_5. For a particular choice of $\mathcal{L}_{\text{int}}^{sc}$, the fermion reducibility can again be used to produce a $B - L$ conservatin law, thereby forbidding

Majorana masses. In this scheme, the fermion mass matrix has an Abelian global family symmetry, $U_1 \times U_1 \times U_1$, which allows for a t-quark mass of the order of 15 GeV.

Finally, the most promising candidate of type III is based on E_6 [7], with the imbedding

$$E_6 \supset SO_{10} \times U_1 \ ,$$

with the fermions in the (complex) $\underset{\sim}{27}$ (see Table III)

$$\underset{\sim}{27} = \underset{\sim}{16} + \underset{\sim}{10} + \underset{\sim}{1} \ .$$

This produces the singlet used in the preceeding model, but introduces for each family ten additional helicities. Interestingly, a feature of the GWSW model that was lost in SU_5 and SO_{10} is regained: all the symmetry breaking needed can occur in the fermion-fermion operator with, for instance, $\underset{\sim}{16} \times \underset{\sim}{1}$ breaking E_6 down to SU_5, and the $\underset{\sim}{10} \times \underset{\sim}{10}$ which contains a $\underset{\sim}{24}$ of SU_5 breaking SU_5 down to $SU_2 \times U_1 \times SU_3^c$. The Higgs structure of the model is found in

$$\underset{\sim}{27} \times \underset{\sim}{27} = (\overline{\underset{\sim}{27}} + \underset{\sim}{351}')_s + \underset{\sim}{351}_A \ .$$

Spinless bosons transforming as the $\underset{\sim}{27}$ are not sufficient to give the particles their requisite masses. At least one $\underset{\sim}{351}'$ of Higgs is needed to get $\underset{\sim}{16} \times \underset{\sim}{1}$ and $\underset{\sim}{10} \times \underset{\sim}{10}$ terms. Then, with these families, the theory is just asymptotically free. Addition of another fermion family, or of another $\underset{\sim}{351}'$ changes the sign of the derivative of the running gauge coupling constant. The consequences of a temporarily free theory are not understood, but one may speculate that the attrative and repulsive vector forces may compete with that of gravity, leading to the "bounces", thereby at least delaying the formation of singularities. Although much work has been done on the mass matrix of the E_6 model, there is no obvious way to reproduce the neutral lepton mass matrix. Still one should not be discouraged as it presents the greatest unification of the fermions. However, none of these attempts provides a reason for the triplication of families.

III) <u>The search for family symmetries</u>

After having presented certain "unified" descriptions of the fermions, we are still not quite unified enough because of the apparent triplication (and possibly infinite xeroxing) of families. As the concept of families arises by looking at the fermion masses, we must search for ways to narrow down $\mathcal{L}_{\text{int}}^Y$. This term is the most arbitrary in the $SU_2 \times U_1$ model in which an enormous number of parameters is introduced. Several attempts have been made to introduce a family symmetry, both discrete[10] and continous. In one of these[25], a gauged family symmetry (called horizontal symmetry) based on SO_3 is introduced at the $SU_2 \times U_1 \times SU_3^c$ level. However, this scheme, because of the symmetries of the fermion mass matrix cannot be extended directly to SU_5 (unless one wants to introduce three 50's of Higgs!). Another, previously mentioned[6], uses a global $U_1 \times U_1 \times \tilde{U}_1$ family symmetry in SO_{10}.

Although it is not clear at what stage one should introduce the family symmetry, we find it convenient to do it at the level of SU_5 because of the regularities in the mass matrix, where we have just the nearly massless e-family and the heavier μ- and τ-families. We propose to use the family group to tell us why the e-family is so light and at the same time reduce the number of parameters in $\mathcal{L}_{\mathrm{int}}^Y$. The family group may or may not be gauged. At least it should be a symmetry of the terms of dimension -4 in $\mathcal{L}$. If it is gauged, we should beware of anomalies since the family spinors are Weyl spinors. Also it must be badly broken to avoid at low energy flavor changing forces. This last aspect is neatly done by having the agent that does the superstrong breaking do at the same time the family breaking. We now present an illustrative example of this type.

Let SU_2 be the family symmetry[26]. Assume that the two lightest SU_5 families form a doublet under it. The fermion content in terms of $SU_2^f \times SU_5$ is then

$$F \equiv \begin{pmatrix} F_e \\ F_\mu \end{pmatrix} \sim (\underset{\sim}{2}, \underset{\sim}{10}) \quad ; \quad f \equiv \begin{pmatrix} f_e \\ f_\mu \end{pmatrix} \sim (\underset{\sim}{2}, \underset{\sim}{\bar{5}})$$
$$F_\tau \sim (\underset{\sim}{1}, \underset{\sim}{10}) \qquad\qquad f_\tau \sim (\underset{\sim}{1}, \underset{\sim}{\bar{5}}) \ .$$

The spinless bosons are taken to be

$$h \sim (\underset{\sim}{1}, \underset{\sim}{5}) \qquad\qquad h' \sim (\underset{\sim}{1}, \underset{\sim}{5})$$
$$H \sim (\underset{\sim}{2}, \underset{\sim}{5}) \qquad\qquad H' \sim (\underset{\sim}{2}, \underset{\sim}{5}) \ .$$

as well as $\Phi \sim (\underset{\sim}{2}, \underset{\sim}{24})$. Φ acquires a very large vacuum expectation value, breaking SU_5 down to $SU_2 \times U_1 \times SU_3^c$ and breaking the family SU_2. The Yukawa interactions are given by the terms $F_\tau F_\tau h$, $F_\tau f_\tau \overline{h}'$, $F_\tau F H$, $F_\tau f \overline{H}'$, $F f_\tau \overline{H}'$, which besides $SU_2^f \times SU_5$, respect three global U_1's. One of those is explicitly broken in $\mathcal{L}_{\mathrm{int}}^{sc}$ by a term of the form $\overline{h}h'\overline{H}H'$; the other two, X and Y are summarized in the table

	h	g'	H	H'	F_τ	f_τ	F	f
X	1	1	1	1	$-1/2$	$3/2$	$-1/2$	$3/2$
Y	1	-1	-1	1	$-1/2$	$-1/2$	$3/2$	$3/2$

Both X and Y are to be broken spontaneously. The breaking of X does not give rise to a Goldstone boson because X-conservation is replaced by $B - L$ conservation. Y-breaking gives rise to a Goldstone boson which acquires mass by QCD instanton effects[27]. In view of the non-existence of a low mass axion[28] it might be necessary to add a Higgs mass of the form $h\overline{h}'$ which explicitly breaks Y (this can be used to obtain the desired sign for the $u - d$ mass difference). One interesting consequence of this model is that the e-family is massless in lowest order but acquires calculable (e.g., finite) corrections, by means of, among others, terms of the form $FFH\overline{H}'h'$, FFH^2h, $Ff H\overline{H}'\overline{h}$, $Ff\overline{H}'^2 h$, consistent with all the symmetries of $\mathcal{L}$. Such a model alters the value of m_e/m_d without affecting the others. Also, it yields small u and d masses, accounting for isospin symmetry. This SU_2 family symmetry looks more natural in an SO_{10} model for there only the SU_2 forbids the e-family from acquiring a mass (whereas here Y is needed) by having Higgs with $(1, \underset{\sim}{10})$ and $(\underset{\sim}{2}, \underset{\sim}{10})$ only. This model is presented here as an illustration of the uses one may make of family symmetries. Note that a gauged SU_3 family symmetry is feasible provided one uses an anomaly-free set of representations. Finally, one may unify the family symmetry with the SU_5. There is a strong candidate is SU_8, with the fermions appearing in $\underset{\sim}{56}_L + \underset{\sim}{\overline{56}}'_L$. We hope to present these models in detail in a later publication.

In closing, let us mention what we have not discussed: the all-important problem of hierarchies[29] for which this survey offers no answers nor provides any clues.

<u>Acknowledgement</u>

The author wishes to thank his collaborators, Professors M. Gell-Mann and R. Slansky for invaluable comments, suggestions and ideas concerning the material presented here.

Table I

$\underline{SU_5}$ Unification Picture

$SU_5 \supset SU_2 \times U_1 \times SU_3^c$

$$\begin{pmatrix} \nu_{eL} \\ e_L \end{pmatrix} \qquad \underset{\sim}{\overline{d}}_L \qquad\qquad \underset{\sim}{\overline{u}}_L \qquad \begin{pmatrix} \underset{\sim}{u}_L \\ \underset{\sim}{d}_L \end{pmatrix} \qquad \overline{e}_L$$

$$\underbrace{(\underset{\sim}{2},\ \underset{\sim}{1}^c) \ + \ (\underset{\sim}{1},\ \underset{\sim}{\overline{3}}^c)}_{\underset{\sim}{\overline{5}}} \qquad\qquad \underbrace{(\underset{\sim}{1},\ \underset{\sim}{\overline{3}}^c) \ + \ (\underset{\sim}{2},\ \underset{\sim}{3}^c) \ + \ (\underset{\sim}{1},\ \underset{\sim}{1}^c)}_{\underset{\sim}{10}}$$

$$\begin{pmatrix} \nu_{\mu L} \\ \mu_L \end{pmatrix} \qquad \underset{\sim}{\overline{s}}_L \qquad\qquad \underset{\sim}{\overline{c}}_L \qquad \begin{pmatrix} \underset{\sim}{c}_L \\ \underset{\sim}{s}_L \end{pmatrix} \qquad \overline{\mu}_L$$

$$\underbrace{}_{\underset{\sim}{\overline{5}}} \qquad\qquad \underbrace{}_{\underset{\sim}{10}}$$

$$\begin{pmatrix} \nu_{\tau L} \\ \tau_L \end{pmatrix} \qquad \underset{\sim}{\overline{b}}_L \qquad\qquad \underset{\sim}{\overline{t}}_L \qquad \begin{pmatrix} \underset{\sim}{t}_L \\ \underset{\sim}{b}_L \end{pmatrix} \qquad \overline{\tau}_L$$

$$\underbrace{}_{\underset{\sim}{\overline{5}}} \qquad\qquad \underbrace{}_{\underset{\sim}{10}}$$

Table II
$\underline{SO_{10}\ \text{Unification}}$

$SO_{10} \supset SU_5 \times U_1$

$$
\underbrace{\begin{pmatrix} \nu_{eL} \\ e_L \end{pmatrix} \quad \bar{d}_{\underset{\sim}{L}}}_{\underset{\sim}{\bar{5}}} \qquad \underbrace{\bar{u}_{\underset{\sim}{L}} \quad \begin{pmatrix} u_{\underset{\sim}{L}} \\ d_{\underset{\sim}{L}} \end{pmatrix} \quad \bar{e}_L}_{\underset{\sim}{10}} \qquad \underbrace{\bar{\nu}_{eL}}_{\underset{\sim}{1}}
$$

$$
\underbrace{}_{\underset{\sim}{16}}
$$

$$
\underbrace{\begin{pmatrix} \nu_{\mu L} \\ \mu_L \end{pmatrix} \quad \bar{s}_{\underset{\sim}{L}}}_{} \qquad \bar{c}_{\underset{\sim}{L}} \quad \underbrace{\begin{pmatrix} c_{\underset{\sim}{L}} \\ s_{\underset{\sim}{L}} \end{pmatrix}} \quad \bar{\mu}_L \qquad \bar{\nu}_{\mu L}
$$

$$
\underbrace{}_{\underset{\sim}{16}}
$$

$$
\underbrace{\begin{pmatrix} \nu_{\tau L} \\ \tau_L \end{pmatrix} \quad \bar{b}_{\underset{\sim}{L}}}_{} \qquad \bar{t}_{\underset{\sim}{L}} \quad \underbrace{\begin{pmatrix} t_{\underset{\sim}{L}} \\ b_{\underset{\sim}{L}} \end{pmatrix}} \quad \bar{\tau}_L \qquad \bar{\nu}_{\tau L}
$$

$$
\underbrace{}_{\underset{\sim}{16}}
$$

Table III
$\underline{E_6}$ Unification

$$E_6 \supset SO_{10} \times U_1$$

$$
\begin{pmatrix} \nu_{eL} \\ e_L \end{pmatrix} \quad \underset{\sim L}{\bar{d}} \quad \underset{\sim L}{\bar{u}} \quad \begin{pmatrix} \underset{\sim L}{u} \\ \underset{\sim L}{d} \end{pmatrix} \quad \bar{e}_L \quad \bar{\nu}_{eL} \qquad \underset{\sim L}{Y_{eL}} \qquad \begin{pmatrix} N_{eL} \\ E_L^- \end{pmatrix} \quad \underset{\sim L}{\bar{h}} \quad \begin{pmatrix} \overline{N}_{eL} \\ \overline{N}_{eL} \end{pmatrix} \quad \underset{\sim L}{h}
$$

$$\underbrace{\qquad\qquad\qquad \underset{\sim}{16} \qquad\qquad\qquad}\qquad \underset{\sim}{1} \qquad\qquad \underset{\sim}{10}$$

$$\underbrace{\qquad\qquad\qquad\qquad \underset{\sim}{27} \qquad\qquad\qquad\qquad}$$

$$
\begin{pmatrix} \nu_{\mu L} \\ \mu_L \end{pmatrix} \quad \underset{\sim L}{\bar{s}} \quad \underset{\sim L}{\bar{c}} \quad \begin{pmatrix} \underset{\sim L}{c} \\ \underset{\sim L}{s} \end{pmatrix} \quad \bar{\mu}_L \quad \bar{\nu}_{\mu L} \qquad \underset{\sim L}{Y_{\mu L}} \qquad \begin{pmatrix} N_{\mu L} \\ M_L^- \end{pmatrix} \quad \underset{\sim L}{\bar{k}} \quad \begin{pmatrix} \overline{M}_{eL} \\ \overline{N}_{\mu L} \end{pmatrix} \quad \underset{\sim L}{k}
$$

$$\underbrace{\qquad\qquad\qquad \underset{\sim}{16} \qquad\qquad\qquad}\qquad \underset{\sim}{1} \qquad\qquad \underset{\sim}{10}$$

$$\underbrace{\qquad\qquad\qquad\qquad \underset{\sim}{27} \qquad\qquad\qquad\qquad}$$

$$
\begin{pmatrix} \nu_{\tau L} \\ \tau_L \end{pmatrix} \quad \underset{\sim L}{\bar{b}} \quad \underset{\sim L}{\bar{t}} \quad \begin{pmatrix} \underset{\sim L}{t} \\ \underset{\sim L}{b} \end{pmatrix} \quad \bar{\tau}_L \quad \bar{\nu}_{\tau L} \qquad \underset{\sim L}{Y_{\tau L}} \qquad \begin{pmatrix} N_{\tau L} \\ T_L^- \end{pmatrix} \quad \underset{\sim L}{\bar{j}} \quad \begin{pmatrix} \overline{T}_{\tau L} \\ \overline{N}_{\tau L} \end{pmatrix} \quad \underset{\sim L}{j}
$$

$$\underbrace{\qquad\qquad\qquad \underset{\sim}{16} \qquad\qquad\qquad}\qquad \underset{\sim}{1} \qquad\qquad \underset{\sim}{10}$$

$$\underbrace{\qquad\qquad\qquad\qquad \underset{\sim}{27} \qquad\qquad\qquad\qquad}$$

References

1. J. C. Pati and A. Salam, Phys. Rev. $\underline{D8}$ (1973) 1240; $\underline{D10}$(1974) 275.
2. H. Georgi and S. L. Glashow, Phys. Rev. Lett. $\underline{32}$ (1974) 438.
3. Y. Nambu in "Preludes in Theoretical Physics," ed. A. de Shalit (North-Holland, Amsterdam, 1966); H. Fritzsch and M. Gell-Mann, Proc. of the XVI International Conference on High Energy Physics, Vol. 2, p. 35 (national Accelerator Laboratory).
4. S. L. Glashow, Ph.D. Thesis, Harvard University 1959, and Nucl. Phys. $\underline{22}$ (1961) 579; S. Weinberg, Phys. Rev. Lett. $\underline{19}$ (1967) 1264; A. Salam, Proc. 8th Nobel Symp., Stockholm, ed. N. Svartholm (Almquist and Wiksells, Stockholm 1968) p. 367; A. Salam and J. C. Ward, Phys. Lett. $\underline{13}$ (1964) 168.
5. H. Fritzsch and P. Minkowski, Ann. of Phys. $\underline{93}$ (1975) 193; H. Georgi, Particles and Fields, 1974 (APS/DPF Williamsburg) ed. C. E. Carlson (AIP New York, 1975) p. 575; M. S. Chanowitz, J. Ellis and M. K. Gaillard, Nucl. Phys. $\underline{B128}$ (1977) 506.
6. H. Georgi and D. V. Nanopoulos, Harvard preprints 1978-1979.
7. F. Gürsey, P. Ramond and P. Sikivie, Phys. Lett. $\underline{B60}$ (1975) 177; F. Gürsey and M. Serdaroğlu, Yale preprint, 1978; Y. Achiman and B. Stech, Phys. Lett. $\underline{77B}$ (1978) 389; Q. Shafi, Univ. of Freiburg preprint 1978.
8. For a review see P. Fayet, S. Ferrara, Phys. Reports $\underline{32C}$ (1977) 249.
9. D. Z. Freedman, P. van Nieuwenhuizen, and S. Ferrara, Phys. Rev. $\underline{D13}$(1976) 3214; S. Deser and B. Zumino, Phys. Lett. $\underline{B62}$ (1976) 335.
10. S. Pakvasa and H. Sugawara, Phys Lett. $\underline{73B}$ (1978) 61, Wisconsin preprint COO-881-66, 1978; H. Sato, University of Tokyo preprint, 1978; E. Derman, Rockefeller preprint, 1978.
11. H. Georgi and S. . Glashow, Phys. Rev. $\underline{D6}$ (1973) 429.
12. N. Kobayashi and K. Maskawa, Progr. Teor. Phys. $\underline{49}$ (1973) 652.
13. K. Symanzik in Coral Gables Conference on Fundamental Interactions at High Energies II. (A. Perlmutter, G. J. Iverson and R. M. Williams eds., Gordon and Breach, 1970.)
14. P. W. Higgs, Phys. Rev. Lett. $\underline{12}$ (1964) 132; F. Englert and R. Brout, Phys. Rev. Lett. $\underline{13}$ (1964) 321; G. S. Guralnik, C. R. Hagen, and T. W. B. Kibble, Phys. Rev. Lett. $\underline{13}$ (1964) 585.
15. J. Goldstone, Nuovo Cimento $\underline{19}$ (1961) 154.
16. S. Coleman and E. Weinberg, Phys. Rev. $\underline{D7}$ (1973) 1888; E. Gildener and S. Weinberg, Phys. Rev. $\underline{D13}$ (1976) 3333.
17. S. Weinberg, Phys. Rev. Lett. $\underline{29}$ (1972) 1698.
18. S. L. Glashow, J. Illiopoulos, L. Maiani, Phys. $\underline{D2}$ (1970) 1285.
19. M. Gell-Mann, P. Ramond, and R. Slansky, Rev. Mod. Phys. $\underline{50}$ (1978)721.
20. A. J. Buras, J. Ellis, M. K. Gaillard, and D. V. Nanopoulos, Nucl. Phys. $\underline{B135}$ (1978) 66.
21. H. Georgi, H. R. Quinn, and S. Weinberg, Phys. Rev. Lett. $\underline{33}$ (1974) 451.
22. T. Goldman and D. Ross, Caltech preprint, CALT-68-704 (1979).
23. M. Yoshimura, Phys. Rev. Lett. $\underline{39}$ (1977) 1385; S. Dimopoulos and L. Susskind, SLAC-PUB-2126 (1978); D. Toussaint, S. B. Treiman, F. Wilczek

and A. Zee, Princeton preprint 1978; J. Ellis, M. K. Gaillard and D. V. Nanopoulos, CERN preprint, TH-2596 (1978); S. Weinberg, Harvard preprint, HUTP-78/A040, 1978.

24. M. Gell-Mann, P. Ramond, and R. Slansky, unpublished.

25. F. Wilczek and A. Zee, Phys. Rev. Lett. $\underline{42}$(1979)421.

26. This concept has been used earlier under the name of M-spin. See for instance F. Gürsey and G. Feinberg, Phys. Rev. $\underline{128}$ (1962) 378; T. D. Lee, Nuovo Cimento $\underline{35}$ (1965) 975; S. Meshkov and S. P. Rosen, Phys. Rev. Lett. $\underline{29}$ (1972) 1764; Phys. Rev. $\underline{D10}$ (1974) 3520.

27. G. 't Hooft, Phys. Rev. Lett. $\underline{37}$ (1976) 8.

28. R. D. Peccei and H. Quinn, Phys. Rev. Lett. $\underline{38}$ (1977) 40; Phys. Rev. $\underline{D16}$(1977)1791; S. Weinberg, Phys. Rev. Lett. $\underline{40}$ (1978) 22; F. Wilczek, Phys. Rev. Lett. $\underline{40}$ (1978) 279.

29. E. Gildener, Phys. $\underline{D14}$ (1976) 1667.